Food From the Sea

Other Titles in This Series

The Changing Economics of World Energy, edited by Bernhard J. Abrahamsson

Indonesia's Oil, Sevinc Carlson

Desertification: Environmental Degradation in and around Arid Lands, edited by Michael H. Glantz

The Nuclear Impact: A Case Study of the Plowshare Program to Produce Natural Gas by Underground Nuclear Stimulation in the Rocky Mountains, Frank Kreith and Catherine B. Wrenn

Natural Resources for a Democratic Society: Public Participation in Decision-Making, edited by Albert E. Utton, W. R. Derrick Sewell, and Timothy O'Riordan

Social Assessment Manual: A Guide to the Preparation of the Social Well-Being Account for Planning Water Resource Projects, Stephen J. Fitzsimmons, Lorrie I. Stuart, and Peter C. Wolff

Water Needs for the Future: Political, Economic, Legal, and Technological Issues in a National and International Framework, edited by Ved P. Nanda

Environmental Effects of Complex River Development: International Experience, edited by Gilbert F. White

The Geopolitics of Energy, Melvin A. Conant and Fern Racine Gold

EARTHCARE: Global Protection of Natural Areas, edited by Edmund A. Schofield

Food, Fuel, and Shelter: A Watershed Analysis of Land-Use Trade-Offs in a Semi-Arid Region, edited by Timothy D. Tregarthen

Water in a Developing World: The Management of a Critical Resource, edited by Albert E. Utton and Ludwik A. Teclaff

Conservation and the Changing Direction of Economic Growth, edited by Bernhard J. Abrahamsson

DISCARD
B.C.H.E. – LIBRARY
00020523

Westview Special Studies in Natural Resources and Energy Management

Food From the Sea: The Economics and Politics of Ocean Fisheries

Frederick W. Bell

Over the last century, the dramatic increase in the world's population and demand for food has attracted substantial international interest in the sea as a food source. In this book, an exploration of the factors at work in determining the past, present, and future use of the sea as an important source of protein, Frederick Bell provides an analysis of the current use of living marine resources and the prospects for and obstacles to obtaining maximum utilization. He covers such topics as the current exploitation levels of fishery stocks, fishery management and the new world order of extended fishery jurisdiction, the conflict between pollution and fishery resources, the potential for fish farming, the plight of the individual fishing firm, and alternative uses of fish resources (e.g., for recreational activities), and includes a critique of the role of the government in coming to grips with fishery problems. Economic and biological principles are integrated to provide a comprehensive analysis of how economic and political forces interact to determine how effectively the countries of the world, especially the United States, are dealing with fishery problems.

The book will serve as a reference, as an adjunct text for courses in the economics of natural resources, and as a text for courses in fishery economics. It will also be of interest to those who seek a better understanding of both the fables and realities of bringing food from the sea.

Frederick W. Bell, professor of economics at Florida State University, was previously chief of economic research for the National Marine Fisheries Services, U.S. Department of Commerce. During his service with the government he was responsible for all economic analysis and policy evaluation concerning the U.S. fishing industry. Dr. Bell has published several books and monographs as well as over 20 articles on resource and environmental economics.

Food From the Sea: The Economics and Politics of Ocean Fisheries

Frederick W. Bell

Westview Press / Boulder, Colorado

Westview Special Studies in Natural Resources and Energy Management

Published in 1978 in the United States of America by
Westview Press, Inc.
5500 Central Avenue
Boulder, Colorado 80301
Frederick A. Praeger, Publisher

Library of Congress Cataloging in Publication Data
Bell, Frederick W.
Food from the sea.
(Westview special studies in natural resources and energy management)
Bibliography: p.
1. Fishery resources. 2. Fisheries. 3. Fishery policy. 4. Fishery management. 5. Food supply.
I. Title.
SH327.5.B44 333.95 77-28756
ISBN 0-89158-403-x (hardcover)
ISBN 0-89158-353-x (paperback)

Printed and bound in the United States of America

To Jane, John, and Kathleen

Contents

Figures

Tables

Preface

Although the United States and other affluent nations have more than an adequate food supply, other nations daily face the specter of starvation. The world now has a critical population/food dilemma of potentially major proportions. Production from the sea and the land is not keeping pace with a world population that is doubling every thirty-five years. Unless this age-old Malthusian problem is solved, millions face starvation and ultimately death.

The situation has stimulated substantial international interest in the sea as a source of food and raw materials. The potential of the sea—not as a panacea, but as an important source of protein to augment the world's food supplies and thereby as a means of mitigating the crises we face—is a continuing theme throughout this book. At present, fish provide approximately 9 percent of the world's protein. Fish are sought not only for food but also for recreation and pleasure. What forces determine the present supply and demand for fishery products? More important, what steps are needed to utilize the full potential of the sea as a source of food and recreation? This book explores these forces and thus provides an insight into food potential from the sea.

The book is directed at two groups in particular: those who want a comprehensive treatment of fishery economics, or more precisely the political economy of food from the sea, as a reference source; and those who can use the book for courses in resource, environmental, and fishery economics. Written for both the general reader and the student, this book requires no previous training in either biology or economics.

Chapter 1 surveys recent trends in the production and value of fishery products. Intercountry comparisons are made. Chapter 2 deals with world food consumption and, in particular, the role of fisheries. The distribution of fishery consumption among the world's countries is detailed. In addition, the determinants of fishery demand—both general and unique—are discussed with numerous empirical examples. Chapter 3 draws attention to the fishery resource crisis, that is, to the fact that the world's oceans cannot support historical levels of production of traditional species. The major fishing powers are identified, and the magnitude and value of their catch are given. Population dynamics models currently in use are discussed with detailed empirical cases. This provides the basis to develop a bioeconomic supply response for individual fishery resources. The overall fishery productivity of the world's oceans is estimated. Chapter 4 deals with the critical problem of why so many of our ocean resources, such as the blue whale, haddock, and sardines, have been overfished and ultimately depleted. The economic reasons for the market system's failure to avoid depletion of valuable resources are detailed. Historical and proposed systems to deal with resource depletion are discussed along with the economic impact of extended fishery jurisdiction, a movement that is virtually sweeping the world. Chapter 5 deals with another obstacle to higher fishery productivity—environmental deterioration. It discusses the reasons for, and the impact on fishery resources of, water-borne pollutants along with possible solutions. Chapter 6 deals with a sleeping giant—the recreational fishing industry—which is by all measures economically more important in the United States than domestic commercial fishing for food! The problems of valuation and conflicts with competing uses are discussed. Although there are surely limits to production from world fishery stocks, aquaculture (fish farming) presents us with new horizons in food fish and shellfish production. This is discussed in chapter 7. Some predict that by the year 2000 more fish will be farmed in coastal and inland areas than will be harvested from the wild stocks of the sea. The economics and biology behind aquaculture are explored. Chapter 8 deals with underutilized or untapped resources such as the Antarctic krill and lanternfish with respect to their potential, both biological and economic. In chapter 9, the plight of the individual fishing enterprise is considered along with the financial rewards and the present financial condition of U.S. fisheries. Finally, chapter 10 addresses the pros and cons of government intervention in the fisheries. Generally, attempts to manage fishery stocks without adequate economic input have been ineffective. In addition, governments throughout the world are trying to outsubsidize each other in a rather futile attempt to gain some competitive advantage and thereby maintain fishing for food and employment. Unfortunately, our conclusions are pessimistic: in general, there is less potential for food from the sea than we might hope, and, in particular, it will be difficult to help the less-developed countries. However, extended fishery jurisdiction, aquaculture transferability to food-starved countries, and the availability of some underutilized species do offer some hope for a solution to the population/food dilemma.

The book should also appeal to world governments interested in formulating fishery policy. It should be a handy reference guide for practitioners in the field—biologists, home economists, economists, oceanographers, and others. More importantly, it is hoped that with increased emphasis by the U.S. Sea Grant Program, more interdisciplinary courses will deal with the political economy of the ocean fisheries. For such courses, this book could serve as a principal or supplementary textbook. At the back of the book are study questions for each chapter to be used by students or others.

Frederick W. Bell
Florida State University

Acknowledgments

This book has been evolving over the last fifteen years—from the author's original study of the economics of the New England fishing industry to his recent inquiries into the impact of coastal water quality on fishery stocks. Thus, there are many colleagues whom I wish to thank for their advice and contribution to my knowledge of this field. Drs. James Crutchfield, Guilio Pontecorvo, Francis Christy, Virgil Norton, Ernest Carlson, Lawrence Van Meir, Fred Smith, Lee Anderson, and the late Frederick V. Waugh have greatly contributed, through intellectual interchange, to the fruition of this book. My colleagues at the National Marine Fisheries Service (NMFS)—William Schaaf, Brian Rothschild, Jerome Pella, William Chapoton, Morton Miller, Bruno Noetzel, Darrel Nash, Donald Cleary, Jack Greenfield, Jack Richards, and Richard Kinoshita—all contributed to my learning process on the economics of food from the sea. The NMFS also supplied financial assistance, data, and other detailed information, for which I would like to express my gratitude.

In any book, of course, certain individuals must be especially mentioned. Fred Olson of the NMFS contributed greatly in his exploratory work on the value of the world's fisheries. My thanks to Dr. Olson extend not just as a colleague, but as a warm personal friend, who was always available when called on. At Florida State University, Manley Johnson, now at George Mason University, aided with many computer runs, which have been incorporated into the book. Bernard Schmitt, now an economist with the U.S. Congress, provided data on the overall food situation in the world today as well as material on nutrition. E. Ray

Canterbery and Philip Sorensen are fellow researchers who contributed both their comments and material to the writing of this book.

Many students at Florida State University have used the book in draft form. They have pointed out errors, difficult sections in need of clarification, and most important, they have provided an audience by which the book could be tested. I express my gratitude for their indulgence of early draft chapters. In addition, Richard H. Stroud of the Sport Fishery Institute reviewed the chapter on recreational fishing and provided many useful comments and references. Ken Roberts of the Office of Sea Grant helped me obtain many documents, and Jack Glick of the USGAO provided many reports on that agency's studies of the U.S. government's role in the fisheries. I am also indebted to the Florida Sea Grant Program, directed by Hugh Pompenoe, for a small educational grant to develop a course in the economics of living marine resources; this was a major factor prompting the writing of this book.

Finally, my chief editorial assistant in this undertaking was Neil Walsh, who not only tried to make the text more readable, but also took time off to ask me several times to make the Beverton and Holt model more understandable. Donna Quick, Gennelle Jordan, Constance Wester, Margaret Crutchfield, and Carol Mahle helped type the manuscript with their unusually high level of competency. William Laird, chairman of the economics department at Florida State, provided moral support in this laborious task. My wife and children knew me only as "Uncle Daddy" during the two years this book was in progress. My wife, Jane, did an excellent job in proofreading and helping prepare the index for this book. I need not say more, except to place at least part of the blame on my colleagues, from whom I learned and who supported me in this effort. No one in his right mind would take full responsibility for the *first* comprehensive textbook and reference source in this field.

F. W. B.

1. Trends in World Fisheries

Over the last century, the world's population has increased dramatically, life-styles and attitudes have changed, and man's accommodation with the environment has become a political, social, and economic problem of the highest priority, one that truly deserves the label crisis. In fact, there is a multitude of crises: the population crisis, the environmental crisis, the raw materials crisis, and the food crisis, to mention but a few. These crises have stimulated substantial international interest in the sea as a source of food and raw materials. As a result, there is an ever-increasing "crisis at sea." We will explore this crisis throughout this book.

The potential of the sea—not as a panacea—but as an important source of protein to augment the world's food supplies and thereby become a means for mitigating the crises man faces will be a continuing theme throughout this book. How important are world fisheries with respect to these crises? In the United States, for example, annual per capita consumption of fish as direct food expressed in edible weight is approximately twelve pounds.[1] Denmark, Iceland, Norway, Portugal, Sweden, Taiwan, Hong Kong, Japan, Cambodia, and the Philippines consume at a minimum more than four times the U.S. per capita consumption of fish. In Japan, Iceland, and many Scandinavian countries, fish provides up to 25 percent of the protein in the diet. It has been estimated that fish either directly or indirectly may currently account for nearly one-tenth of the world's protein supply. This will be discussed in some detail in chapter 2. In addition, many countries, such as Peru, Iceland, Japan, Canada, and Denmark, critically depend upon fish exports for foreign exchange. Over 90 percent of the value of

Iceland's exports, for example, is fishery products. Without this trade, these economies would suffer greatly. The three leading fishing powers—Japan, China, and the USSR—derived over $2.5 billion each in 1973 from their fish catch measured at dockside. The retail value probably approaches $7–8 billion for each of these countries. Sysoev (1970) reported that in recent years the Soviet fishing industry has employed approximately 11 percent of all workers in their food industry nationally and contributed over 7 percent to gross food output. Furthermore, the common property nature of fishery resources opens up a "Pandora's box" for the overexploitation and, for all practical purposes, the extinction of a valuable renewable natural resource.[2] Jurisdictional disputes over world fisheries have led to open warfare on the high seas; a recent dispute between Iceland and England even threatened the NATO alliance! Moreover, overexploitation depresses national incomes by diverting valuable labor and capital into unproductive pursuits. Jacques Cousteau, the famous marine biologist, has indicated that the combination of fishery overexploitation and ocean pollution could have a significant impact on the interrelationship between the oceans and the land, an impact that could reduce agricultural production with catastrophic consequences. Finally, although fishing for wild stock fish may soon reach the limits of production potential on sustainable basis, aquaculture (or fish-farming) may offer a significant potential for developing nations faced with food shortages (Bell and Canterbery 1976).

In essence, the pressure of a growing world population on natural resources in general and the fisheries in particular has given rise to the specter of Malthusian stagnation and even the possibility of a collapse of civilized life on our planet.[3] A world population that currently doubles every thirty-five years will raise more and more disputes over such common property resources as fisheries.

For these reasons, the world's fisheries should be viewed within the context of a world population/food imbalance that, according to Meadows et al. (1972) portends a collapse in per capita food production early in the twenty-first century.[4] Except for a few economists, social scientists have been only vaguely aware of the fisheries both as an international combat zone for scarce resources and as a vast wasteland of economic inefficiency. Physical scientists need to know more about the economic, social, and political implications of their work on the marine environment. In 1950–1970, the world fishery catch expanded rapidly, increasing from 21 million to 70 million metric tons (there are 2200 pounds to a metric ton and only 2000 pounds to a "short ton"). This amazing growth in the catch—nearly 12 percent annually—greatly increased the per capita supply of fish protein. Since 1970, however, the world catch has either declined or remained constant. But population continues to grow! What role can the fishery resources of the ocean play in augmenting future food supplies?

The purpose of this book is to provide the reader with the foundations of the economics of the fisheries so that these and similar questions may be answered. Although helpful, no previous training in economics is required. The book integrates economic principles into the various subject areas. This chapter will provide some descriptive data on recent trends in world fisheries. Chapter 2 will deal directly with the world food crisis and the past and potential role of fisheries. In addition, we shall discuss the demand analysis so critical in evaluating the future demand for food in general and fishery products in particular. Chapter 3 will deal with a corresponding crisis on the supply side, or the growing depletion of fishery stocks. Chapter 4 asks, why manage the fisheries? As we shall see, the need for management of the fisheries is a direct outgrowth of the nature of fishery supply: this resource is common property in nature—owned by everyone and at the same time by no one. The economic aspects of the Law of the Sea and the recent wave of extended fishery jurisdictions throughout the world will be analyzed in terms of the economic, social, and political impact on the various countries of the world. Chapter 5 examines *a crisis within a crisis,* or the growing threat of environmental pollution and its impact on fishery resources. Chapter 6 deals with what many have called the "sleeping giant" or the use of fishery resources for recreational purposes. This aspect of the fisheries has been greatly underemphasized, principally because of the lack of information on the economic importance of recreational fishing.[5] Chapter 7 discusses aquaculture—fish-farming—which many feel will one day dwarf the catch from the ocean fisheries' wild stocks. What can we reasonably believe about aquaculture's potential to increase world food production? Chapter 8 takes a look at underutilized wild stocks of fish and the economic feasibility of expanding the total fishery supply through their use. Since most of our data will be drawn from the experience of the United States, we shall look at the financial rewards and risks in the U.S. fishing industry. This will be the subject of chapter 9. In general, we shall take a brief look at the fishing firm as a financial venture. Finally, chapter 10 gives a brief history of government participation in the fisheries and its effectiveness in solving economic problems. Throughout the book, our emphasis will be on marine, or saltwater, fisheries. However, the principles are equally applicable to freshwater species, and these will be discussed in varying places in this book.

The Measurement of Trends in Fisheries: The Apples and Oranges Dilemma

Before one launches into statistics on world fishery catches, a fundamental problem should be recognized. The United States and other countries of the world have been adding "apples and oranges" for years in reporting the total tonnage of the fish and shellfish they have caught. Adding tons of lobsters to tons of menhaden is as meaningful as adding pounds of beef to pounds of horsemeat.[6] Yet the practice of adding up tons of widely differing fish pervades the literature and greatly distorts economic evaluations. We have done it above in citing increases in *total* world catch. So before we get entrenched in talking about tons of "fish," let us begin on the right foot! Value is the usual measure used in "adding things up." For example, our gross national product is the total value of all goods and services produced, not the total *weight* of everything produced.

We have tabulated the trend in the estimated *ex vessel value* (i.e., dockside) for twenty-nine major fishing nations and the "rest of the world" over 1967–1973 using data from Olson and Bell (1976). Table 1.1 indicates that the value of the world's fisheries (in U.S. dollars) increased from approximately $6.3 billion to $18.0 billion from 1963 to 1973, or 288 percent (unadjusted for inflation). Figure 1.1 compares the trend in value and catch for the world over 1963–1973. Japan, China (People's Republic of China and Republic of China), the USSR, and the United States were consistently the leading fishing countries in terms of *value of catch.* It is interesting that these estimated rankings have not changed appreciably over the eleven-year period analyzed. Table 1.2 is of great interest: it indicates the *value per metric ton* landed for each of the twenty-nine countries. This reflects the mix of fishery products and its trend for each country in terms of *unit value.* As is evident, many Asian countries, such as Bangladesh, Malaysia, and Australia, have some of the highest values per metric ton because of their heavy landings of shrimp and lobster, which command a higher per unit price on the international market.

Table 1.1: Estimated Exvessel Value of World Fishery Landings by Selected Countries, 1963-73 1/

Country	1963	1964	1965	1966	1967	1968	1969	1970	1971	1972	1973
					Million U.S. dollars						
1. Japan	1,299.8	1,343.3	1,552.4	1,748.6	1,963.1	2,083.0	2,371.8	2,713.3	3.135.0	2/(2,049.2)	2,515.0
2. China 3/	(802.0)	(1,008.0)	(1,223.0)	(1,375.0)	(1,276.0)	(1,447.0)	(1,585.0)	(2,005.0)	(2,417.0)	(2,598.0)	3,060.0
3. USSR	(619.9)	(706.0)	(964.0)	(954.4)	(1,039.9)	(1,130.0)	(1,250.0)	(1,520.0)	(1,709.5)	(1,931.0)	2,646.0
4. USA	376.6	388.8	445.1	471.9	439.1	496.9	526.2	613.0	643.2	703.6	907.4
5. Spain	194.2	212.1	261.7	282.8	325.5	316.7	326.8	378.0	(490.1)	602.1	316.0
6. France	218.2	223.7	236.7	249.0	265.4	265.2	249.4	286.7	315.4	374.2	505.5
7. Philippines	200.0	218.2	193.6	236.6	271.4	369.5	332.5	434.1	650.8	359.2	513.3
8. India	(125.6)	(182.4)	(185.0)	(187.3)	(211.4)	186.2	275.3	341.3	341.9	356.7	443.3
9. Thailand	92.9	119.7	118.7	122.6	146.4	201.9	230.3	286.2	301.1	342.4	357.1
10. Italy	132.4	132.4	155.2	181.0	186.9	186.9	203.8	222.1	255.8	311.4	308.6
11. United Kingdom	155.8	167.5	177.6	180.6	174.7	154.3	165.9	188.0	248.6	283.3	383.8
12. Viet-Nam, S.	77.1	80.9	129.4	122.2	(135.2)	(134.3)	(161.0)	(203.3)	(236.2)	(290.7)	314.4
13. Rep. of Korea	61.7	63.8	73.4	89.5	112.5	143.8	171.9	210.6	254.9	264.3	383.9
14. Canada	119.0	134.8	145.9	160.5	149.5	169.6	169.6	193.4	199.8	236.0	300.1
15. Norway	98.3	109.9	155.1	187.2	166.2	146.3	148.6	196.8	237.9	235.0	355.5
16. Bangladesh	(85.5)	(90.2)	(100.6)	(108.7)	(113.5)	116.7	122.5	155.7	(169.0)	(181.0)	191.6
17. Brazil	(54.4)	(53.9)	(65.7)	67.9	77.0	87.6	98.2	110.1	160.4	(217.1)	250.1
18. Denmark	68.6	80.3	94.2	91.3	91.3	100.5	99.6	114.0	141.7	160.9	238.9
19. Malaysia	58.5	61.6	66.7	77.5	88.6	93.2	90.8	91.0	112.6	151.1	274.7
20. Poland	(31.1)	(36.2)	(42.2)	(48.9)	(50.5)	(62.2)	(66.1)	(82.6)	(107.7)	(119.7)	170.4
21. Germany (FR)	75.3	79.3	89.1	92.5	92.8	94.7	99.9	106.2	122.9	116.5	180.7
22. Netherlands	39.4	44.9	52.6	61.4	61.9	62.6	66.9	77.5	93.6	108.7	152.9
23. Australia	35.6	37.0	43.3	47.2	49.7	61.6	64.8	64.8	83.2	102.2	121.0
24. Peru	70.7	95.1	(88.8)	113.6	124.0	(131.9)	117.4	187.2	173.9	90.8	85.9
25. Burma	63.8	63.8	63.8	63.8	72.8	77.6	82.3	87.5	80.1	81.4	93.1
26. Portugal	60.4	68.7	63.4	67.4	74.2	74.2	74.0	77.4	95.9	185.9	136.0
27. Mexico	57.3	56.7	58.3	66.3	75.8	75.5	75.3	90.8	95.1	128.1	145.0
28. Iceland	42.4	52.2	65.9	69.5	52.1	35.2	52.7	67.7	72.1	71.2	69.2
29. Sweden	42.1	46.8	58.9	54.7	50.8	49.1	46.3	47.6	48.5	49.6	54.2
Total	5,358.6	5,958.2	6,970.3	7,579.9	7,938.2	8,554.2	9,324.9	11,151.9	13,003.9	12,755.3	15,525.4
Other Countries	907.4	1,247.7	1,496.4	1,629.1	1,676.6	1,880.9	1,877.9	1,966.9	2,113.7	1,787.8	2,510.3
World Total	6,266.0	7,205.9	8,466.7	9,209.0	9,614.8	10,435.1	11,202.8	13,118.8	15,117.6	14,543.1	18,035.7

1/ Food and Agriculture Organization of the United Nations (FAO), Yearbook of Fishery Statistics, Catches and Landings, Vol. 36.
2/ Numbers in parentheses are estimates by the National Marine Fisheries Service.
3/ FAO includes both People's Republic of China (Mainland) and Republic of China (Taiwan)

Figure 1.1: World Trend in Ex Vessel Value and Metric Tons of Fishery Landings, 1963-73

Table 1.2: Estimated Exvessel Value Per Metric Ton of Fish Landings by Country, 1963-73

Country	1963	1964	1965	1966	1967	1968	1969	1970	1971	1972	1973
					U.S. dollars per metric ton						
1. Japan	194	212	224	245	248	240	275	290	315	1/199	235
2. China	(119)	(158)	(214)	(227)	(226)	(244)	(260)	(292)	(321)	(343)	(404)
3. USSR	(156)	(158)	(189)	(178)	(180)	(186)	(192)	(210)	(233)	(249)	307
4. USA	136	147	165	188	183	203	211	221	228	266	340
5. Spain	173	176	193	205	223	207	215	246	326	372	201
6. France	294	287	308	309	324	330	324	375	425	478	634
7. Philippines	354	350	282	326	353	391	340	438	620	317	411
8. India	(120)	(138)	(139)	(137)	(151)	122	171	194	185	218	226
9. Thailand	222	208	189	167	173	185	181	198	190	204	211
10. Italy	456	478	439	169	501	514	549	574	657	752	792
11. United Kingdom	162	172	170	169	170	148	153	171	225	262	335
12. Viet-Nam, S.	204	204	345	321	(327)	(337)	(347)	(393)	(402)	(429)	441
13. Rep. of Korea	117	106	115	127	148	(171)	196	226	237	197	232
14. Canada	99	111	116	119	115	113	121	139	155	202	261
15. Norway	71	68	67	65	51	51	60	66	77	74	120
16. Bangladesh	(407)	(392)	(389)	(420)	(428)	438	442	630	(684)	(732)	775
17. Brazil	(157)	(157)	(169)	173	183	177	200	213	276	(368)	424
18. Denmark	81	92	112	112	85	69	78	93	101	112	163
19. Malaysia	253	256	294	286	259	245	262	267	306	421	618
20. Poland	(137)	(137)	(142)	(146)	(149)	(153)	(162)	(176)	(208)	(220)	294
21. Germany (FR)	116	127	141	141	140	139	153	173	242	278	380
22. Netherlands	109	116	140	174	197	194	207	258	291	312	445
23. Australia	503	482	547	532	540	607	706	632	746	868	980
24. Peru	10	10	(12)	13	12	(13)	13	15	16	19	37
25. Burma	177	177	177	177	191	196	199	202	181	180	201
26. Portugal	112	114	114	133	133	147	162	155	207	243	302
27. Mexico	236	228	227	231	217	207	213	235	224	279	301
28. Iceland	54	54	55	56	58	59	76	92	95	98	131
29. Sweden	124	125	155	166	145	150	167	162	203	219	239
Weighted Average	131.0	134.9	156.0	159.3	157.5	161.6	179.8	189.5	221.1	235.8	287.7

Source: Food and Agriculture Organization of the United Nations, Yearbook of Fishery Statistics, Catches and landings, Vols. 26 and 36. Computed from total value and quantity of landings.

1/ Numbers in parentheses are estimates by the National Marine Fisheries Service.

Table 1.3 is the aggregate fish catch by country expressed in metric tons and reflects the process of adding apples and oranges. However, these statistics are used throughout the fisheries literature, and we present them here to illustrate their shortcomings. The most striking thing about these three tables—total value, value per metric ton, and quantity by country—is the wide dispersion in ex vessel value per metric ton: from $37 for Peru to $980 for Australia in 1973. The value of a ton of fish landed in Australia was twenty-seven times greater than a ton of fish landed in Peru. In table 1.3 Peru was the leading fishing nation in 1971, landing over 10.6 million metric tons. For many analysts, this figure has often been a key indicator of the importance of a nation's fishing success. Peru's catch yielded nearly $0.174 billion dollars, which ranked it sixteenth among major fishing countries. The major reason for the difference in ranking is the kind of fish caught. In Peru, nearly all of the catch was anchovies, which have a low value per metric ton and which are used for the production of fish meal and oil. In 1971, the United States landed over 2.8 million metric tons (about 27 percent of Peruvian landings) but ranked fifth in the world by value.

The confusion between pounds and values is a simple error in logic that has caused governments to accuse fishery officials of failing to raise the total metric tons of fish caught. For example, irate fishermen and others interested in fisheries have written letters condemning a U.S. policy that has allowed our rank to slip over the past ten years in terms of total metric tons of fish landed. Examiners of fishery agencies from both the executive and congressional branches of government have repeatedly cited the fact that the U.S. catch in metric tons has remained unchanged from 1963 to 1973. Some of those at the highest levels of government have accepted the fallacy that you apparently can add apples and oranges. Table 1.1 shows a more incisive trend: the *value* of the U.S. catch increased by approximately 141 percent over the 1963–1973 period, or an average of 12.8 percent annually. Inflation, or the increase in the U.S. wholesale price index (WPI), increased by approximately 46 percent over the same period, or an average of 4.2 percent a year.[7] Thus, the real value of the U.S. fishery catch increased by an average of 8.6 percent annually, even though the total physical catch remained constant.[8] This means that the "real contribution or value" of the U.S. fish-harvesting sector was increasing considerably faster than the average for all other goods and services in the economy.

Table 1.3: World Fishery Catch by Selected Countries, Expressed in Thousands of Metric Tons, 1963-73

Country	1963	1964	1965	1966	1967	1968	1969	1970	1971	1972	1973
1. Japan	6,698.5	6,350.7	6,928.8	7,131.6	7,901.6	8,694.2	8,638.5	9,366.4	9,949.6	10,272.6	10,701.9
2. China 2/	6,715.9	6,372.4	5,714.7	6,056.0	5,645.2	5,932.4	6,095.9	6,868.0	7,530.2	7,574.0	7,574.0
3. USSR	3,977.2	4,475.7	5,099.9	5,348.8	5,777.2	6,082.1	6,498.4	7,252.2	7,337.0	7,756.9	8,619.0
4. USA	2,777.0	2,647.1	2,696.2	2,515.3	2,405.5	2,451.7	2,489.1	2,776.5	2,819.5	2,649.5	2,669.9
5. Spain	1,125.3	1,203.5	1,355.0	1,379.7	1,459.8	1,533.3	1,522.0	1,538.8	2/1,505.1	1,616.9	1,570.4
6. France	742.3	730.4	767.6	804.8	819.7	803.1	770.5	764.4	741.7	783.0	796.8
7. Philippines	565.6	623.5	685.8	726.0	769.2	944.6	978.1	992.0	1,049.7	1,131.9	1,248.5
8. India	1,046.3	1,320.0	1,331.3	1,367.3	1,400.4	1,525.6	1,606.8	1,756.1	1,841.6	1,637.3	1,958.0
9. Thailand	418.7	577.0	626.7	732.6	846.6	1,088.9	1,269.9	1,447.7	1,587.1	1,678.9	1,692.3
10. Italy	290.4	324.6	353.8	369.0	373.1	363.8	370.9	386.7	389.5	414.4	389.7
11. United Kingdom	960.0	974.3	1,047.1	1,068.8	1,026.2	1,040.2	1,083.0	1,099.0	1,107.3	1,081.5	1,144.4
12. Viet-Nam, S.	378.6	397.0	375.0	380.5	410.7	410.0	463.8	517.4	587.5	677.7	713.5
13. Rep. of Korea	529.6	599.5	640.4	704.1	760.7	841.1	879.1	933.6	1,073.7	1,338.6	1,654.6
14. Canada	1,197.0	1,211.0	1,262.3	1,346.0	1,295.7	1,498.7	1,404.8	1,389.0	1,289.8	1,169.1	1,151.6
15. Norway	1,387.9	1,608.1	2,311.8	2,872.3	3,265.7	2,855.7	2,490.7	2,980.4	3,074.9	3.162.9	2,974.5
16. Bangladesh	3/(210.0)	(230.0)	2/258. 7	258.7	265.1	266.4	277.3	247.2	2/247.2	2/247.2	247.2
17. Brazil	346.2	343.6	388.8	393.1	419.7	495.1	492.2	517.3	580.7	2/589.9	589.9
18. Denmark	847.9	871.1	840.8	850.8	1,070.4	1,466.8	1,275.4	1,226.5	1,400.9	1,442.0	1,464.7
19. Malaysia	230.8	241.1	226.9	270.9	341.7	380.9	346.7	340.5	367.8	358.7	444.7
20. Poland	226.7	264.3	297.5	334.9	338.9	406.7	408.1	469.3	517.7	544.0	579.6
21. Germany (FR)	647.2	624.3	632.7	657.3	661.5	682.3	651.6	612.9	507.6	418.8	475.2
22. Netherlands	361.0	387.8	377.1	353.1	314.5	323.3	323.2	300.7	321.2	348.3	343.8
23. Australia	70.8	76.7	79.2	88.8	92.1	101.4	91.8	102.6	111.5	117.8	123.5
24. Peru	6,899.0	9,116.5	7,631.9	8,844.5	10,198.6	10,555.5	9,243.6	12,612.9	10,606.1	4,768.3	2,299.3
25. Burma	360.0	360.0	360.0	360.0	380.7	396.1	413.9	432.4	442.7	453.3	463.4
26. Portugal	539.7	603.7	553.6	506.0	559.8	506.3	457.0	498.4	462.7	436.7	452.7
27. Mexico	243.0	249.0	256.4	287.1	349.8	364.3	353.2	386.8	425.2	459.2	482.1
28. Iceland	784.5	972.7	1,198.9	1,240.3	897.6	600.6	689.5	733.8	684.9	726.5	906 2
29. Sweden	340.5	375.5	379.6	329.6	350.9	328.1	278.0	294.8	238.5	226.7	226.9
Total	40,918.5	44,181.1	4,4678.5	47,577.9	50,398.6	52,939.2	51,863.0	58,844.1	58,808.9	54,083.5	53,958.3
Other Countries	5,681.5	7,718.9	8,521.5	9,722.1	10,001.4	10,960.8	10,837.0	11,155.9	11,391.1	11,416.5	11,741.7
World total	46,600.0	51,900.0	53,200.0	57,300.0	60,400.0	63,900.0	62,700.0	70,000.0	70,200.0	65,500.0	65,700.0

1/ Food and Agriculture Organization of the United Nations (FAO). Yearbook of Fishery Statistics, Catches and Landings, Vol. 36

2/ Estimated by FAO

3/ Numbers in parentheses are estimates by the National Marine Fisheries Service.

Physical and Dollar Value of the World Fishery Catch: Sorting Out the Apples and Oranges

Table 1.4 ranks all twenty-nine countries by *value* of fish landings by year from 1963 to 1973. Japan, China, and the USSR were ranked in the top three during all eleven years. The United States was fourth in ten out of the eleven years. The ranks of the other countries were fairly stable on the whole. However, Poland and Brazil demonstrated a consistent increase in rank. In 1973, the first five fishing powers caught nearly 54 percent of the total value of the world's catch. Figure 1.2 shows how each country's share of the value of the world catch changed over the 1963–1973 period. Note that the dominance of the big three (Japan, China, USSR) increased from 43 percent of the value of the catch to 46 percent in 1973. Japan and the USSR greatly expanded their deepwater fishing fleets, although Japan's share declined over 1963–1973. Both the Soviets and Japanese can be seen off the coast of most countries that have abundant fishery resources. Because of their presence, in fact, many countries have extended and will extend their national fishery jurisdiction unilaterally.

Figure 1.3 shows the 1973 ranking of the value of the catch per metric ton for selected countries. The weighted Food and Agriculture Organization (FAO) average is $288 per ton. The range is from $980 per metric ton in Australia to $37 in Peru. Hence, it is not just the *volume* of the catch but also the *composition* (in terms of unit values) of the catch that is important in producing revenue for a country. Judging by figure 1.3, Australia, Italy, Bangladesh, and France land high-value species. Note that the USSR, China, and the United States have above average value per metric ton but that Japan is somewhat below the world weighted average of $288. Hence, for these four countries, a combination of relatively high value per metric ton and overall volume accounts for their world leadership in the fisheries. The U.S. fish catch is so diverse—from lobsters to menhaden—that its overall value per metric ton is only slightly above the world average. Two factors are important here. First, some countries are naturally endowed with high-unit-value fish near their shores, for example, Australia and Italy. Second, many countries, such as the USSR and Japan, actively seek to improve on their own coastal fishery endowments by fishing off the coasts of foreign countries. This has led to international conflict and in some cases open warfare.

We have thus far been looking primarily at comparisons among countries for individual years. Adjusted for inflation, the real

Table 1.4:Ranking of 29 Countries by Value of Fish Landings, 1963 - 1973 1/

Rank	1963	1964	1965	1966	1967	1968	1969	1970	1971	1972	1973
1	Japan	Japan	Japan	Japan	Japan	Japan	Japan	Japan	Japan	China	China
2	China	China	China	China	China	China	China	China	China	Japan	USSR
3	USSR	USSR	USSR	USSR	USSR	USSR	USSR	USSR	USSR	USSR	Japan
4	USA	USA	USA	USA	USA	USA	USA	USA	Phil.	USA	USA
5	France	France	Spain	Spain	Spain	Phil.	Phil.	Phil.	USA	Spain	Phil.
6	Phil.	Phil.	France	France	Phil.	Spain	Spain	Spain	Spain	France	France
7	Spain	Spain	Phil.	Phil.	France	France	India	India	India	Phil	India
8	UK	India	India	India	India	Thail.	France	France	France	India	ROK
9	Italy	UK	UK	Norway	Italy	Italy	Thail.	Thail.	Thail.	Thail.	UK
10	India	Italy	Italy	Italy	UK	India	Italy	Italy	Italy	Italy	S.V. Nam
11	Canada	Canada	Norway	UK	Norway	Canada	ROK	ROK	ROK	UK	Thail
12	Norway	Thailand	Canada	Canada	Canada	UK	Canada	S.V. Nam	UK	S.V. Nam	Norway
13	Thail.	Norway	S.V. Nam	Thail.	Thail.	Norway	UK	Norway	Norway	ROK	Spain
14	Bangla.	Peru	Thailand	S.V. Nam	S.V. Nam	ROK	S.V. Nam	Canada	S.V. Nam	Canada	Italy
15	S.V.Nam	Bangla.	Bangla.	Peru	Peru	Peru	Norway	UK	Canada	Norway	Canada
16	Germany	Denmark	Denmark	Bangla.	Bangla.	S.V. Nam	Bangla.	Peru	Peru	Bangla.	Bangla.
17	Peru	Germany	Germany	Germany	ROK	Bangla.	Peru	Bangla.	Bangla.	Brazil	Malaysia
18	Denmark	Peru	Peru	Denmark	Germany	Denmark	Germany	Denmark	Brazil	Denmark	Brazil
19	Burma	ROK	ROK	ROK	Denmark	Germany	Denmark	Brazil	Denmark	Malaysia	Denmark
20	ROK	Malaysia	Malaysia	Malaysia	Malaysia	Malaysia	Brazil	Germany	Germany	Mexico	Germany
21	Portugal	Iceland	Iceland	Iceland	Brazil	Brazil	Malaysia	Malaysia	Malaysia	Poland	Poland
22	Malaysia	Brazil	Brazil	Brazil	Mexico	Burma	Burma	Mexico	Poland	Germany	Neths.
23	Mexico	Burma	Burma	Portugal	Portugal	Mexico	Mexico	Burma	Portugal	Neths.	Mexico
24	Brazil	Portugal	Portugal	Mexico	Burma	Portugal	Portugal	Poland	Mexico	Portugal	Portugal
25	Iceland	Sweden	Sweden	Burma	Neths.	Neths.	Neths.	Neths.	Neths.	Australia	Australia
26	Sweden	Mexico	Mexico	Neths.	Iceland	Poland	Poland	Portugal	Australia	Cuba	Cuba
27	Neths.	Neth.	Neth.	Sweden	Sweden	Australia	Australia	Iceland	Burma	Peru	Burma
28	Greece	Australia	Australia	Poland	Poland	Sweden	Iceland	Australia	Iceland	Burma	Senegal
29	Australia	Poland	Poland	Australia	Australia	Hungary	Sweden	Cuba	Cuba	Iceland	Peru

1/ The 29 countries with the highest value of fish landings in 1973 were first ranked. The same 29 countries were then ranked by value for each year from 1963 to 1972. Thus, the same countries were ranked each year, but it does not necessarily mean that these countries were always in the top 29 (1963-1972).

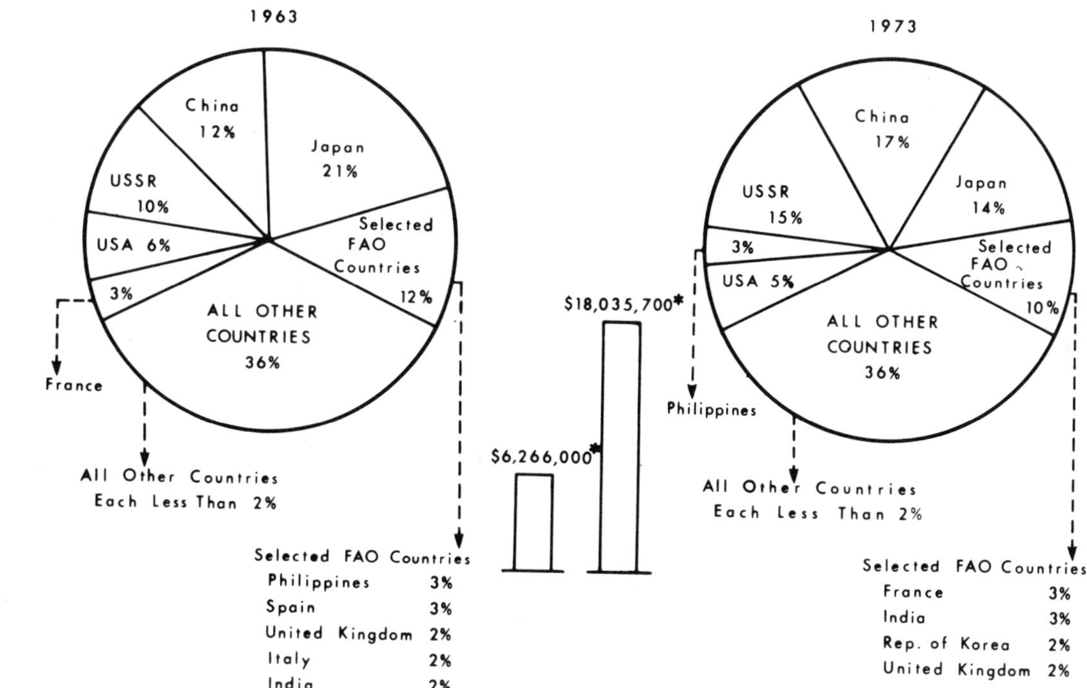

Figure 1.2: Ex Vessel Value of World Fishery Landings by Selected Countries as Reported by FAO, 1963 and 1973

*thousands of dollars

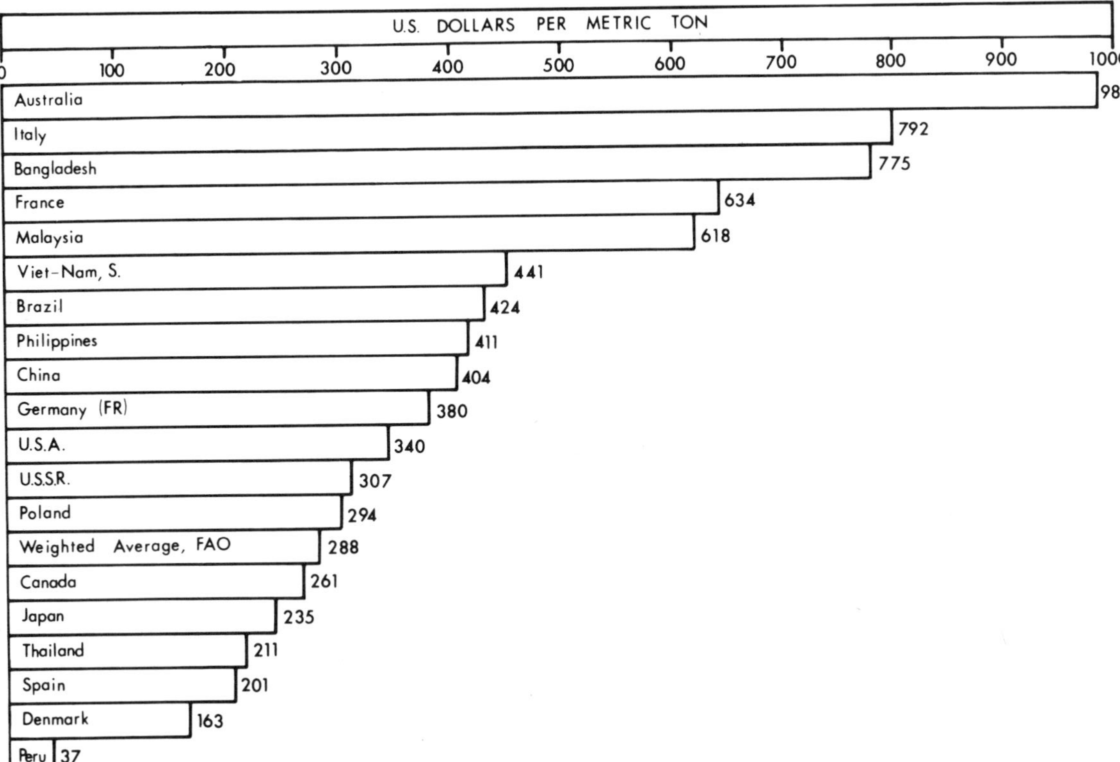

Figure 1.3: Ex Vessel Value Per Metric Ton of Fish Landings By Selected Country, 1973

value of the world fish catch has increased by 6.8 percent annually over the 1963–1973 period.[9] Table 1.5 shows the *annual compound rate of growth* of the real value (i.e., constant dollars) and quantity (i.e., metric tons) for twenty-nine nations over 1963–1973. The Republic of Korea, Poland, Brazil, and Thailand have experienced compound growth rates of 12 percent or greater. Two of the big three, China and the USSR, did considerably better than the world average, although Japan slipped below the world average growth rate. Although below the world average, the U.S. real value landings increased by 4.8 percent a year. Countries such as the Netherlands, Bangladesh, the United States, Canada, and France showed no increase in their *physical catch,* but all showed significant increases in the real value of their fishery catch. This illustrates the importance of distinguishing between real value and physical quantities. Of the twenty-nine countries, twenty showed a positive and statistically significant trend in their real value per metric ton of fish landed. The Netherlands, the Federal Republic

Table 1.5: Annual Compound Rates of Growth of Fishery Landings in Terms of Real Value, Real Value per Metric Ton and Quantity by Selected Countries with FAO Average, 1963-73

Rank (By Value)	Country	Value of Landings (1967 Prices)	Value Per Metric Ton (1967 Prices)	Quantity (Metric Tons)
1	S. Korea	17.0	5.7	10.7
2	Poland	13.5	3.6	9.6
3	Brazil	13.2	6.5	6.3
4	Thailand	12.0	3.1	15.5
5	S. Viet Nam	11.2	4.0	6.9
6	U.S.S.R.	10.2	2.1	7.5
7	China	9.6	7.2	2.2
8	Malaysia	9.3	2.8*	6.4
9	Australia	9.1	3.5	5.4
10	Netherlands	8.7	10.1	-1.3*
11	India	8.4	3.0	5.2
12	Philippines	7.9	-.2*	7.8
13	Norway	6.9	.1*	6.7
14	Denmark	6.9	.1*	6.7
15	World Total, FAO	6.8	3.9	3.7
16	Mexico	6.1	-1.3*	7.6
17	Italy	5.8	4.5*	2.6
18	Portugal	5.7	6.4	-2.7
19	Spain	5.3	2.0*	3.1
20	Bangladesh	5.3	4.5	1.1*
21	U.S.A.	4.8	4.6	.2*
22	Canada	4.5	4.7	-.1*
23	Japan	4.4	-1.2*	5.5
24	U.K.	3.9	2.4*	1.4
25	France	3.6	3.5	.9*
26	Germany (FR)	3.5	7.3	7.1
27	Sweden	-2.7	2.7	-5.5
28	Burma	.8*	-2.2	3.0
29	Iceland	.4*	5.9	3.1*
30	Peru	.3*	6.5	-6.1*

*Not statistically significant from zero at the 5% level.

Source: Tables 1.1, 1.2 and 1.3

of Germany, China, Peru, Brazil, and Portugal all showed annual compound growth rates of 6.4 percent or more; the FAO world average was 3.9 percent. The increase in the weighted average price of fish (value per metric ton) in the world, in constant dollars, indicates that the cost of catching fish and shellfish is increasing faster than it is for other commodities. Thus, fishery products are becoming relatively more valuable than most other commodities. *Fishermen experience rising real value, while consumers find that fish are becoming relatively more expensive.*

Finally, table 1.6 gives some idea of the trends in fisheries in certain areas of the world. The USSR, Oceania, and South America have experienced the most rapid growth in the real value of landings, although the physical landings in South America showed no statistically significant trend. In this last case, the real value per metric ton increased more rapidly than in any other continent or region. As we have indicated before, a country can increase the value of its landings by unit value—value per metric ton—or by physical volume or by both. Africa and Asia showed no statistically significant trend in value per metric ton; both, however, showed growth in the real value of their landings owing to increases in physical production.

Summary

The purpose of this chapter is twofold: (1) to give a general idea of the coverage and scope of this book, and (2) to acquaint the reader with the trends in world fishery value and production. There is no question that world fishery resources, both marine and freshwater, are part of the population/food crisis that is prevalent today and quite likely to get worse over the next fifty to seventy-five years. The fisheries play a major role in supplying protein to the world and have increased in real value over the

Table 1.6: Annual Compound Rates of Fishery Landings in Terms of Real Value, Real Value Per Metric Ton and Quantity by Continent or Region, 1963-73

Rank (By Value)	Continent	Value of Landings (1967 Prices)	Value Per Metric Ton (1967 Prices)	Quantity (Metric Tons)
1	U.S.S.R.	10.2	2.1	7.5
2	Oceania	9.1	2.9	6.0
3	South America	7.6	11.1	-3.5*
4	Asia	6.4	1.5*	4.8
5	North and Central America	5.9	4.2	1.6
6	Africa	5.1	-.4*	5.2
7	Europe	4.3	1.5*	2.7

*Not statistically significant from zero at the 5% level.

1963–1973 period. Fishery production has been concentrated in Japan, China, the USSR, and the United States in terms of real value. Poland and Brazil have consistently moved up in the rankings over the 1963–1973 period. Although below the FAO compound growth of 6.8 percent in real value, the real value of the U.S. catch has grown at a respectable rate of 4.8 percent.

We have consistently pointed out the difficulty of comparing one country's total physical catch to that of any other. The "apples and oranges" problem is so ingrained in the fisheries literature that it may take years for people working in this area to think in monetary terms. We hope this chapter has made a contribution toward this end, even though the data base is weak and even though heroic assumptions have sometimes had to be made.

For the world as a whole, physical fishery production leveled off after 1970 and has not changed for five consecutive years. Because of increased demand pressures, the real value of these resources has risen. The real value per metric ton for the world grew at an annual compound growth rate of 3.9 percent. This is probably indicative of the inability of world supply to keep pace with the demands placed upon a renewable, but limited, fishery resource. We must give reasons for these trends, which are largely descriptive in nature. Chapter 2 will explore factors that influence the demand for fishery products; one of the most important is the population crisis. Chapter 3 will survey the resource crisis, or the supply side of the problem. The trends discussed in this chapter will be given greater meaning and substance after we have developed some analytical insight into the "crisis at sea."

Notes

1. Direct fish consumption is different from indirect consumption: the former is restricted to fish that are eaten by the consumer after some degree of processing; the latter include fish sold for industrial purposes and converted to fish meal and oil, which are used for animal feed and as an input for other products.

2. Ocean fisheries are "owned" by no country; therefore, they are common property as opposed to private property. This important distinction and its implications will be discussed in chapter 4.

3. In 1798, Thomas Malthus published his now famous *Essay on Population*, which theorized that although population would grow geometrically, food supplies could only be increased arithmetically. Hence, food production per individual would fall to a subsistence level, thereby producing stagnation in both output and population growth.

4. In April 1968, a group of thirty individuals from many countries—

scientists, educators, economists, humanists, industrialists, and national and international civil servants—gathered in the Accademia dei Lincei in Rome. They met to discuss a subject of staggering scope—the present and future predicament of man. Out of this meeting grew the Club of Rome, whose purpose is to foster an understanding of our global socioeconomic system and to call attention to policy-makers and the public complex problems facing man over the next seventy-five years.

5. The major use of fishery resources has been as a source of food. However, in many affluent countries, they are becoming more economically important as a source of recreation.

6. Menhaden is a sardinelike fish caught not for human but industrial consumption; it ultimately becomes animal feed.

7. We have used the wholesale price index, which indicates inflation at the producer's level (i.e., fishermen, manufacturers), since our fishery value is not retail value. Hence, we did not use the consumer price index. However, both indexes are highly correlated.

8. The *compound* growth in the *real* value of the U.S. catch was 4.8 percent a year. A compound rate of growth is lower than *average* annual growth and is considered a more valid measure of growth rates.

9. The purpose of the deflation was to measure growth rates in constant or real dollars (without inflation). All countries' value of landings was divided by the U.S. wholesale price index (WPI) to adjust for inflation. Although inflation rates differ among countries, fishery products must compete in an international market where a world average inflation is involved. No index of overall world inflation is available, so the U.S. rate of inflation was selected, since the United States is a major producer and consumer of fishery products. In addition, all values are in U.S. dollars (adjusted by exchange rates, which reflect different rates of inflation), which is an additional reason for using the U.S. index for all countries.

References

Bell, Frederick W., and Canterbery, E. Ray. 1976. *Aquaculture for developing countries.* Cambridge, Mass.: Ballinger Publishing Co.

Food and Agriculture Organization. 1973. *Yearbook of fishery statistics.* Vol. 36.

Meadows, Donella H.; Meadows, Dennis L.; Randers, Jorgen; Behrens, William W., III. 1972. *The limits to growth.* New York: Universe Books.

Olson, Fred L., and Bell, Frederick W. 1976. Apples, oranges and fish: The value of the world's fisheries. Unpublished manuscript for the National Marine Fisheries Service.

Sysoev, N. P. 1970. *Economics of the Soviet fishing industry.* Translated for the National Marine Fisheries Service.

2. *The Food Crisis: The Demand for Fishery Products*

Famine threatens the continents of Asia and Africa. According to some experts, within a year 10 to 30 million people are expected to die of starvation or of diseases made fatal by malnutrition. Several West African nations and India, Pakistan, Indonesia, and the Philippines have been hit by torrential floods and by blistering droughts that have caused crop failures. The terrible tragedy underlying today's population/food dilemma is that even if we could feed the famished, we would only defer the starvation of far greater numbers—*unless* food supplies keep pace with, or exceed, population growth.

The human population is now increasing at about 2 percent a year. As Malthus suggested, geometric population growth is clear, especially over the last two hundred years. At the beginning of the Christian era, the death rate was nearly fifty per 1000. Today, it stands around ten per 1000. Improvements in medical science and dietary habits have been principally responsible for the acceleration in population growth as the death rate has declined. The world's human population is now doubling every thirty-five years. Even more important, the areas of the world where food supplies are scarce are precisely those areas where population growth rates exceed the world average, especially in Central America, South America, and North-Central Africa.

According to Meadows et al. (1972), the present rate of resource exploitation and exponential population growth will result in a collapse of the world's economic system, that is, a collapse in food production per capita.[1]

World Food Supplies and Demand Pressures

There are more controversies, even among experts, about the potential for food production than about any other factor in the population problem. In the United States, for example, agricultural productivity has been among the highest for any economic sector. Moreover, only Canada, Australia, New Zealand, Argentina, and Uruguay can match the U.S. standard of diet. Yet these countries constitute only 9 percent of the world population. Most experts agree that North America, as the only major food surplus area, would have to contribute heavily to the have-not countries for a considerable time to avoid mass starvation. At present, 80–93 percent of the world's acceptable agricultural land is being used. Worldwide expansion of agricultural hectarage was a principal source of increased production until 1950, but during the past decade the production increment from this source fell off to only one-fifth on a worldwide basis. Figure 2.1 shows the historical trends in food consumption for the more-developed and less-developed countries of the world. The more-developed countries have experienced substantial increases both in *aggregate* and *per capita* food consumption. The less-developed countries have matched the developed areas in increasing *aggregate* food production, but they failed to increase *per capita* food consumption over the 1961–1973 period.

The same demand pressures of population growth have placed added emphasis upon increasing production from the sea. The worldwide growth in the real value of fishery production has been increasing at a compound annual rate of 6.8 percent over 1963–1973. Over the same period, the world's landings expressed in metric tons have increased by 3.7 percent a year. At best, total physical catch is a crude indicator of the supply of protein from the world's oceans. More important, the rate of increase of the physical catch has been *declining,* which explains the rise in the real value of fishery products. This will be discussed at some length in chapter 3. What is the present role of fish in world food consumption?

The Role of Fish in Food Consumption

Although fish production continued up to 1970 to respond fairly well to the food needs of an expanding world population, one should place fish within the general scope of food consumption. This will give the reader an overall perspective. Table 2.1 shows the commodities the world consumed and their caloric and protein yield averaged over 1969–1971. In terms of protein, *directly edible* fish and shellfish supply about 5.4 percent of the world's protein and, when adjusted for protein quality, as much

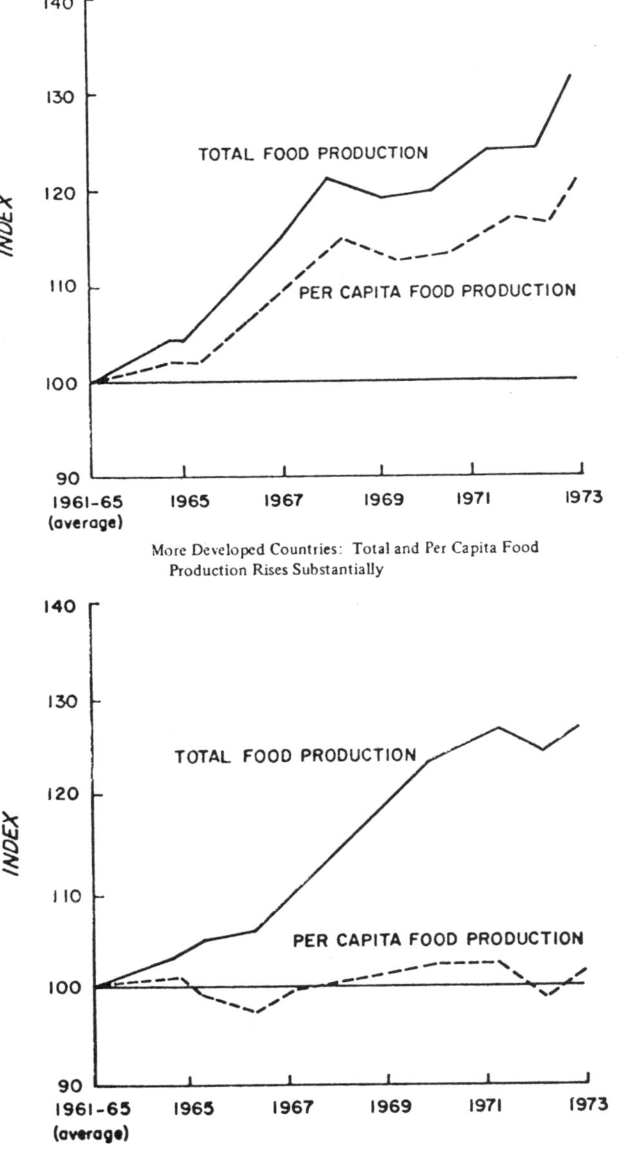

More Developed Countries: Total and Per Capita Food Production Rises Substantially

Less Developed Countries (excluding Communist Asia): Population Growth Absorbs Food Production Increases

Figure 2.1: Comparison of Aggregate and Per Capita World Food Consumption Between More and Less Developed Countries

Source: U.S. Department of Agriculture

Table 2.1: World Consumption by Major Commodity Groups

Commodity	Consumption 1969-71[a] (Million Metric Tons)	Share of Total Quantity (%)	Calorie Consumption 1969-71 (Trillion Calories)	Share of Total Calories (%)	Protein Consumption 1969-71 (Million Metric Tons)	Share of Total Protein (%)	Adjusted Protein Consumption[b] 1969-71 (Million Metric Tons)	Share of Total Adjusted Protein (%)
Cereals[c]	787	36.3	2800	62.5	75.8	55.8	42.9	46.3
Wheat[c]	332	15.3	1162	25.9	40.5	29.8	21.9	23.2
Rice[c]	310	14.3	1116	24.9	20.8	15.3	13.7	14.8
Coarse grains[c]	145	6.7	522	11.7	14.5	10.7	7.7	8.3
Starchy roots	279	12.9	254	5.7	4.5	3.3	2.6	2.8
Sugar, centrifugal (raw basis)	70	3.2	271	6.1	–	–	–	–
Sugar, noncentrifugal	12	0.6	42	0.9	0.1	–	–	–
Pulses, nuts and oilseeds	52	2.4	181[f]	4.0	13.0[g]	9.6	7.4	8.0
Vegetables	223	10.3	49	1.1	2.2	1.6	0.8	0.9
Fruits	158	7.3	58[g]	1.3	1.3	1.0	0.7	0.8
Meat	107	4.9	251	5.6	15.5	11.4	15.5	16.7
Beef and veal	39	1.8	78	1.7	6.9	5.1	6.9	7.5
Mutton and lamb	7	0.3	17	0.4	1.1	0.8	1.1	1.2
Pigmeat	36	1.7	135	3.0	4.3	3.2	4.3	4.6
Poultry	16	0.7	21	.0.5	3.2	2.4	3.2	3.5
Eggs	19	0.9	27	0.6	2.4	1.8	2.4	2.6
Fish	41	1.9	28	0.6	7.4	5.4	7.3	7.9
Whole milk, including butter[d]	389	17.9	226	5.0	13.6	10.0	13.0	14.0
Fats and oils[e]	33	1.5	291	6.5	–	–	–	–
Butter (fat content)	5	0.2	44	1.0	–	–	–	–
Vegetable oils	22	1.0	158	3.5	–	–	–	–
TOTALS	2170		4478		135.8		92.6	

Sources: Consumption figures from FAO, *Assessment of the World Food Situation*, Rome, November 1974, p. 80. Calorie content from FAO, *Agricultural Commodities – Projections for 1975 and 1985*, Rome, 1967, p. 55. Protein content and data to calculate protein adjustments from FAO, *Amino Acid Content of Foods and Biological Data on Proteins*, Rome, 1970.

[a]1969-71 averages. [b]Protein consumption for each food type is adjusted for quality using the relevant chemical score calculated in accordance with accepted FAO procedures. [c]Including milk products in liquid milk equivalent. [d]Including animal fats. [e]Calorie content used is an average of various pulses, nuts and oilseeds. [f]Protein content used is an average of various pulses, nuts and oilseeds. [g]Calorie content used is an average of various fruits. [h]Protein content used is an average of various fruits.

as 8.0 percent.[2] The industrial uses of fish manifest themselves as edible protein through poultry and pigs, which together supply 5.6 percent of the world's protein; of this perhaps as much as 1 percent is due to fish used as feed. Thus, fish may supply as much as 9.0 percent of the world's protein adjusted for protein quality.

In contrast to the protein yield, fish supplied only 0.6 percent of the world's calories. The balance between calories and protein will be discussed later in this chapter. The high protein yield of fish is readily apparent. Note that the conversion of all commodities to common denominators—calories and protein—enables comparison of their relative importance in the diet. The first column of table 2.1 is meaningless for intercommodity comparisons. The "apples and oranges" problem thus continues on a somewhat different plane.

To avoid the "apples and oranges" problem, we can disaggregate the world fish catch into species categories. The UN Food and Agriculture Organization continues to express per capita consumption of all fish in terms of kilograms (or pounds) by country. Hence, intercountry comparisons are, at best, only rough indicators of fish consumption. More meaningful comparisons can be made by disaggregation.

We have divided the aggregate catch into eleven major *food* fish categories and an industrial category called "fish meal."[3] The latter is not used for human consumption but is fed to livestock and other animals for indirect production of food, whether for humans or pets. Note the term *round weight,* which is equivalent to *live weight,* or the weight of the fish when first caught. International statistics are usually expressed in round weight. The edible or "meat weight" will vary from species to species depending on the species' particular makeup (i.e., biological structure). For example, scallops are caught "live" and usually shucked aboard ship. Approximately 14 percent of the live or landed scallops are edible or are sold as scallop meat. However, the term *edible weight* refers to what the consumer actually eats. A more important reason for expressing fish in round weight is that consumption can be related directly to the resource, or what is known as the biomass. The biomass is the total quantity of a particular species living in a particular area. In chapter 3, the resource base (i.e., available supply) is always expressed as round, or live, weight since biological behavior deals with live fish. Hence, the consumption-supply relation can be expressed in a common term—live, or round, weight.

What does table 2.2 tell us about the consumers of fishery products? The species categories in table 2.2 account for approximately two-thirds of the world's fishery catch expressed in physical weight. We have restricted our analysis to the major marine, or saltwater, fish. Although our primary emphasis, as noted in chapter 1, will be on marine fisheries, the study of fisheries economics includes freshwater fish as well. In terms of value, the marine species in table 2.2 account for at least 90 percent of the world's ocean catch. Of the total world catch in 1973, 9.76 million metric tons were from fresh water. Thus, ocean, or marine, fisheries accounted for 55.94 million metric tons. Table 2.2 accounts for 42.07 million metric tons, or almost 76 percent of the total *ocean* catch. The remaining marine catch is distributed among many hundreds of species categories. In the food fish categories, the United States is the leading consumer in seven out of the eleven categories. Japan and the USSR are also major consumers in practically every fishery category. The four major consuming countries account for over two-thirds of the world consumption of groundfish, tuna, salmon, halibut, and crabs, and over ninety percent of the clams, scallops, and oysters.[4] The dominance of the USSR, Japan, and the United States in the consumption of marine fish is striking. In many cases, the U.S. per capita consumption is not

Table 2.2: Rank of Four Leading Countries in the Consumption of Selected Fish Products, 1973 (Round of Live Weight)

Species Category	Country	Total Consumption (mil. lbs.)	Percent of Total	Per Capita Consumption (lbs.)
1. Groundfish[1]	U.S.S.R.	7,841	30	31.4
	Japan	7,071	27	65.6
	U.S.A.	2,064	8 [2]	9.8
	U.K.	1,539	6 (68)	27.5
	World Total	26,050	100	6.8
2. Tuna	U.S.A.	1,379	37	6.6
	Japan	839	22	7.8
	S. Korea	266	7	8.0
	France	126	3 (63)	2.4
	World Total	3,763	100	1.0
3. Salmon	Japan	371	36	3.4
	U.S.S.R.	179	17	.7
	U.S.A.	174	17	.8
	Canada	129	12 (82)	5.8
	World Total	1,044	100	.3
4. Halibut	U.S.S.R.	82	31	.3
	U.S.A.	41	15	.2
	Japan	33	12	.3
	Canada	217	10 (68)	.1
	World Total.	265	100	.05
5. Sardines/Herring[3]	U.S.S.R.	1,855	16	7.4
	Denmark	1,131	10	226.2
	Japan	1,058	9	9.8
	S. Africa	872	8 (43)	36.8
	World Total	11,450	100	3.0
6. Shrimp	U.S.A.	699.5	29	3.3
	India	368	15	.6
	Thailand	198	8	5.0
	Japan	195	8 (60)	1.8
	World Total	2,441	100	.6
7. Lobster	U.S.A.	205	48	1.0
	Chile	56	13	.5
	U.K.	34	8	.6
	France	29	7 (76)	.6
	World Total	424	100	.1
8. Crabs	U.S.A.	238.9	30	1.1
	Japan	170	21	1.6
	France	44	5	.8
	S. Korea	32	4 (60)	1.0
	World Total	805	100	.2
9. Clams	U.S.A.	641.3	47	3.1
	Japan	450	33	4.2
	Malaysia	88	6	7.6
	S. Korea	77	6 (92)	2.3
	World Total	1,374	100	.4
10. Scallops	U.S.A.	246.4	51	1.2
	Japan	120	27	1.1
	France	66	15	1.3
	U.K.	19	4 (97)	.3
	World Total	480	100	.1
11. Oysters	U.S.A.	955	53	4.5
	Japan	292	16	2.7
	France	236	13	4.5
	S. Korea	194	11 (93)	5.8
	World Total	1,812	100	.5
12. Fish Meal (Industrial)	Japan	8,773	21	81.4
	U.S.S.R.	5,262	12	21.1
	U.S.A.	4,067	10	19.3
	U.K.	2,810	7 (50)	50.3
	World Total	42,659	100	11.2

1. Includes Flounders, soles, cods, hakes, haddock, etc.
2. Number in parentheses is cumulate percentage of world total for the four leading consumers
3. This category unfortunately overlaps with fishmeal. FAO does not break down food-fish and industrial use of landings. Sardine/Herring category (excluding Anchovy) is a combination food-industrial category and per capita calculations are indicative of this.

Source: FAO Yearbook of Fishery Statistics, 1973

high (e.g., groundfish); however, its large population makes the United States a large market for the major fishery products of the world.

These figures are merely descriptive and tell us very little about the factors influencing the trend in consumption from year to year. To learn more about the factors that influence fishery consumption, we must review some basic economics on the demand for food in general and fisheries in particular. This will give us an analytical insight into the reasons why demand for one fishery product differs from the demand for another.

The Demand for Fishery Products: General Theoretical Concepts

To analyze changes and trends in the consumption of fishery products (or, for that matter, most consumer products from food to automobiles), we must first specify the theoretical model or relationship between consumption and the determinants of consumption. To do this, we start with one consumer who typifies the average consumer. It should be recognized that this typical consumer may differ significantly from many in the "market"; however, we shall show that all consumers generally adhere to what are called the "general laws of demand." The consumer has almost an infinite number of wants, conditioned and created by physiological need, personal characteristics, and the social and physical environment. A typical consumer has only limited income with which to satisfy his many wants. Therefore, he is faced with the problem of choosing which wants he wishes to satisfy within the limits of his personal income. Let us take a concrete example of the *hypothesized* factors a consumer considers when choosing how much fish to consume relative to all other goods that he may want.

For example, if one looks at the per capita consumption for the average U.S. consumer of canned tuna over the last two decades, one generally sees a persistent rise in this series (from .86 lbs in 1947 to 2.42 lbs in 1971). What determines how fast the per capita consumption of canned tuna has risen? Why has it not remained constant? Why did the average American consumer consume 2.42 lbs (expressed in edible weight) of canned tuna in 1971? We can attempt to answer these questions by hypothesizing what is called a demand "function," or relationship for canned tuna, as

$$q_T = f(P_T, \frac{Y}{N}, P_1, \ldots, P_n), \tag{2.1}$$

where

q_T = per capita consumption of tuna (canned)

P_T = price per unit (usually a pound) of tuna (canned)

$\frac{Y}{N}$ = real per capita income (i.e., real aggregate consumer income Y divided by population N)

$P_1,\ldots,P_n$ = Price per unit for goods that are substitutes for canned tuna in consumption.[5]

Hence, we have postulated three major determinants of the per capita consumption of tuna for a typical individual: the price of tuna, real per capita income, and the price of close substitutes. It is important in economic analysis that we be able to tell the separate impact of each demand determinant. The reason for this is that any theoretical model must stand the test of whether one can predict the consequences of, for example, a rapid rise in tuna prices on per capita tuna consumption, while *holding all other demand determinants constant.* If this is done for each demand determinant, we can then easily identify which determinants have been the most important, and to what degree, in any observed changes in per capita consumption. Let us now analyze the separate effects of each demand determinant. We shall use canned tuna as an example.

The Price of Tuna

In order to look at the independent relation between the price of tuna and how much a "typical" consumer buys, we must hold all other demand determinants constant, or

$$q_T = f[\, p_T,\ \overline{(\tfrac{Y}{N})}, \overline{P}_1, \ldots,\ \overline{P}_n\,] \qquad (2.2)$$

In equation (2.2) the bars over the demand determinants mean that they are held constant so that we can look at the partial relationship of q_T to P_T. One fundamental law of demand states that there is an inverse or negative relationship between quantity consumed and the price of that commodity. This is graphically displayed in figure 2.2. The relationship is only hypothetical and not based on real market data. (The reader should note that economists place price on the *y*-axis and quantity consumed on the *x*-axis, although normal mathematical convention usually places the causal variable on the *x*-axis.) What makes consumers purchase more of a commodity such as tuna at a lower price or restrict consumption to smaller quantities at higher prices? The funda-

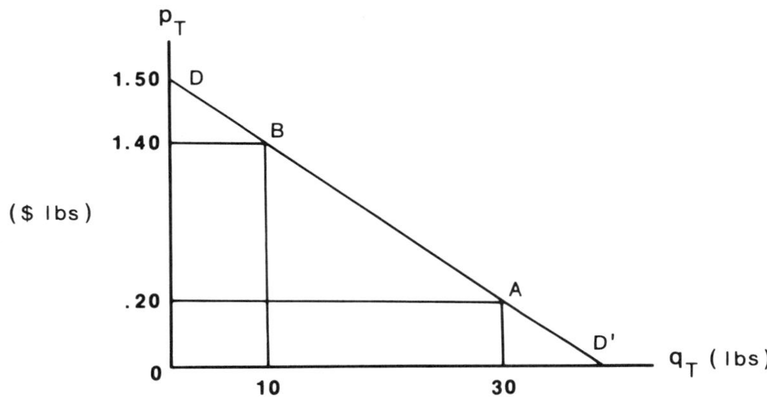

Figure 2.2: Law of Demand for Canned Tuna

mental reason is the existence of substitutes! But, we have already included very close substitutes in equations (2.1) and (2.2). They have been held constant. However, we have identified but a few close substitutes. Tuna is but one food item among many and competes with *close* as well as distant substitutes for the consumer's dollar. Thus, as the price of tuna rises, it becomes more expensive relative to all substitutes, both close and distant. Hence, other goods are substituted for tuna.[6] Note that we have held the consumer's income constant at some level. That is, his income is limited. If the level of income were set, say, at $8000 a year and if tuna prices rose to $10 a pound, the consumer would try to maximize his use of the $8000 by substituting *less* costly foods (such as eggs, vegetables, and potatoes) for tuna.[7] Hence, we can see that consumer will allocate his fixed income so as to maximize his satisfaction. It would indeed be silly for the consumer to pay $10 a pound for tuna if there are cheaper substitutes. The savings from the substitution can be used to purchase other products. In figure 2.2, the consumer would actually cease to purchase any canned tuna at $1.50 per pound. Tuna has "priced itself out of the market"! The relation shown in figure 2.2 is called a *demand curve* (although it is drawn as a straight line).[8] The term *demand* usually refers to the entire demand curve, and as prices rise (fall), the "quantity demanded" falls (increases). There is practically no good under the sun that does not adhere to the law of demand because of the pervasiveness of substitutes. Exceptions to the law of demand are rare.

As indicated above, the consumer will not buy tuna at $1.50 a pound—the sacrifice is too high. Therefore, substitutes are found. Remember that only the price of tuna is increased (decreased)—not any other price (i.e., the price of eggs, vegetables, and other substitutes remains fixed). The *relative* cost of consuming tuna is increased (decreased) when its price rises (falls). When we consider fishery products, substitutes that are usually *not* considered close substitutes are numerous, such as rice, macaroni, beans, and fruit.

If we knew the slope and intercept of the demand curve in figure 2.2, we could easily predict how much tuna would be consumed at every price. Another very useful concept when dealing with the price-quantity relationship is *price elasticity,* which is defined as follows:

$$e_p = - \frac{\text{percent change in quantity demanded}}{\text{percent change in price}} \tag{2.3}$$

Equation (2.3) tells us the "sensitivity" of quantity demanded to changes in price. For example, if $e_p = -2.0$, we can predict the *percentage* decline in quantity demanded as a result of a 10 percent increase in price, or

$$e_p = - 2.0 = \frac{x}{+10\%}, \tag{2.4}$$

or

$$x = -20\%.$$

The percent change in quantity indicates that consumer response to price changes is apparently very sensitive, or more precisely, elastic. What makes demand curves elastic or inelastic? The availability of substitutes is clearly an important factor. For example, salt has *few* substitutes, and its price-quantity relation is therefore usually considered very inelastic. Another factor is the importance of the item in the budget of the purchaser. If the expenditure on some good is *small* relative to your total income, you are likely to be less sensitive to a change in its price (i.e., price inelastic).

Figure 2.3 shows the aggregate demand curve in the United

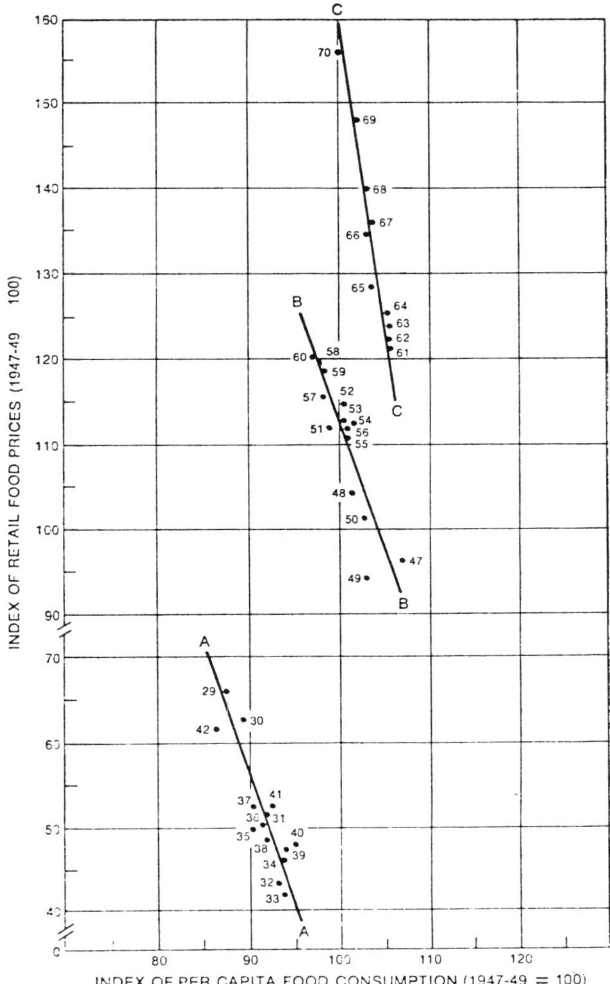

Figure 2.3: The Aggregate Demand Curve for Food

Source: Wilcox et al (1974)

States for food. It is highly inelastic (–0.17 to –0.32). That is, consumers change the quantity of food consumed very little as food prices vary relative to nonfood prices. Should this surprise you? Think of your own behavior. The human stomach has a limited capacity, and you probably fill yours several times a day. When food prices are rising, we do not deprive our bodies of a necessity of life. When your budget is squeezed, nonessential items like TV sets, CB radios, and dishwashers are some of the first consumption items to be dropped, but we keep eating. Substitutes for food (i.e., alternatives to eating) are not quite as pervasive as substitutes for refrigerators, automobiles, TV sets, or chairs. Price elasticity is thus a rough barometer of the substitutability of the product in question.

Price elasticity also allows us to *predict* changes in total sales as a result of changes in price. You may think that the higher the price, the greater the sales. What are sales? Sales volume is the result of multiplying price times quantity sold. Economists use the term *total revenue,* or *TR,* for sales, or

$$TR = pq. \tag{2.5}$$

In (2.5), if price goes up by 10 percent, we might expect total revenue (sales) to rise by 10 percent, but that was before we were acquainted with the law of demand. A rise of 10 percent in the price of canned tuna may prompt consumers to substitute eggs and macaroni and a whole host of other items for canned tuna; therefore, quantity consumed will fall. What will happen to sales volume, or *TR?* That depends on the price elasticity. If the price elasticity for tuna is –0.666, a 10 percent increase in price will result in a 6.66 percent *fall* in quantity sold. Therefore, total revenue (*TR*), or sales, will rise: the percentage change in p is greater than the percentage decline in q. If price elasticity is greater than unity (one), total revenue will actually fall in the face of a price increase, since the percentage increase in price is less than the percentage drop in quantity demanded. Table 2.3 illustrates the impact of price elasticity on *TR.*

We have explored the impact of price upon consumption, or quantity demanded. Let us now look at the role of per capita income in our demand function (equation [2.1]).

Real Per Capita Income

As with price, we shall look at the independent influence of *real* per capita income. What does the word *real* mean? Real income

Table 2.3: Relation of p and TR

Price Change	$e_p < 1$	$e_p = 1$	$e_p > 1$
p ↑	TR ↑	$\overline{TR}$	TR↓
p↓	TR↓	$\overline{TR}$	TR ↑

is quite different from "money" income—the difference is that all the income or dollars one receives (i.e., money income) is subject to inflation, or a general rise in the overall level of prices.[9] If your money income—take-home pay—went up by 10 percent, would that increase your "purchasing power," or ability to buy more goods to satisfy those insatiable wants? That depends on whether all prices have gone up by, say, 10 percent. If so, your real income has not changed, or expressed another way—you really cannot purchase increased *amounts* of existing goods. Thus, real per capita income represents purchasing power, and an increase will enable the consumer to increase his overall consumption if he so decides. Equation (2.6) expresses the fact that we are looking at the relation between q_T and Y/N while holding all other demand determinants constant:

$$q_T = f(\overline{p}_T, \frac{Y}{N}, \overline{P}_1, \ldots, \overline{P}_n). \quad (2.6)$$

What will be the impact on consumer purchases of canned tuna if real per capita income changes? Figure 2.4 illustrates the influence of income upon quantity consumed. *DD* represents the demand curve before the increase in real per capita income. The market price is 20 cents a pound, and our typical consumer is willing to purchase 30 pounds per year. This consumer has a real per capita income of $5000 per year and has just received a $1000 increase in real income. The increase might have been more in "money" terms, but we have adjusted for inflation. In our hypothetical situation, the demand curve shifts out and to the right, or *D′D′*. At every point on the q_T axis, the consumer is willing to pay more for tuna than he was before his raise. The consumer's response is to increase his per capita consumption of tuna by fifteen pounds a year, which costs him an additional $3.00. This will come out of his raise of $1000. An increase in per capita

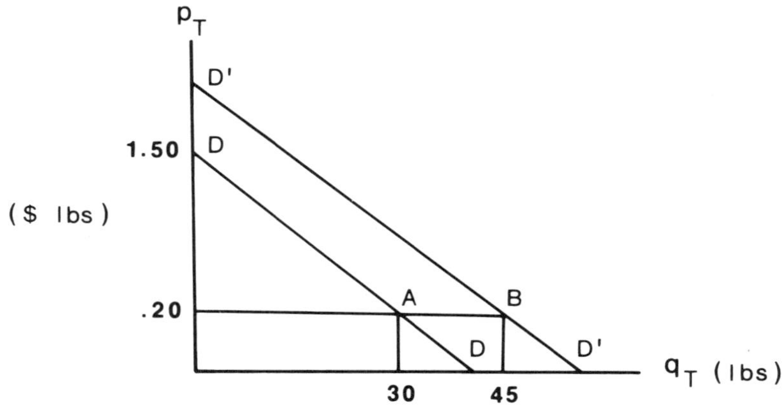

Figure 2.4: Impact of Increase in Real Per Capita Income on Tuna Consumption

income due to a real wage increase, for example, has made our typical individual more affluent. What he does with his increase in income depends upon a number of factors. If he is relatively poor, he may spend his additional income exclusively on food. That is, the more likely situation is that this incremental income will be spent on the necessities of life.

But does the consumer go on and on purchasing more canned tuna as his real income increases? Well, it all depends. It depends, in the case of food products, on *Engel's law.* This is the second law we have encountered in this chapter. Engel observed that as real per capita income increases, a reduced percentage of income is devoted to food. Can Engel's law be quantified? We have two sets of statistics in table 2.4 to illustrate Engel's law. As United States family income rises, the percent spent on food drops from 29 to 20 percent.[10] At low levels of per capita income ($20–$30) in developing countries, 70 percent is spent on food; however, at an income per capita level of $2,000, only 20 percent is spent on food. The data are consistent with Engel's law. This "law," as we shall shortly see, has an important bearing upon the response of food consumption in general and fish consumption in particular to changes in real per capita income. The point of this law is that as a consumer gets more and more affluent, he will in most cases

Table 2.4: Percent of Total Expenditures Devoted to Food

Family Income Level	U.S. Non Farm[1]	Per Capita Income	Developing Economies[2]
Under $3000	29	$20-$30	70
$3000-5000	26	$100-$200	50
$5000-7500	25	$500	35
$7500-10,000	24	$1000	25
$10,000-15,000	23	$2000	20
Over $15,000	20		

[1] 1961 data taken from U.S. Department of Labor (1965).

[2] Burk and Ezekiel (1967)

not be expected to increase his consumption of *any* food item in the same proportion as his increase in income. The consumer will become satiated by food and drink and at some point may spend no more money for tuna or any other food items out of his increases in income. Engel's law can be expressed in terms of the consumer's *income elasticity,* or

$$e_{Y/N} = \frac{\text{percent change in consumption}}{\text{percent change in real income}} . \qquad (2.7)$$

Income elasticity is a concept analogous to price elasticity. It shows the sensitivity (in percentages) of changes in consumption (i.e., per capita) to changes in income. Generally, as real income (Y/N) increases, the income elasticity for most food products will decline and approach zero, especially at high levels of affluence. In *theory,* the absolute value of the income elasticity should be less than unity for all food and all fish as categories. However, the current estimated value of the income elasticity may vary widely from fishery product to fishery product, as we shall see below. Put differently, Engel's law should result in a *less* than proportional increase in food consumption in response to a given change in real income.

For most food products, consumption *does* increase absolutely with increases in income. Economists call these "normal goods." However, for some goods, such as white flour, consumption actually declines *absolutely* with increases in income. Economists call white flour an "inferior good."

Close Substitutes

Finally, if we hold the price of tuna and real per capita income constant, what is the role of close substitutes for tuna (i.e., $p_1, \ldots, p_n$)? Let us assume that p_1 is the price of chicken—and not from the sea! If the price of chicken increases, tuna will become relatively cheaper; therefore, the consumer may attempt to maximize the use of his fixed income by consuming less chicken and *more* of the relatively less expensive tuna. Per capita consumption of tuna will rise. In our demand relation, close substitutes will therefore exhibit a positive sign (i.e., an increase in the price of a substitute will increase the per capita consumption of the good in question—tuna). Graphically, if p_1 increases, the demand curve will shift to the right for tuna, as shown in figure 2.5.

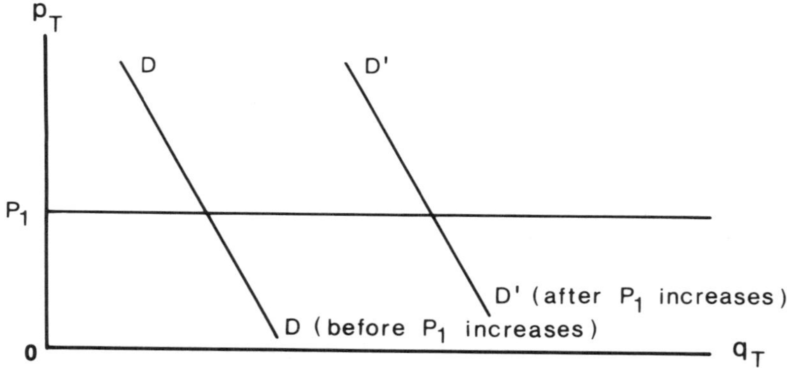

Figure 2.5: Impact of an Increase in the Price of a Close Substitute on the Demand Curve for Tuna

One may theorize that tuna, for example, will have many close substitutes, such as canned salmon, chicken, and beef.[11] As we shall see, this is only a hypothesis; it needs empirical verification. The answer to this question will come later in this chapter. In contrast to competitive products in consumption (i.e., substitutes), corned beef and cabbage are *complementary* products in consumption; consumption of more of one is usually accompanied by greater consumption of the other. Or, for example, if the price of scallops or shrimp drops, there will be a greater demand for tartar sauce. These are likewise complementary products. The complementarity may only work in one direction, since a drop in the price of tartar sauce is not likely to stimulate more fish consumption. Why? Because tartar sauce is not that important in the consumption of scallops or shrimp. Therefore, complementarity may be rigid or loose and may work in only one direction, depending on the goods in question.

Population

We have so far been talking about the average consumer. Statistically, the per capita consumption of tuna would be derived by dividing total consumption, say for the United States, by population. We have used this approach since it coincides with the statistical analysis of demand relations. For purposes of illustration, let us assume the following hypothetical demand function has been found for tuna (where p_C = price of chicken per pound):

$$q_T = -5P_T + .001\left(\frac{Y}{N}\right) + 4p_C, \tag{2.8}$$

and given $P_T = \$0.20$ (i.e., 20 cents a pound)

$\left(\frac{Y}{N}\right) = \$3000$

$p_C = \$0.50$ (i.e., 50 cents a pound).

Substituting into (2.8), we obtain a per capita consumption of tuna of four pounds. But what about the role of population? At the beginning of this chapter, we stressed the "population bomb" that was increasing the demand for food on the planet. Well, it is still a bomb! Equation (2.8) tells us the per capita consumption, but we must multiply through by the size of the population to obtain the aggregate demand for tuna. If there are 220 million Americans, this means that 880 million pounds of tuna are demanded by all consumers in the United States. Thus, even if tuna prices, incomes, and the price of close substitutes remain constant, the aggregate demand for tuna will grow at the population growth rate. World population, as indicated earlier, is doubling every thirty-four years. If this rate of population growth continues to be about 2 percent annually, the size of the tuna market will also double every thirty-four years. Thus, population is a major determinant of demand.[12] Now you can see that the food crisis on the demand side depends not just on population, but also on the behavior of prices, income, and the availability of close substitutes. Now let us look at some special aspects of the demand for fish.

Unique Factors in the Demand for Fishery Products

As with most products, the demand for fish is influenced by certain unique features in addition to the general factors discussed above. Some of these factors apply to all fishery products, but some apply only to certain species.

Size of Fish. It is well known that consumers are willing to pay premium prices for fish or shellfish of a certain size. Why is this so? Well, you would really have to read the mind of the consumer. Economists usually express this as a difference in "tastes and preferences" and let it go at that. An example should clarify the situation. As a practical example, the more shrimp there are per pound (i.e., obviously the greater the number per pound, the smaller each shrimp), the lower the price per pound. Let us look at the ex vessel price of shrimp landed in the western Gulf of Mexico by size (i.e., number of shrimp in a pound), as shown in table 2.5. These data reflect the premium that the consumer places on large shrimp. For example, a pound of shrimp (heads off) con-

Table 2.5: Ex Vessel Price Per Pound, heads-off Shrimp*
December, 1975

Size	Price Per Pound
Un 15	$3.65
15/20	3.25
21/25	3.04
26/30	2.88
31/35	2.69
36/42	2.33
43/50	1.94

*Source: National Marine Fisheries Service (1976)
Un means under while slashes are used in trade to indicate intervals.

taining only twelve shrimp sells for $3.65, but one containing forty-five shrimp sells for $1.95. The premium is truly startling! The consumer is willing to pay 87 percent more per pound for large shrimp. We can, of course, only speculate as to why this psychological response exists. After all, the consumer still gets the same amount of shrimp—one pound—no matter what the "size count" is. (The phrase *size count* is used in the trade to express the number of shrimp [or intervals of sizes] per pound.) "Tastes and preferences" are being expressed, and obviously the consumer feels the 87 percent price differential is worth it. (It *is* easier to shell twelve big shrimp than to shell forth-five little ones. This may be a factor.) Although no definitive studies have been done, there is apparently no appreciable variation in *quality* (i.e., protein yield and taste) or *cost of harvesting* from the Un 15 size count to the 43–50 size count. Shrimp is not the only species that has this price differential per pound. Within certain limits, lobsters and yellowtail flounder (Gates 1974) show the same general relationship.

Figure 2.6 shows two different demand curves: one for a size count under 15 and the other for 43/50. The line *AB* indicates the price differential, as we have observed above. Put quite plainly, the consumer's preference for larger shrimp, as expressed in the price differential, shifts the demand curve out and to the right. This is not just an interesting exposition of consumer preference; it actually governs the economic behavior of fishermen. Go down to any Gulf of Mexico shrimping community and ask the fishermen about their fishing strategy. They usually say that they know where you can catch the larger shrimp (i.e., low size count) and will plan their fishing trip with this in mind.

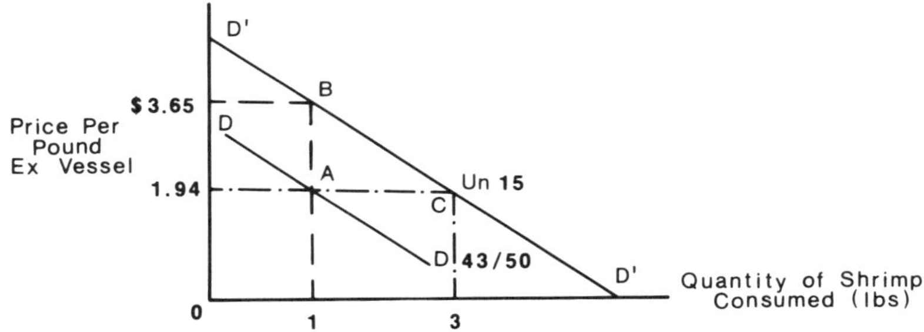

Figure 2.6: Relation between Shrimp Size Count and Price Per Pound

Fresh vs. Frozen Fish. There is a considerable difference in price between fresh and frozen fish. Fish tend to deteriorate immediately after being caught. Autolysis takes place, during which enzymes digest the tissues and thus cause a softening of the tissue and a change in flavor and odor. The speed of deterioration depends primarily on temperature. Even if fish are frozen after being caught, however, there is apparently a loss in flavor, although there are many advantages to freezing, such as storage and transportation efficiency.

Gaede and Storey (1969), in their study of consumer purchases of seafood in Springfield, Massachusetts, developed table 2.6.

Table 2.6: Household Opinions about Characteristics of Fresh Seafood Compared to Frozen Unprepared Seafood

Seafood Characteristics	Household Opinion: Fresh Better than Frozen	Fresh Same as Frozen	Fresh Worse than Frozen	No Opinion
	(percent of households indicated below)			
Taste	70	13	1	16
Nutritional Value	45	32	-	22
Quality, general	49	29	3	19
Cooking Qualities	41	35	5	19
Availability	21	42	16	21
Price	16	25	24	35

Source: Gaede and Storey, 1969.

When comparing fresh seafood to frozen unprepared seafood, nearly all households that had an opinion rated fresh seafood better with respect to taste, nutritional value, general quality, and cooking qualities. Fresh seafood was generally judged much less available in marketing outlets than frozen seafood. Finally, table 2.6 indicates that more households thought frozen seafood was cheaper than fresh.

Table 2.7 shows a direct comparison of fresh versus frozen fillets for haddock, cod, and flounder over the 1970–1975 period.

Table 2.7: Relation Between the Retail Price of Fresh and Frozen Fish Fillets in Cents*

Species	Year: 1970	1971	1972	1973	1974	1975
Haddock						
(1) Fresh[1]	115.6	115.3	132.2	209.0	214.6	207.0
(2) Frozen[2]	87.9	100.0	106.3	131.4	149.4	151.5
(1)÷(2)	1.32	1.25	1.24	1.59	1.44	1.37
Cod						
(1) Fresh[1]	101.2	126.0	134.3	180.6	195.7	219.1
(2) Frozen[2]	66.7	78.5	89.7	111.1	140.3	146.0
(1)÷(2)	1.52	1.61	1.50	1.63	1.39	1.50
Flounder						
(1) Fresh[1]	129.1	144.7	166.5	221.2	238.2	279.2
(2) Frozen[2]	97.6	96.3	105.9	131.1	154.7	166.0
(1)÷(2)	1.32	1.50	1.57	1.69	1.54	1.68

*fillet is a fully eviscerated fish, scaled with head off. Fresh and frozen fillets are physically identical except for the state of preservation

[1]1-lb package, U.S. average price reported by the U.S. Bureau of Labor Statistics (BLS)

[2]Estimated retail food prices (BLS) at Boston, Massachusetts

Source: National Marine Fisheries Service, Food Fish Market Review and Outlook, CEA F-23, (March, 1976)

Over this period, fresh haddock fillets sold, on the average, at a retail price 37 percent higher than frozen fillets. Even more sizable price differentials were observed between the fresh versus the frozen cod and flounder. In terms of economic analysis, the preference for fresh over frozen would be handled the same way as an apparent consumer preference for large over small shrimp. That is, if we hold all other factors constant (including price), consumers will consume more fresh than frozen fish, thereby increasing per capita consumption (see figure 2.6).

Distance from the Sea. An important factor influencing fish consumption is the consumer's familiarity with the product. Does he see it in stores? Does he see it being landed while he is boating? Is it served widely in restaurants? In other words, do the eating habits of the region respond to its widespread availability in coastal areas? For the moment, let us consider marine or saltwater fish as an example. A relatively recent survey by the National Marine Fisheries Service under the direction of Miller and Nash (1971) for the United States indicated that some fishery products have achieved the status of a "national food" but that others are strictly regional. Table 2.8 is a rough measure by species and more aggregated categories of the difference in the per capita consumption of fish between principally coastal areas (data on individual coastal states were not available) and noncoastal areas. The differential for *shellfish* and the *finfish* indicates that coastal areas consume 83 percent more per capita than noncoastal areas. These fish apparently do not enjoy the same popularity as food in the hinterlands of the United States. This is partly a function of transportation cost, which raises the delivered price and thereby reduces per capita consumption. However, the differential is much too large to be explained by higher prices. A fairly widespread hypothesis is that these inland markets have not developed because of consumer unfamiliarity with these products. In essence, a cultural factor is involved.

Finally, Bell (1968a) demonstrated through statistical analysis that *dollar* expenditures per family on all fish and shellfish products decline with an increase in linear distance from the sea. The data base was for 1950 and involved forty-eight cities. For every hundred miles from the coast, annual expenditures per family on fishery products fell \$7.30. Table 2.8 does seem to indicate that canned fishery products are more evenly distributed between coastal and noncoastal areas—the differential is only 17 percent. The *nonperishability* of canned products increases the potential

Table 2.8: A Comparison of Fishery Per Capita Consumption Between Coastal and Non-Coastal Areas of the United States for Selected Fishery Products, 1969* (pounds)

Species	Coastal Area[1] (1)	Non-Coastal Area[2] (2)	1÷2 (3)
1. Shrimp	1.053	.826	1.27
2. Oysters	.237	.189	1.25
3. Crabs	.223	.058	3.84
4. Lobsters	.383	.016	23.94
5. Clams	.145	.013	11.15
6. Scallops	.096	.102	.94
7. Total Shellfish	2.275	1.342	1.70
8. Total Finfish	6.063	3.224	1.88
9. Total Shellfish and Finfish	8.338	4.566	1.83
10. Canned Fish[3]	5.807	4.957	1.17
11. Shellfish, Finfish and Canned Fish	15.646	10.722	1.46

*Estimated home consumption (i.e., excludes away from home meals) and includes fish both purchased and caught by the consumer.

[1]New England, Middle Atlantic, South Atlantic, E. South Central, W. South Central, and Pacific region (i.e., all have ocean coast line).

[2]E.N. Central, W. North Central, and Mountain regions (no marine coastal area).

[3]Principally tuna, salmon, shrimp and crab.

Source: Miller and Nash, 1971.

for exposure of such items as salmon and tuna in the midwestern United States, thereby increasing the potential for consumer awareness, experimentation, and ultimate acceptance. Hence, distance from the sea may shift the demand curve down and to the left as we get farther away from coastal areas. Innovations in marketing may reduce such cultural barriers.

Demographic Factors. Demographic factors refer to such population characteristics as *occupation, education, race, age structure,* and *marriages.* These characteristics change over time, and questions naturally arise in demand analysis as to whether such factors are related to fish consumption. This has not been well studied for fishery products throughout the world. As indicated above, the growth in the *total* population of the world has been of overriding importance in the demand for fishery products. However, population characteristics have been linked to fish consumption. For example, as Purcell and Raunikar (1968) state in their cross-section study of households in Atlanta, Georgia,[13]

> The number of persons in each of the five age classifications showed considerable differences in the effect on quantity and expenditures for the fish and shellfish categories. Persons in the 6-10 and 11-18 year-old age classification had the greatest effect on *quantity* purchased of total fish and shellfish; however, the number of persons over 18 years old had the greatest effect on *expenditures* for total fish and shellfish (1968, p. 33).

They go on to state:

> Tuna was the only category of fish and shellfish for which it was estimated that there was no significant difference due to race in the quantity purchased and expenditures per quarter. The white household had lower quantity purchased and expenditures than the nonwhite household for fresh fish, salmon, and fish and shellfish. In addition, the quantity of sardines in oil purchased by the white household was lower than in the nonwhite household (1968, p. 25).

The study by Purcell and Raunikar included most of the other demand determinants discussed above, and their conclusions therefore refer to the separate (or independent) influence of age and race.

A more comprehensive consumer panel was developed by the National Marine Fisheries Service (Nash 1970). *Without* controlling other economic variables, table 2.9 shows some of the results from this NMFS study. The table might lead the casual observer to conclude that blacks eat more shellfish (73 percent more), finfish (189 percent more), and canned fish (46 percent more) than whites. If the data are accurate, this is a true statement.

But does race really make a difference? The answer lies in the general approach to demand theory as presented in this chapter. The point is that all other variables *have not been held constant.* Blacks, especially near the coast, obtain fish at lower prices. Moreover, this survey asked what was consumed, not what was bought. As we shall see in chapter 4, recreational fishing or just plain fishing for food might explain part of the differential between black and white fish consumption. So it is very important to analyze data such as that presented in table 2.9 with some skepticism. Does race—a cultural factor—really make a difference, or does it mask economic variables such as price, distance from the sea, or the price of substitutes? A statistical analysis of the same consumer panel by Fullenbaum (1971) indicated that blacks do in

Table 2.9: Relation Between Per Capita Consumption of Selected Fishery Products and Race, Occupation and Education, Feb. 1969-Jan. 1970, United States*

	Shellfish, Fresh and Frozen			Finfish Fresh and Frozen		Canned Fish			
	Shrimp	Oysters	All Shellfish	Flounder	All Finfish	Light Tuna	Pink Salmon	All Canned	Grand Total
Race									
White	.909	.366	1.731	.54	3.947	1.730	.841	5.163	12.264
Black	1.988	.194	3.003	.70	11.426	2.210	1.796	7.544	23.054
Occupation									
Professional	.755	.127	1.311	.470	3.419	1.517	.352	3.719	9.437
Managers	.812	.249	1.612	.475	3.521	1.572	.772	4.903	11.429
Clerical-Sales	1.231	.195	2.265	.817	4.432	1.904	.831	5.515	14.059
Craftsman	1.110	.199	1.970	.462	3.760	1.937	.801	5.199	12.282
Operative	.867	.111	1.335	.352	3.946	1.233	.844	4.149	10.154
Education									
Less than 4 years high school	.977	.249	2.102	.634	5.833	2.062	1.521	7.244	15.958
Less than 4 years college	1.103	.227	1.916	.572	6.999	1.856	.805	5.135	15.595
College Graduates	.875	.131	1.558	.518	3.676	1.492	.333	3.975	10.318

*Grand total includes specialty items such as clam chowder and T.V. Dinners that are not shown. Also, total consimption includes both fish commercially bought and that caught by the individual.

Source: Darrel A. Nash (1970)

general consume more of all kinds of fish. Fullenbaum held many economic and social variables constant. This finding agrees with that of the Purcell and Raunikar study. This kind of information can be very important for the efficient marketing of fish and for predicting future demand for fish. This leads us directly to a fascinating area of demand determinants—social institutions.

Social Institutions. Various aspects of a culture may sometimes encourage or discourage the consumption of fish. This may take the form of established laws—either formal or informal—customs, or other socially acceptable practices. In the Hamito-Semitic cultures of North Africa, for example, there is a general dietary prohibition against nonscale fish of any sort, including shellfish.[14] This prohibition is associated with the Judeo-Islamic tradition and usually extends to all areas where Islam is found, whether these include specifically Arab cultures or not.

Culturally, black Africa has very definite attitudes about the consumption of fish. Although they vary greatly by tribe and degree of urbanization, tribal taboos against fish are somewhat less than religious, but more than just custom. In some instances, these taboos are directed against women or children—the reasoning being that delicacies such as fish should be reserved for adults or men. The Hindu and Buddhist religions stress vegetarianism; therefore, fish is not widely consumed in the Indian subcontinent or in parts of Indo-Asia.

For over a thousand years, the Catholic church required its members to abstain from meat on Friday in the spirit of penance. This religious practice was always considered favorable to the consumption of fish. Since the renunciation of meat is not always the most effective means of practicing penance and since meat is no longer an exceptional food, the Catholic church in November 1966 abolished meatless Fridays throughout the United States. Bell (1968b) studied how this affected the price of fish in New England (population approximately 45 percent Catholic). Table 2.10 shows the impact on prices paid to New England fishermen and revenue losses. Thus, we may conclude that this doctrinal change shifted the demand curve down and to the left (i.e., consumers purchased less at each price level).

In another survey of the impact of religion on consumption, the National Marine Fisheries Service survey (Nash 1970) indicated the following per capita consumption (pounds) by religion from February 1969 to January 1970:

	Shellfish[1]	*Finfish*[1]	*Canned*	*Total*[2]
Jewish	2.230	10.087	10.081	27.254
Catholic	2.193	3.981	5.280	13.061
Protestant	1.662	4.142	5.105	12.322

1. Fresh and frozen.
2. Includes specialty items such as TV dinners, chowders, and smoked fish not shown elsewhere.

Although these statistics must be judged with caution, it would appear that Jews consume considerably more finfish and canned fish than Catholics or Protestants. However, the margin narrows for shellfish, as might be expected since the Jewish religion generally considers shellfish to be nonkosher. It is also interesting that three to four years after the abolition of meatless fridays, Catholics consumed more fish per capita than Protestants did. Although meatless Fridays no longer exist, Catholics may have developed a greater preference for fish owing to hundreds of years of conditioning.

Table 2.10: Impact of the Abolition of Meatless Friday on New England Landing Prices for Seven Fish and Shellfish Species (Dec. 1966-Aug. 1967)[1]

Species	Percent Change in Price Due to Abolition of Meatless Friday	Revenue Loss (thousand of dollars)
Sea Scallops	-17	603
Yellowtail Flounder	-14	487
Small Haddock (scrod)	- 2	86
Large Haddock	-21	668
Cod	-10	217
Perch	- 8	161
Whiting	-20	312
All Species	-12.5	2,534

1. Excludes Lent months of February and March
Source: Bell (1968)

In summary, although the main determinants of fish consumption are principally *price, real per capita income,* and *population,* we must be very mindful of special demand determinants such as those surveyed in this section.

Industrial Demand for Fish. This discussion of the theory behind the demand for fishery products has so far applied only to food fish. However, large quantities of fish harvested from the ocean are not *directly* eaten by the consumer. Most of the fish that are caught primarily for reduction (i.e., made into meal and oil) come from the herring group and include the herrings, menhadens, anchovies, and sardines. The most important use of fish meal is as a high-protein concentrate feed for poultry and pigs.

The consumer, especially in Western countries, eats quite a bit of chicken and pork. Therefore, economists have developed a concept called "derived demand" for fish meal through the consumption of chicken, bacon, and pork chops. Hence, we are really interested in the consumer demand for foods that depend on feed, specifically fish meal. The formula feeds make use of high-protein concentrates, such as fish meal, soybean meal, and other animal meal. Colonel Sanders' "finger lickin' good" chicken is really, in part, a fish in disguise! According to the FAO (1961), fish meal contains an unidentified factor called UGF, which stimulates rapid growth of up to 2.5 percent of the total feed ration of commercial broilers. However, Colonel Sanders wants no more than about 10 percent of fish meal in the feed ration. Why? After this percentage is reached, a fishy flavor appears in the poultry (and meat) although the actual percentage varies depending on the kind of fish in the meal and on processing methods. It is quite apparent from this discussion that we can formulate a demand relation for fish meal in the following manner:

$$Q = f(p_F, C, p_{SB}), \tag{2.9}$$

where

Q = utilization (quantity) of fish meal
p_F = price of fish meal
C = aggregate consumption of chickens
p_{SB} = price of soybean meal.

For fish meal, in contrast to food fish, we are not dealing in per capita figures. Theoretically, there is no reason why we could not divide Q and C by population; however, the literature in this area usually deals in aggregate values. So Q is obtained by multiplying the per capita utilization by population. The term *utilization* is used, since fish meal is not consumed directly but is utilized to produce a *consumed* product. Following our reasoning above, Q and p_F would obey the law of demand and have an inverse relation. If the price of fish meal *increases,* producers may substitute meat scraps and tankage (slaughterhouse waste), feather meal, and blood meal for fish meal. C, or aggregate chicken consumption, is the demand determinant that is most critical to the overall fish meal demand. Like food fish, chicken has its own demand function and corresponding demand determinants such as income and population. Finally, the feed ration is usually a combination of soybean and fish meal, whose proportions can be varied. The price of soybean meal *relative* to fish meal (holding all other factors constant) determines the feed ration proportions. In our later sections, we shall have more to say about the world fish meal market and its role in the population/food crisis. Now let us look at some empirical testing of the theory developed above.

Consistency of Demand Theory with Market Data

An Empirical Example

The U.S. shrimp market expanded very rapidly over the 1947–1971 period. Expressed as *edible weight,* aggregate consumption increased from 99.4 million pounds in 1947 to 310.8 million pounds in 1971. Over this period, per capita consumption rose from 0.69 lbs. to 1.52 pounds. These figures depict an expanding aggregate market and a rising use of shrimp by the average U.S. consumer. To explain why this has taken place, we shall draw upon our theoretical tools developed above. In essence, we are about to conduct an experiment similar to any performed in chemistry, physics, marine biology, or oceanography. Our laboratory is the U.S. market for shrimp. And we hope to work under controlled conditions—even though in the social sciences, "controlled conditions" are less rigorous than in the physical sciences, for the social sciences deal with people who react to changes in economic variables under less than optimally controlled conditions. We cannot, of course, bring 220 million Americans into a laboratory and ask them how much shrimp they would purchase if they were given a certain income or certain prices. But even the biologist or oceanographer often has difficulty in duplicating actual environmental conditions in the laboratory. In many cases, the marine biologist must go to the natural environment itself, where he attempts to control important variables, although this is less optimal. The same problem confronts the economist.

In the case of shrimp, let us hypothesize that the demand function is

$$\left(\frac{Q}{N}\right)_{Sh} = a - b(p_{Sh}) + c\frac{Y}{N} + d(p_M) + h(p_P) + m(p_{SF}), \quad (2.10)$$

where *demand* is

$\left(\frac{Q}{N}\right)_{Sh}$ = U.S. shrimp per capita consumption (edible weight) and *demand determinants* are

(p_{Sh}) = real price of shrimp (ex vessel price divided by the consumer price index, CPI)[15]

$(\frac{Y}{N})$ = U.S. real per capita disposable income (money per capita income divided by CPI)

(p_M) = real price of meat (index of retail meat prices divided by CPI)

(p_P) = real price of poultry (index of retail poultry prices divided by CPI)

(p_{SF}) = real price of selected shellfish (average ex vessel price of lobsters, scallops, and clams divided by CPI).

Note the similarity between equation (2.10) and our theoretical demand function as specified in equation (2.1). The demand determinants are the price of shrimp, per capita income, and three possible close substitutes—meat, poultry, and other shellfish. The data for this experiment are shown in table 2.11. The reader may

Table 2.11: Per Capita Consumption of Shrimp and Theoretical Demand Determinants, U.S.* (1947-71)

Year	Demand	Demand Determinants				
	Per Capita Consumption[1]	Real Shrimp Price ($)	Real Per Capita Income ($)	Real Meat Price (Index)[2]	Real Poultry Price (Index)	Real Shellfish Price ($)
1947	.69	.5785	1,513	1.07	2.12	.5949
1948	.69	.4827	1,567	1.12	2.18	.5922
1949	.71	.5350	1,547	1.07	2.07	.4874
1950	.72	.5284	1,646	1.11	1.97	.5354
1951	.84	.5000	1,657	1.17	1.91	.4949
1952	.89	.5119	1,678	1.13	1.88	.5975
1953	.89	.6167	1,726	1.05	1.82	.5243
1954	.91	.4733	1,714	1.04	1.63	.5242
1955	.95	.5299	1,795	.96	1.70	.5511
1956	.90	.6523	1,839	.92	1.47	.5749
1957	.81	.7141	1,844	.98	1.39	.5089
1958	.87	.6617	1,831	1.06	1.33	.5416
1959	1.05	.4651	1,881	1.02	1.21	.5624
1960	1.09	.5085	1,883	.98	1.21	.4803
1961	1.05	.5558	1,909	.99	1.08	.5190
1962	1.06	.6788	1,969	.99	1.13	.5221
1963	1.15	.5060	2,015	.97	1.09	.5802
1964	1.18	.5694	2,126	.94	1.06	.6502
1965	1.26	.5725	2,239	.99	1.07	.7291
1966	1.25	.6685	2,335	1.06	1.10	.6677
1967	1.35	.5450	2,403	1.00	1.00	.7690
1968	1.43	.5988	2,486	.98	.99	.8704
1969	1.40	.5811	2,534	1.01	.99	.8342
1970	1.58	.4978	2,610	1.01	.93	.9046
1971	1.52	.5812	2,683	.96	.90	.9629

*All money values divided by the Consumer Price Index.

[1]Edible weight, lbs.

[2]An index rather than actual price, 1967=100.

Source: National Marine Fisheries Service, Basic Economic Indicators, Shrimp, 1947-72, Current Fisheries Statistics No. 6131, (June 1973); Economic Report of the President, 1975; U.S. Dept. of Labor, BLS, The Consumer Price Index, 1947-72.

be curious to know why we have not included all those "unique demand determinants" for fish. For example, the size count of shrimp is very important as a demand determinant. The racial and religious composition of the population may also influence shrimp consumption. But over the period to be analyzed (1947–1971), there has been little change in these unique determinants of shrimp demand. Hence, they have been omitted from equation (2.10). And let us be quite clear on one point! We are not saying that these unique demand determinants have no bearing on the level of shrimp per capita consumption. They really do. However, if there is no appreciable change in these unique determinants over our period of analysis, they cannot help explain the rise (or change) in the per capita consumption of shrimp.

Table 2.11 shows a progressive rise in the per capita consumption of shrimp (fresh, frozen, and canned); however, there have been periods of decline, such as from 1955 (0.95 pounds per capita) to 1957 (0.81 pounds per capita). This approximately 15 percent decline in per capita consumption had a significant impact on aggregate consumption. Despite population increases from 1955–1957, aggregate shrimp consumption in the United States declined by 18.1 million pounds. Why? This can and will be explained shortly. But first let us look at the behavior of the demand determinants over the 1947–1971 period. The money price (ex vessel) of shrimp kept pace with inflation and showed no appreciable trend, although great fluctuations from year to year were evident. The average American became more affluent, as his real disposable personal income increased 77.3 percent.[16] The candidates for close substitutes showed interesting trends. Real meat prices showed an upward trend, but real poultry prices declined considerably. If chicken and other poultry products were close substitutes for shrimp, consumers would have switched from shrimp to poultry over the 1947–1971 period. Finally, the average real price of lobsters, scallops, and clams showed a strong upward trend, indicating a possible switch from these products to shrimp.

Are the market data consistent with our theory? Remember that statistical analysis can never prove causality—all we can say is that the data are consistent or conflict with our hypothesis. The parameters (*a, b, c, d, h,* and *m*) of our demand equation were estimated using the technique of multiple regression and the market data in table 2.11.[17] Preliminary results indicated that *d, h,* and *m* all were negative and not statistically significant (i.e., statistically, they had no appreciable impact on the per capita consumption of shrimp). (To be close substitutes, the parameters should have positive signs.) That is, the price of meat, poultry, and other shellfish was not statistically related to shrimp consumption. However, *b* exhibited a negative sign and *c* a positive sign. These parameters (for the real price of shrimp and real per capita income) were highly statistically significant. Has this experiment been done before? Yes, but with somewhat different data, time period, and techniques. Doll stated that "shrimp undoubtedly do have complementary or substitution effects with other seafood and meat but none could be isolated in preliminary specifications" (1972, p. 433). Gillespie, Hite, and Lytle (1969) found that meat was not a close substitute for shrimp. Cleary (1969) discovered that meat, poultry, soft clams, blue crab, scallops, and lobster could not be statistically linked to shrimp consumption. Apparently, shrimp does not have any close substitutes that can be statistically isolated.

In light of the preliminary findings, the substitution variables (p_M, P_P, and P_{SF}) were dropped from the equation. The final demand equation for shrimp is the following:

$$\left(\frac{Q}{N}\right)_{Sh} = -.1245 - .5514p_{Sh} + .00075\left(\frac{Y}{N}\right). \qquad (2.11)$$

Equation (2.11) explains over 98 percent of the variation in the per capita consumption of shrimp over the 1947–1971 period. The explanatory power of the equation is outstanding. Figure 2.7 shows the *actual* and predicted per capita shrimp consumption over the period of analysis. The word *predicted* is used to indicate the level of per capita consumption obtained by using equation (2.11) and the actual values of the demand determinants for each year. Table 2.12 shows the actual and predicted per capita shrimp consumption, population, and the actual and predicted *aggregate* consumption of shrimp in the United States from 1947 to 1971. Remember the question we posed about the decline in shrimp consumption during the 1955–1957 period. Our equation predicted an aggregate decline of only 4.8 million pounds; however, it did predict the decline. If we look at table 2.11, we find that real shrimp prices increased from 52.99 cents in 1955 to 71.41 cents in 1957, about a 35 percent increase. Real per capita income was not much of a factor, increasing from $1,839 in 1955 to $1,844 in 1957, an insignificant change. Using equation (2.11), the rise in the real shrimp prices coupled with virtually no change in real per capita income (i.e., this was actually a recessionary period)

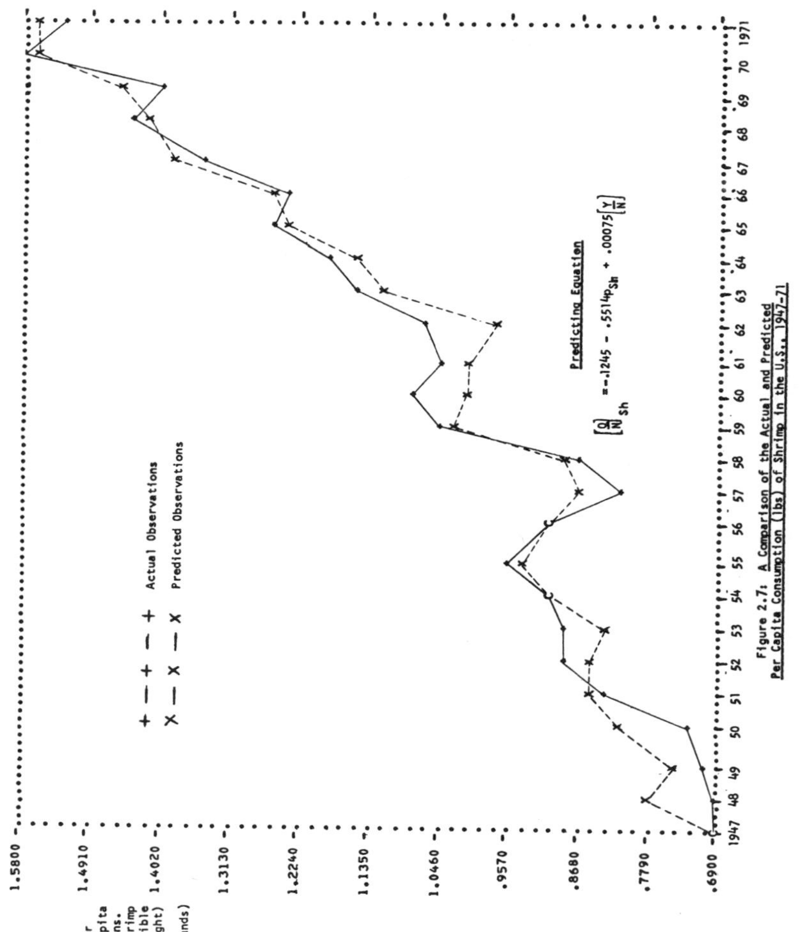

Figure 2.7: A Comparison of the Actual and Predicted Per Capita Consumption (lbs) of Shrimp in the U.S., 1947-71

Table 2.12: Actual and Predicted Per Capita Shrimp Consumption, Population, and Actual and Predicted Aggregate Shrimp Consumption, U.S., 1947-71

Year	Per Capita Consumption (lbs) (1) Actual	(2) Predicted	U.S. Population[1] (3) (Millions)	Aggregate Consumption[2] (4) Actual (1)x(3)	(5) Predicted (2)x(3)	Ratio of Predicted to Actual (5)÷(4)
1947	.69	.69	144.1	99.4	99.4	1.00
1948	.69	.78	146.7	101.2	114.4	1.13
1949	.71	.74	149.3	106.0	110.5	1.04
1950	.72	.82	151.5	109.1	124.2	1.14
1951	.84	.84	154.6	129.9	129.9	1.00
1952	.89	.85	157.1	139.8	133.5	.95
1953	.89	.83	159.0	141.5	131.9	.93
1954	.91	.90	161.9	147.3	145.7	.99
1955	.95	.93	165.0	156.8	153.5	.98
1956	.90	.90	168.2	151.4	151.4	1.00
1957	.81	.87	171.2	138.7	148.9	1.07
1958	.87	.88	175.0	152.2	154.0	1.01
1959	1.05	1.03	177.2	186.1	182.5	.98
1960	1.09	1.01	179.7	195.9	181.5	.93
1961	1.05	1.00	183.1	192.3	183.1	.95
1962	1.06	.98	183.3	194.3	179.6	.93
1963	1.15	1.11	189.5	217.9	210.3	.97
1964	1.18	1.16	191.4	225.8	222.0	.98
1965	1.26	1.24	194.7	245.3	241.4	.98
1966	1.25	1.26	195.6	244.5	246.5	1.01
1967	1.35	1.38	197.8	267.0	272.9	1.02
1968	1.43	1.41	199.7	285.5	281.6	.99
1969	1.40	1.46	201.2	281.7	293.8	1.04
1970	1.58	1.56	203.9	322.1	318.1	.99
1971	1.52	1.57	204.5	310.8	321.1	1.03

1. Total resident population; (2) millions of pounds (edible weight).

Source: Equations (15) and Basic Economic Indicators: Shrimp (1973).

explains the downturn in aggregate shrimp consumption. At last we have blended theory with reality. It actually works!

Let us now take a brief look at the demand curve for shrimp. How can we determine the demand curve? The answer is that there are many demand curves, depending on the level of consumer income. That is, as real per capita disposable income expands, the demand curve for shrimp shifts up and to the right. We have chosen 1951 and 1970 to illustrate the shift in the demand curve for shrimp. Figure 2.8 illustrates the demand curve for each year. They were derived in the following manner. The level of real per capita income was $1,657 and $2,610 in 1951 and 1970, respectively. As noted in our section on the theory of demand, increases in real income will shift the demand curve. These real per capita income figures were inserted in equation (2.11), which had the effect of increasing the constant term. Let us take 1951 as an example.

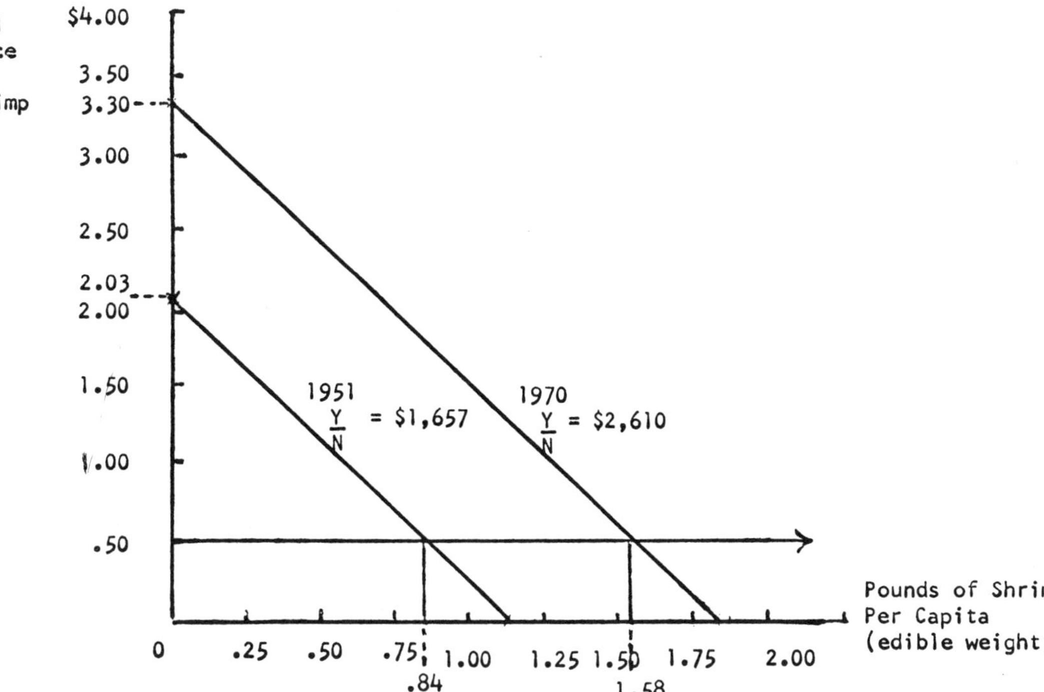

Figure 2.8: Demand Curves for Shrimp, 1951 and 1970

$$\left(\frac{Q}{N}\right)_{Sh} = -.1245 - .5514p_{Sh} + .00075\ (\$1{,}657), \qquad (2.12)$$

or adding 1.2449 to -.1245, we have

$$\left(\frac{Q}{N}\right)_{Sh} = 1.1204 - .5514p_{Sh}\ . \qquad (2.13)$$

However, economists like to place price on the *y*-axis and quantity consumed on the *x*-axis. Don't ask me why! We may solve equation (2.13) in terms of price, or

$$p_{Sh} = 2.032 - 1.8136\left(\frac{Q}{N}\right)_{Sh}. \qquad [1951] \qquad (2.14)$$

Inserting \$2,610, or per capita real income for 1970, in equation (2.12) and repeating the steps above gives the following demand curve for 1970:

$$p_{Sh} = 3.300 - 1.8136\left(\frac{Q}{N}\right)_{Sh}. \qquad [1970] \qquad (2.15)$$

Note that the intercept is higher for 1970 than 1951 (3.300 versus 2.032). This represents the power of real per capita income to shift the demand curve. Figure 2.8 illustrates the impact of income. But what about per capita consumption and real prices for these years? We chose 1951 and 1970 for certain reasons: (1) the real prices of shrimp were almost identical at \$0.50 per pound, and (2) the predicting equation was extremely accurate in each year (see table 2.12).

In figure 2.8, we see that at a constant real price, per capita consumption of shrimp increased from 0.84 to 1.58 lbs over the 1951-1970 period. Except for year-to-year fluctuations in price, which influenced fluctuations in per capita consumption of shrimp, real prices showed no real trend over the period of analysis. The data indicate, therefore, that real per capita income explains (statistically) the rise in per capita consumption of shrimp. As far as the *aggregate* consumption of shrimp is concerned, we have found that its rise over the period of study was presumably due to (1) rising affluence and (2) population growth. We should again be cautious about causality. How much did each demand factor increase the aggregate consumption of shrimp in the United States? Per capita consumption more than doubled over the period (2.2 times), but population increased 1.42 times. Multiplying these increases together (2.2 x 1.42), we obtain 3.13—or aggregate consumption of shrimp more than tripled over the 1947–1971 period. Roughly 60 percent of the increase was due to a rise in per capita consumption stimulated by rising affluence, and the balance was due to population growth.

Finally, what can be said about the price and income elasticities for shrimp? The price elasticity e_p is important for predicting change in sales, and the income elasticity $e_{Y/N}$ is important for evaluating the response of consumption to income. In 1951 and 1970, the per capita shrimp consumption was 0.56 and 0.92 lbs, respectively. Furthermore, per capita consumption of shrimp increased by over 80 percent in response to a 58 percent increase in income (holding the price constant at $0.50). This is indicative of a reasonably high income elasticity. Within the range of the market data, we can estimate the average price and income elasticity over the 1947–1971 period:

$$e_p = -.30$$

$$e_{Y/N} = 1.37.$$

The implications of these elasticities are quite clear. Over the period of study, increases (decreases) in price have generally meant a rise (decline) in revenue for the shrimp market (see table 2.3), since prices were (on average) in the inelastic portion of the demand curve. For this period, the income elasticity was substantially greater than one and showed no sign of declining. Thus over this range of income per capita, Engel's law with respect to shrimp was not validated by the data.

Previous Studies: Fishery Demand Analyses

There have been several studies of the critical price, income, and cross elasticities for groundfish (cod, haddock, flounder) both in the United States and the United Kingdom. Bell (1968a) and Farrell and Lampe (1965) have studied the New England groundfish market. They generally found that various species of groundfish are price elastic, are income inelastic, and are influenced by the price of meat, poultry, and other fish prices. A contrasting study of haddock, cod, plaice, and other demersal (groundfish) species by the White Fish Authority (1971) indicated price *inelasticity* for the same species marketed in Britain. We might be led tentatively to conclude that New Englanders have more substitutes at hand than Englishmen. Here we see a substantial difference between two countries in consumption behavior for almost identical products. Crutchfield and Zellner (1962) indicated in their study of the Pacific halibut, a flat groundfish, that the demand was very price elastic.

Table 2.13 is a survey of what is probably the best estimate now available on price, income, and cross elasticities. Note the tendency for prices to be highly elastic within the groundfish group. This group also exhibits strong cross elasticities with meat and poultry. Except for yellowtail flounder and halibut, the income elasticities are quite low. Canned tuna is a notable exception to the finfish group: it is price inelastic. Apparently, the consumer substitutes other commodities with some reluctance when tuna prices rise. Over the last twenty years, canned salmon has become less and less a competitor for the consumer's dollar. Statistical results indicate that salmon has no association with increasing affluence. However, fresh and frozen salmon steaks have a highly positive income elasticity, and more and more salmon is being diverted to this market. Like white flour, canned sardines are apparently an "inferior good." However, Purcell and Raunikar's (1968) cross-sectional study indicates a positive income elasticity for "sardines in oil." The results on sardines are far from conclusive.

Suttor and Aryan-Nejad (1969) confirm our general results on shellfish. Shellfish are consistently price inelastic. Shrimp, lobsters, and crabs consistently exhibit income elasticities greater than one, and scallops, clams, and oysters have inelastic income responses. Fullenbaum (1971) and Purcell-Raunikar (1968) could find no relation between income and the per capita consumption of either oysters or clams; this lends further support for our income elasticity for oysters.

Finally, fish meal utilization in the United States is price elastic. Over the period 1950–1971, a 1 percent increase in chicken consumption produced a "derived demand" for fish meal, the consumption of the latter increasing by 1.054 percent. There is a very strong cross elasticity between the demand for fish meal and the price of soybeans. According to table 2.13, a 10 percent increase in the price of soybean meal will result in a 12.3 percent increase in the aggregate demand for fish meal.

There have been a few studies of price and income elasticities for all fish and shellfish. Nash (1967) reported a price elasticity in the United States for *all* fish and shellfish of -0.45. For fresh and frozen fish and shellfish (no canned items), he found the income

Table 2.13: Estimated Price, Income and Cross Elasticities for Selected Fishery Products, United States

Fishery Product	Price Elasticity	Income Elasticity	Cross Elasticity	Period	Area
Food Fish					
A. Finfish					
1. Groundfish					
(a) Large Haddock[1]	-2.17	.46	1.91[2]	1957-67	New England
(b) Small Haddock[1]	-2.19	.33	4.89[2]	1957-67	New England
(c) Cod[1]	-3.15	.10	5.48[2]	1957-67	New England
(d) Yellowtail Flounder[1]	-2.29	1.97	4.01[2]	1957-67	New England
(e) Halibut	-10.00[3]	N/A	N/A	1953[3]	Seattle
	-5.00[4]	1.21	N/A	1947-71	Pacific Coast
2. Canned Tuna[5]	-.70	1.21	None	1947-71	U.S.
3. Canned Salmon[6]	-1.00	0.0	None	1947-71	U.S.
4. Fresh and Frozen Salmon[6]	-1.30	1.62	None	1947-71	U.S.
5. Canned Sardines[7]	-0.93	-.29	None	1950-68	U.S.
B. Shellfish					
1. Shrimp (Canned, fresh and frozen)[6]	-.30	1.37	None	1947-71	U.S.
2. Lobster (American and Spiny)[6]	-.65	1.95	None	1947-71	U.S.
3. Crabs (blue, dungeness and king)[6]	-.31	1.21	None	1947-71	U.S.
4. Scallops[6]	-.65	.74	None	1947-71	U.S.
5. Clams[7]	-.61	.25	None	1948-67	U.S.
6. Oysters[7]	-.67	0.0	None	1948-67	U.S.
Industrial					
C. Fish Meal (Utilization)[6]	-1.30	1.054[8]	1.23[9]	1950-71	U.S.

1. Bell (1963); 2. Meat and Poultry; 3. Crutchfield and Zellner (1962); 4. NMFS-BEI (1972); 5. NMFS-BEI (1973); 6. (Chapter 2 in this book); 7. Bell et al.(1975); 8. Percentage change in fish meal utilization in response to a percentage change in U.S. chicken consumption; 9. Price of soybean meal

elasticity to be in the range of 0.65 to 1.00. Robinson and Crispoldi (1971) place the aggregate income elasticity for the United States at 0.28.

Our study of price, income, and cross elasticities for various fishery products is extremely important. First, it provides a basis for demand projections or forecasts. When these projections are compared to available supplies (chapter 3), we shall get a better idea of the sea's potential as a source of protein. Second, price elasticities indicate the direction of the change in sales when prices rise or fall. Farrell and Lampe state that "If a decline in haddock landings were to occur in consequence of long-term factors such as an increase in fishing intensity in Georges Bank, gross revenue to producers [fishermen] would decrease. This follows directly from the fact that demand at the port for fish is highly [price] elastic" (1967, p. 50). In chapter 4, we shall look at demand projections on a world basis for the various kinds of fish. We have deferred these demand projections until we can develop the supply side of the problem, which is covered in chapter 3. Let us now look at the demand determinants for fish in "centrally planned economies."

Demand Determination in Centrally Planned Economies

Although we have not specifically stated that consumers are free to make individual choices in what is called a "market economy," we have strongly implied it. A market economy is strongly associated with the term *capitalism.* Capitalism is a form of economic organization that allows producers and consumers wide latitude, or what is sometimes called freedom of choice, in their economic decision making. Individuals are free to enter business at their own discretion, and consumers freely choose among products on the market.

The opposite end of the spectrum is a centrally planned economy. Decision making is less individualistic and more centralized among a few government leaders. Whether capitalistic or centrally planned, an economic system still must deal with the fundamental problem: scarce resources. That is, there is only so much labor, capital (machinery, plants, etc.), and raw materials to use in producing products for a growing population. In a capitalistic system, a decentralized process of individual decision making—through competitive offers to buy and sell products—determines what is produced. In centrally planned economies, resources—capital, labor, and raw materials—are allocated by a small group of individuals who determine output targets. For example, the USSR has gone through the process of allocating *scarce* resources to develop heavy industry, such as steel, farm equipment, and electrical machinery. Because of scarce resources, consumer products such as autos, refrigerators, toasters, and TV sets have not been produced in great abundance.

But what does this all have to do with fish? Robinson and Crispoldi (1971) estimated that in 1970 centrally planned economies consumed almost 32 percent of the world's fish catch. N. P. Sysoev's *Economics of the Soviet Fishing Industry* (1970) describes the process by which fish consumption is determined in the Soviet Union—a centrally planned economy. First, the Institute of Nutrition of the USSR Academy of Sciences has set the standard yearly consumption of fish products at 18.2 kg per capita (approximately forty pounds). According to Sysoev, this physiological standard for the consumption of fish products is not immutable. On the supply side, the Soviets establish targets for the fishing industry, which must supply the population with a limited assortment of high-grade fish products and supply stockbreeding with fish meal. Remember that the state owns the means of production (i.e., fishing vessels and equipment). Thus, Sysoev states that "it will be necessary to put on the market a wide range of food products able to satisfy the requirements of even the most fastidious consumer" (1970, p. 29). But under Soviet socialism the aim of production is to provide a "correct" diet. The output of the fishing industry is important to a rational diet since it is a major source of animal protein. In a planned economy, the consumer's choice is narrow and is founded more on basic nutritional needs rather than on a choice among a variety of products offered by hundreds of thousands of individual producers. In a planned economy such as that of the USSR, the state is the only producer. The production program is the basic part of the development plan of the industrial sector. To quote Sysoev once again, "It [development plan] is a system of targets for the volume of production and sales, their growth rates, the nomenclature, assortment, and quality of fish products over a definite period" (1970, p. 57).

But what if the consumer wants a pizza, hamburger, fishburger, submarine sandwich, or a foot-long hot dog? Here is the rub! The state decides what is produced in the "best interest" of the public. And here is the basic difference between planned and capitalistic economic systems: the means of production do not respond to consumer preferences, although some consideration is given to foods that are culturally acceptable to the Soviet consumer at various stages in the planning process. That is, although the Soviet people are given some variety in their diet (determined mainly by the Institute of Nutrition) and allocate their income according to their preference as in a capitalistic society, their choices are circumscribed by the objectives of the state, which are determined by relatively few individuals.

The Marketing and Distribution of Fish

Nutritional Value of Fish

The marketing and distribution channels are charged with the job of convincing the public of the value of fish. Why should one eat fish? The physical and nutritional condition of our bodies depends mainly on the type of protein we eat. For example, scientists have found that people living along the northeast coast of Brazil are tall, healthy, and vigorous. Most of their protein comes from fish. Where vegetables are the main source of protein, people tend to be shorter and malnourished. The answer lies in the *quality* of protein. This was discussed briefly earlier in this chapter. Many grains and vegetables lack one or more of the eight amino acids *essential* for good nutrition. Animal protein, including fish, provides a good balance of amino acids. In addition, fish is more digestible than meat.

To look at the *quality* of protein from fish, we must first learn how much raw (unadjusted) protein is obtained. This is called the conversion ratio. Examples for various foods are given in table 2.14. That is, for a kilogram of the commodities listed below, fish yields a considerably higher percentage of raw protein than, for example, meat does. The chemical score is equal to the quantity of the most limiting amino acid in the food protein expressed as a percentage of the same amino acid in the amino acid scoring pattern.[18] Table 2.15 shows the protein-efficiency conversion (or the chemical score, CS).

Table 2.14: Raw Protein in Selected Foods

Item	Percent of Moisture-Free Flesh
Red meats	40-50
Shrimp	88
Flounder	80
Rice	7

Source: Lovell (1973)

Table 2.15: Protein Efficiency Conversions or CS

Food	CS (Egg=100)	Limiting Amino Acids
Fish Muscle	83	Tryptophan
Beef Muscle	80	Methionine + Cystine
Rice	57	Lysine
Navy Beans	47	Methionine + Cystine

Source: World Health Organization (1965)

Each food has an amino acid profile or pattern. Of all animal foods, fish is the best source of lysine, which is the amino acid most limiting in cereal proteins. Lysine is more than 10 percent of the total protein in fish and only 2.8 percent in rice. In table 2.15, eggs are equal to 100. The chemical score of all other foods is expressed in terms of eggs. The chemical score of a food can then be used to adjust raw protein (table 2.14) for quality. In a study by Bell and Canterbery (1976), the chemical score for catfish, mussels, and carp was equal to 100. Eels and rainbow trout scored 89 and 80, respectively. This index can be multiplied by the raw protein yield to obtain quality-adjusted protein. For example, we can contrast fish and rice.

	Weight	*Raw Protein Yield*	*Quality Adjusted Protein Yield*
Rice	1 lb	.07 lbs[1]	.04 lbs[3]
Fish	1 lb	.84 lbs[2]	.70 lbs[3]

1. Table 2.14.
2. Average of shrimp and flounder in table 2.14.
3. Table 2.15.

This shows the dramatic difference between the quality of protein obtained from rice, which is the staple of most of the world, and the quality of protein obtained from fish.

Protein requirements are usually expressed in grams/day/kg of body weight. The ratio is much higher for children than adults because protein is required for growth. In *undernutrition,* the structure of diet is nutritionally satisfactory, but *insufficient quantity* is being consumed. In *malnutrition,* the structure of the diet is nutritionally unsatisfactory. Many studies indicate that there is a protein deficiency in many countries. Is this malnutrition or undernutrition?

According to Sukhatme (1972),

> If one examines the available data the conclusion is clear that what diets lack is adequate energy food which would enable active utilization of protein people eat. . . . protein malnutrition is the indirect result of inadequate energy intake. . . . when food intake is sufficient in amount the protein intake is usually satisfactory. Apparently people can have enough and more of the protein they need from the cereal/pulse diet they normally consume if only they eat enough to meet their energy requirement.

In essence, Sukhatme maintains that if one fails to consume more that 1.5 times the calories needed for basal metabolism (BMR), calorie and protein deficiency will result, no matter what the protein intake.

The controversy stems from the fact that Scrimshaw et al. (1969) saw protein as the *primary limiting factor* (not calories). Hence, the UN report on "International Action to Avert the Impending Protein Crisis" (1968) recommends the production of protein foods such as fish. The Sukhatme-Scrimshaw debate can lead to quite different policy conclusions.

The Distribution and Sale of Fish: Changes in Marketing

Fresh Fish. The major problem with the marketing of fresh fish is perishability. As we have seen, it is in the interest of producers to sell fresh fish because of the price differential between fresh and frozen fish. In countries such as the United States, Italy, and the United Kingdom, which have extensive coastlines, fresh fish can be marketed in cities close to landing areas. In a study of two cities (Quincy, Massachusetts, and Binghamton, New York) near the northeastern coast of the United States, Krebs and Storey (1969) found that 25 percent of the consumers had occasionally or frequently had bad experiences with taste or odor for *fresh* haddock, flounder, and cod. In Quincy, a coastal town, fresh fish was purchased more frequently than frozen or canned. In Binghamton, an inland city, frozen and canned fish were purchased more frequently than fresh. According to Krebs and Storey, inland households buy less fresh fish simply because the products were not continuously available and quality was not optimal. As the price of fresh fish increases, greater incentives may induce suppliers to transport by air. A study by Schary et al. indicated that the physical distribution of fresh salmon from the northwestern United States to the east was "dominated by transportation mode choice decisions" (1971, p. 2). And, as we saw in our section on income elasticities, the market for fresh salmon is growing rapidly. On a world basis, fresh fish is relatively less important by value, but it is still the most important by volume (42 percent of food fish consumption).

Frozen Fish. Because freezing overcomes perishability, frozen fish have steadily grown in importance, especially in fairly affluent countries. According to Bell, the marketing of frozen fish has changed drastically in the United States and many Western countries: "The lower landing prices have greatly contributed to the success of the Canadians in the U.S. market for frozen fillets. The Canadian success in capturing the vast U.S. domestic market for groundfish might not have been so *dramatic* and *extensive* except for the advent of the 'fish stick revolution.' In the past decade [1955–1965] this revolution has transformed the Nation's fish consumption habits" (1966, pp. 33–36). Thus, the frozen groundfish product has changed from fillets to fish sticks and frozen fish dinners that can be marketed through supermarkets and convenience food outlets; frozen groundfish under various brand names have been able to achieve national appeal. The widening of the frozen fish market has increased the demand on the groundfish resources of the North Atlantic; this will be discussed in chapter 3. In addition, freezing has permitted the marketing of new species such as ocean perch in the midwestern United States. Developing areas traditionally have had poor distribution channels. They are clustered, with few exceptions, near the equator, where fish must be eaten immediately after being caught or cured. Since they have not had freezing equipment and retail outlets with proper refrigerating facilities, frozen fish have not been marketed on a large scale.

In the shellfish category, shrimp in the United States are distributed from plants in the Gulf to market centers throughout the nation. New York appears to be the only major market area outside the Gulf that sells fresh shrimp. Miller and Nash (1969a, 1969b) have investigated radiation processing to eliminate spoilage organisms as well as bacterial pathogens and animal parasites. This technique promises to be relatively inexpensive, yet it does not compromise the taste or texture of the product. These economic feasibility studies were carried out for West Coast flounder and Gulf of Mexico shrimp. It was found that low-dosage irradiation can add two weeks to the shelf life of Pacific flounder without altering the "fresh" quality of the fish. At the time of these studies, irradiation preservation of Pacific flounder was not economically feasible but it was feasible for Gulf shrimp. Unfortunately, these studies were not followed up by government (i.e., U.S. Atomic Energy Commission and Department of Agriculture) or industry. Moreover, the problem of consumer acceptance of irradiated products remains (frozen and cured account for 26 and 17 percent of world food-fish consumption, respectively).

Canned Fish. Canned fishery products—primarily salmon, tuna, and sardines—have a mass national and international market. The volume of canned salmon is not limited by the acceptance of the product in the market, but rather by the limitations of the resource. The canned tuna market, as indicated above, has expanded very rapidly in the United States, primarily through the importation of raw tuna from Japan. The consumption of canned fishery products is primarily a characteristic of high-income countries. In a survey of canned fish acceptance, the major reason given by respondents for disliking this product—canned salmon, tuna, sardines, and shrimp—was that they do not consider it appetizing. Only for canned shrimp was there a strong preference for fresh or frozen product. Nutrition and convenience are among the positive factors contributing to the demand for canned fish. Canned fish was among the original convenience foods. The "power of advertising" is used only for canned fish that have an ample resource base—such as tuna. Increasing the volume of salmon sold, through advertising, would be restrained by the limited resource base (canned fish is 15 percent of world food-fish consumption).

Industrial Fish. As we shall see later, the greatest increase in the use of fish has been in the production of fish meal and oil. World production of fish meal and solubles has increased from 877 million pounds (product weight) in 1947 to 11,993 million pounds in 1970, or a whopping 1267 percent. As noted above, fish meal is a protein concentrate that is extracted from raw fish by cooking, pressing, and drying and that is used as feed, primarily for poultry. Is this an inefficient way of delivering protein to the world? The answer is an unqualified yes! However, the Peruvian anchoveta has been both a success story and a major source of fish for reduction. We must again go back to the individuality of consumer choice. Although consumer choice is limited considerably in planned economies, few people would show a great preference for an "anchoveta dinner." To many the idea would be repulsive! Menhaden, which is caught in the Atlantic and Gulf of Mexico, has no food fish appeal, even at bargain basement prices. The major users of fish meal are the United States, the USSR, Japan, the United Kingdom, and West Germany. In recent years, the Peruvian anchoveta has been in trouble as a resource base. In 1972, an oceanographic phenomenon associated with water temperature—termed "El Niño"—has apparently contributed to a serious recruitment failure.[19] Thus, the entire stock of Peruvian anchoveta is at a low level. In chapter 1, we saw the Peruvian catch drop from over 10 million metric tons in 1971 to 2.3 million metric tons in 1973.

Marketing Margins

The differences between the prices charged by the producer and those paid by the consumer can be explained by a price spread. For a fish product, the price spread is the difference between the price paid for the final product by the consumer and the dockside value of an equivalent weight of the product. This difference includes the payments to all agents who perform services in moving fish products to the consumer. These services include handling, processing, storage, transportation, wholesaling, and retailing.

The whole realm of fish processing and marketing is usually overlooked in the literature on the fisheries, but it has a profound impact on the resource base, as was indicated above. Penn (1975) has done the most exhaustive study of marketing margins in the United States. Why are these margins so important? First, the producers, or fisherman, depend upon a resource base that is, in many cases, seriously overfished; therefore, his productivity declines, and the cost of harvesting increases. However, a fish processor may introduce machinery or different production methods to increase productivity and thereby lower cost. This may keep fish competitive at the retail level. Second, there has recently arisen a controversy over who is responsible for price increases for food in general. Fish is naturally involved. The fishermen blame the so-called middlemen, and the consumer is usually confused about the entire issue. Table 2.16 shows for the major fishery products marketed in the United States, just exactly how the consumer's dollar was spent in each of the various marketing levels. Note that for American lobsters and frozen sea scallops, the fisherman received 45.38 and 62.6 percent of the retail value of these shellfish. The reason is obvious. Lobsters are marketed live (no processing), and scallops are shucked before they are landed (i.e., the fisherman does all the processing). Processing costs (as a percent of the consumer dollar) are highest for canned items such as tuna, salmon,

Table 2.16: Distribution of Consumer's Dollar Spent on Various Fish Products in the United States According to the Average Prices of 1972-74, by Marketing Functions and Cost Items

	1 Fresh flounder fillets	2 Fresh cod fillets	3 Fresh haddock fillets	4 Frozen ocean perch fillets	5 Fresh salmon steaks	6 Frozen halibut steaks	7 Canned tuna (light meat)
	percent						
By marketing functions:							
Retailing	29.94	30.71	26.20	33.31	17.93	23.67	24.33
Wholesaling	21.50	19.98	11.32	13.43	24.02	12.86	35.41
Processing	15.83	16.56	14.58	28.20	12.62	16.65	
Harvesting	32.74	32.74	47.90	25.06	45.43	46.82	40.26
Total	100.00	100.00	100.00	100.00	100.00	100.00	100.00

	8 Canned pink salmon	9 Canned sardines	10 Fresh peeled gulf shrimp	11 Fresh blue crab meats	12 Frozen sea scallop meats	13 Live American lobsters
	percent					
By marketing functions:						
Retailing	24.77	26.13	18.74	30.97	28.80	27.06
Wholesaling		1.48	14.16	18.26	8.60	27.56
Processing	42.94	66.89	13.64	30.88	-	-
Harvesting	32.29	5.50	53.46	19.89	62.60	45.38
Total	100.00	100.00	100.00	100.00	100.00	100.00

Source: Penn 1975.

and sardines. Retail margins varied from 17.93 to 33.31 percent of the consumer's dollar for fresh salmon steaks to frozen ocean perch. Ex vessel prices for ocean perch are the lowest among the groundfish. This reflects the greater productivity of the ocean perch resource base (this will be extensively discussed in chapter 3). Hence, the processor's and retailer's margins are proportionally higher than those of other groundfish, since labor and overhead expenses are fixed for all products regardless of their differences in value. Salmon steaks and lobster are shipped by airfreight, and the wholesaler incurs a much larger transportation cost.

Penn (1974) has looked at these margins over varying periods. Figure 2.9 illustrates the trend in retail, wholesale, processing, and harvesting prices for selected species. Two features are immediately noticeable: (1) all the price series move fairly well together, and (2) the *absolute* price differential between marketing levels appears to be maintained. Penn did find that the fisherman's share (ex vessel price divided by retail price) *increased* over the 1959–1971 period for fresh haddock, fresh and frozen halibut, canned tuna, frozen shrimp, lobster, and sea scallops and decreased for fresh flounder, fresh cod, frozen ocean perch, canned salmon, and fresh blue crab. These rather mixed results seem to indicate that unique factors operate in the market for each product. However, the information is quite valuable; for any period of time, it gives the principal reason for retail price increases. That is, which level of distribution is contributing to the increase, or are they all contributing? Once this is ascertained, one can dig deeper into reasons, such as changes in input costs or productivity (i.e., output per unit of inputs such as capital or labor).

Are We Keeping Up?

This chapter began with a discussion of the food crisis confronting the world. As indicated in chapter 1, we view the fishery resources of the sea as an increasingly valuable product. As our theory and empirical experimentation seem to indicate, increases in population and affluence are the main determinants of demand, but a rise in fishery prices acts as a rationing device in reducing demand. The FAO indicates that "the gap between the nutritional requirements and the actual consumption of protein by the greater part of the population in the developing countries is widening. Protein deficiency already has serious consequences for the health and working efficiency of the population of develop-

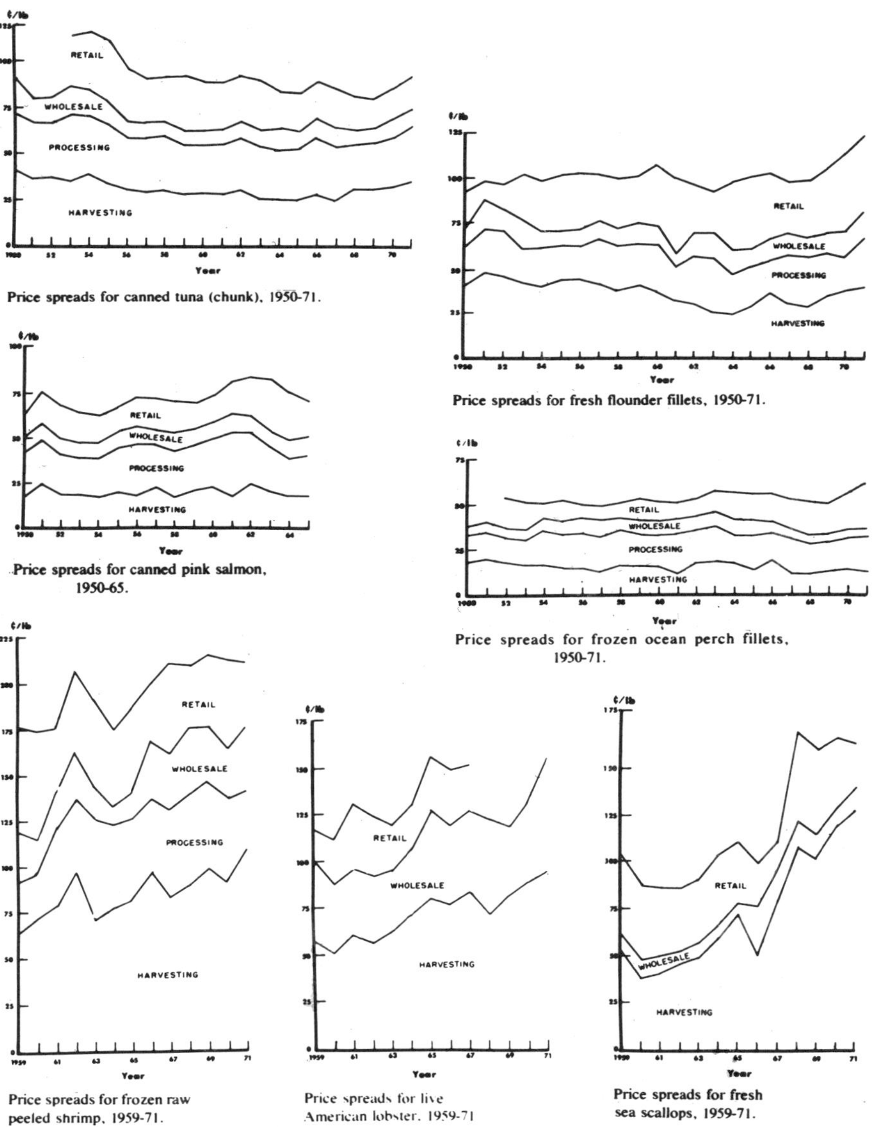

Price spreads for canned tuna (chunk), 1950-71.

Price spreads for fresh flounder fillets, 1950-71.

Price spreads for canned pink salmon, 1950-65.

Price spreads for frozen ocean perch fillets, 1950-71.

Price spreads for frozen raw peeled shrimp, 1959-71.

Price spreads for live American lobster, 1959-71

Price spreads for fresh sea scallops, 1959-71.

Figure 2.9: Retail, Wholesale and Ex Vessel Prices for Various Fishery Products, U.S. Source: Penn (1974)

ing countries (1968, p. 4). FAO goes on to say that "this dependence on the developing world on food shipments from the industrialized countries is increasing steadily and seems likely to continue" (1968, p. 5). One of FAO's specific proposals is to "increase protein production from marine and fresh water fisheries sources" (1968, p. 11).

Let us look at some aggregate trends by species. Figure 2.10 shows an index of the world per capita production/consumption for selected shellfish. It neglects the distribution problem and focuses upon whether production is keeping pace with demand pressures. Clam, crab, and, to a lesser extent, shrimp production are more than keeping up with population growth; their trends are generally up over the 1953–1970 period. Oysters and lobsters are holding their own (i.e., per capita production/consumption has remained stable over the period). Scallops are subject to wild fluctuations but show no pronounced downward trend.

In chapter 1, we pointed out that aggregate fish production had grown at a rate well in excess of population, but began to flounder (excuse the pun, please) after 1970. Figure 2.11 shows the same statistics for finfish and fish meal. As indicated above, the utilization of fish meal per capita increased by 277 percent from 1958 to 1970. However, "El Niño" has drastically reduced this increase. Groundfish and tuna showed slight upward trends, but the salmon index declined almost continuously. As we shall see in chapter 3, a resource crisis is upon us and will drastically alter these trends. However, table 2.2 indicates that the major fish consumers are affluent, not developing, countries. Thus, the per capita distribution of the major fishery products does not provide the developing countries with fish protein. What fish protein they are able to get is largely through labor-intensive (i.e., small vessels or boats with large crews) coastal and freshwater fishing.

If anything, the fishery resources of the world may be analogous to the oil situation: the flow is from developing to developed countries. The international markets are competitive for fishery products (as opposed to the OPEC cartel) but indicate that our discussion on demand determinants explains the flow of valuable fishery resources to countries with the ability to pay. Although the United States is more than self-sufficient in food, its citizens are willing to pay the price for the massive fish imports that now supply around 75 percent of domestic consumption. The fishery resources

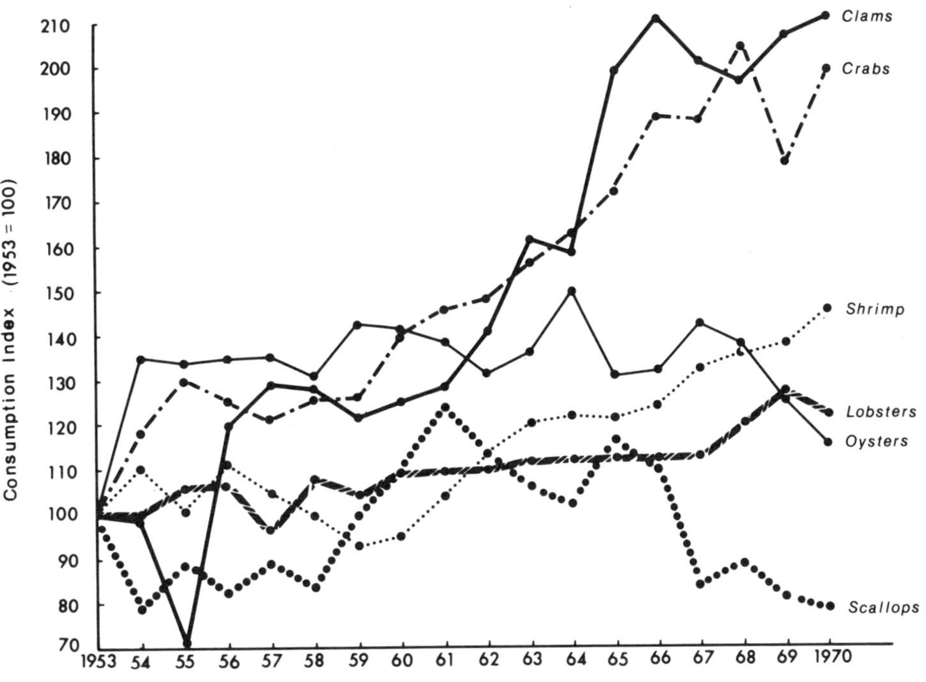

Figure 2.10: Index of the Trend in World Per Capita Production/Consumption for Selected Shellfish, 1953=100

Source: Basic Economic Indicators (1973-75) and United Nations

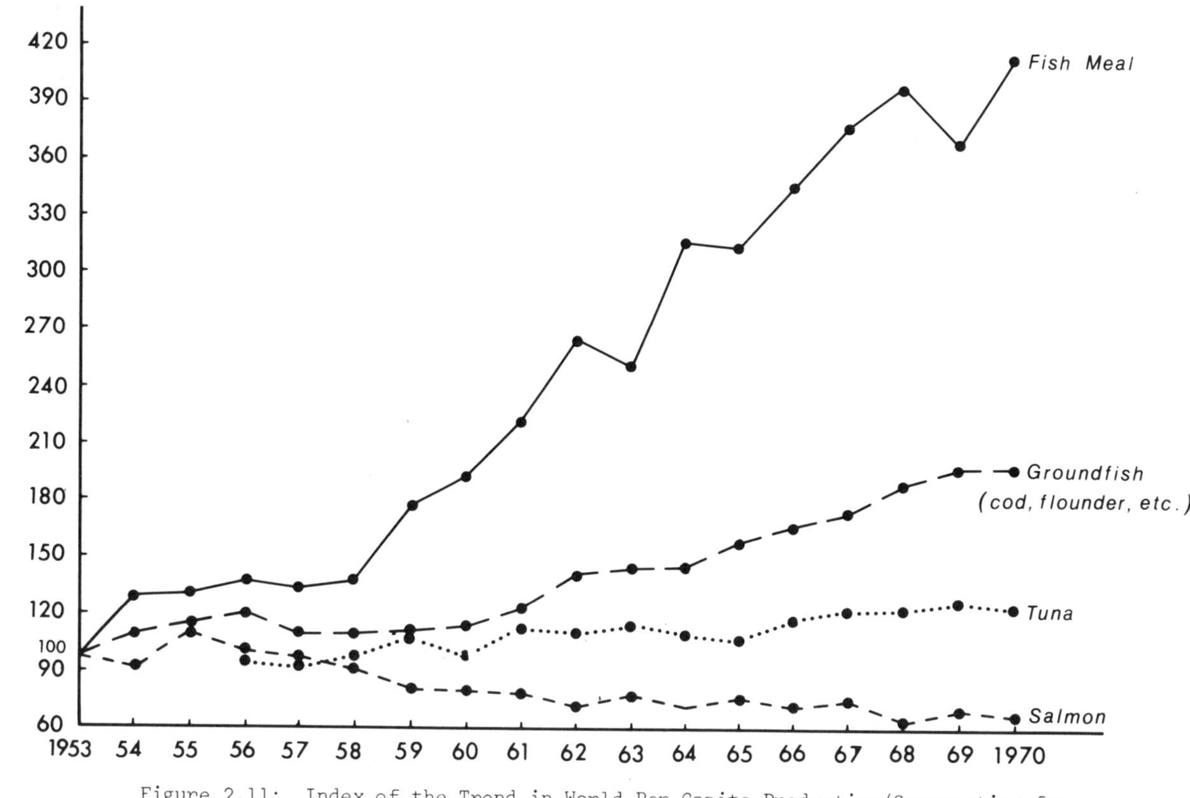

Figure 2.11: Index of the Trend in World Per Capita Production/Consumption for Selected Finfish and Fish Meal, 1953=100

Source: Basic Economic Indicators (1973-75) and United Nations

of the sea are polarizing the nations of the world into those that have abundant food and those that do not.

Summary

We have seen that the demand for fish is growing quite rapidly throughout the world. The population/food problem has brought greater demands for protein from the sea. The major fish-consuming nations are generally affluent and can afford to pay for an increasingly expensive seafood item in their diet. Fish now supplies up to 9 percent of the world's protein adjusted for quality.

In our theoretical discussion of demand for fishery products, we hypothesized that price, real per capita income, and the price of close substitutes determined the level of per capita fish consumption. It is important to know the corresponding elasticities (price, income, etc.) because of their use in prediction and changes in sales. Engel's law, while generally valid for food, is not immutable for selected fishery products, as revealed by our statistical studies. Of course, population growth is another demand determinant that influences the *size* of the market. Other demand factors—the size of the fish, fresh versus frozen, distance from the sea, and demographic characteristics—are also important for selected species.

Using the U.S. market for shrimp, we showed that the theory of demand is consistent with market experience. That is, price, real per capita income, and population allowed us to trace the changes in the level of aggregate consumption for shrimp. No close substitutes were found.

Generally, a survey of the literature indicated that finfish (with the exception of tuna) were price elastic, especially groundfish. Thus, downward pressure on groundfish prices would increase sales to the fisherman. The consumption of cod, haddock, and flounder was appreciably affected by the price of competing products such as meat and poultry. Shellfish, in general, were price inelastic, which implies that the consumer has a strong preference for these products and that his consumption is apparently relatively insensitive to price changes. Income elasticities were greater than one for luxury shellfish such as lobsters, crabs, and shrimp, as well as for canned tuna. Finally, the fish meal market was heavily influenced by price, chicken consumption, and the price of soybean meal, a close substitute.

About one-third of the world's fish is consumed by planned economies (as opposed to market economies). Our market analysis differs greatly by economic organization. The consumer has less individual choice in planned economies since the means of production are state, not private, property. We discussed the consumption of fish in the USSR as an example of a planned economy.

An important element of demand analysis is marketing research and distribution channels. We discussed the change in the marketing of fish from limited regional markets to national convenience store outlets. Fresh, frozen, and canned fish have increased from 1968 to 1973 for the world. We also discussed the protein value of fish and the controversy surrounding FAO recommendations for stimulating the production of fish for protein. Marketing channels and margins were studied using U.S. data as an example.

Finally, although the world per capita production of many species is still rising, the ocean's fishery resources are distributed primarily among the affluent nations. In many instances, developing countries supply fish in return for foreign exchange. These export earnings may or may not help developing countries with respect to food, depending on their political regimes. Thus, the pressure of world demand is more of a crisis involving the world distribution of protein from the sea. Even for affluent nations, however, potential of the sea is limited, and fishery resources will become increasingly more valuable. Therefore, prices will act as a mechanism to divert most of the fishery resources of the oceans to a small group of affluent nations. Chapter 3 will be concerned with the "crisis of supply," which has received much publicity in recent years.

Notes

1. In chapter 1, we mentioned the Club of Rome. One of the first publications of this organization was *The Limits to Growth* by Meadows et al. (1972).

2. The quality of protein varies from food to food (i.e., milk vs. beef). That is, protein is a combination of amino acids, each varying in its potential to supply strength and energy to the human body. Thus the particular combination of amino acids in foods determines protein quality. This topic will be discussed in greater detail later in the chapter.

3. What is a major fish category will obviously vary by area of the world, since preferences differ considerably from one culture or nation to another. We have selected these categories because of their size and importance in international trade.

4. Our data are taken from the *FAO Yearbook of Fishery Statistics*, 1973. However, many communist countries were not members of the United Nations (such as North Vietnam and the People's Republic of China) and

therefore did not report their fishery catch in great detail, but only totals. If adequate data existed, we might expect China to replace many of the countries that we have labeled the four leading countries in fish consumption.

5. We could also include the prices of complementary products. Complementary products are the opposite of substitutes. In the food area, steak goes well with steak sauce or mushrooms. As discussed below, fish may have many complementary products, but their prices do not have an appreciable impact on fish consumption.

6. The simplifying assumption is that all other food items are substituted to some degree for tuna as its price rises to the exclusion of close substitutes because their prices remain constant by assumption. In reality, there is an interaction: a rise in price of tuna may stimulate the demand for salmon, for example, and the price of salmon will increase. This is the difference between partial and general relations. We shall stay with the simplified partial relation.

7. In this example, we are assuming that the consumer does not save any of his income, but spends the entire amount on various products. At higher levels of income, we would subtract out savings leaving income that is allocated among many products the consumer may choose. In addition, the lower price of tuna has an income effect or produces more income available to purchase tuna. This is also responsible for increased tuna consumption.

8. The relation between P_T and q_T may be either linear or curvilinear. It would depend entirely on the behavior of the consumer in terms of his consumption response to a change in price. That is, the law of demand states that there should be an inverse relation between q_T and P_T, but the exact form is an empirical question.

9. You have probably heard or read in the newspaper that the dollar is only worth thirty-five cents today. This statement means that inflation has reduced the purchasing power of the dollar: it will only buy thirty-five cents worth of goods today compared to the same goods several years ago.

10. Income is often expressed in terms of the family unit (usually consisting of two adults and one or more children). For a family of four, its per capita income in the $10,000–$15,000 interval would be roughly $12,500 (the midpoint of the interval) divided by four, or $3,125.

11. There is a third elasticity that measures the extent to which goods are complements or substitutes. Take cod (c) and ground beef (g). If ground beef prices rise, this may affect the quantity of cod consumed or $E_{cg} = [\Delta Q_c/Q_c] \div [\Delta p_g/p_g]$. This *cross elasticity* has a positive sign for substitutes and a negative sign for complementary goods. We would expect a positive sign between cod and ground beef.

12. Traditional theoretical treatments of demand theory hypothesize an individual demand function for each person in the population. These are then added up (i.e., in the case of the demand curve, it is a horizontal summation of all individual demand curves) to obtain an aggregate demand. This only shows that people are different. Our approach is mathematically identical to this and, as indicated above, makes the statistical results of estimating demand functions for fish easier to understand.

13. A cross-section study is one that explores the relation between demand determinants and consumption using a panel or sample of individuals of diverse characteristics at one point in time.

14. Hamito designates a group of African languages related to the Semitic language and including ancient Egyptian, ancient Libyan, modern Berber, and Ethiopian dialects. Semitic designates a major group of languages of southwestern Asia and northern Africa.

15. Ordinarily, we would use consumer prices for shrimp. Consumer prices for headless shrimp, 26–30 count, were available at New York from 1950–1963. From 1964, the U.S. Bureau of Labor Statistics started collecting more representative data from cities around the nation. Ex vessel prices are available from 1947 and represent a more reliable series. Unless marketing margins change considerably over the period, ex vessel prices are an excellent reflection of the behavior of retail prices. The use of ex vessel prices also allows us to focus on the producer as opposed to the retailer, although the same theoretical base applies to both.

16. Disposable personal income is what the consumer has available for consumption or savings. Expressed succinctly, it is his "take-home pay."

17. Multiple regression is a statistical technique designed to find the right combination of parameters and is therefore the demand equation that will maximize the explanatory power of the demand determinants. A simple exposition of this technique is in Brennan (1960, chapter 19).

18. Of the eight amino acids, the smallest quality required by the body is tryptophan. This amino acid is assigned a value of one. The other seven essential amino acids are expressed relative to tryptophan as far as body requirements are concerned. Then, an ideal proportionality scoring pattern of amino acids is constructed, against which the amino acid ratios in different foods may be measured. According to this method, eggs and milk ranked highest as a reference protein against which to measure other foods.

19. Recruitment refers to those fish that enter the fishery and are available to be caught. Nonrecruits would be smaller or juvenile fish. This is discussed in greater detail in chapter 3.

References

Bell, Frederick W. 1966. *The economics of the New England fishing industry: the role of technological change and government aid.* Federal Reserve Bank research report no. 31. Boston.

——. 1968a. Economic and institutional factors affecting the demand for fish and shellfish. In *The future of the fishing industry of the United States,* ed. DeWitt Gilbert. University of Washington New Series, vol. 4. Seattle: University of Washington Press.

——. 1968b. The Pope and the price of fish. *American Economic Review* 51: 1346–1350.

——. 1969. Forecasting world demand for tuna to the year 1990. *Commercial Fisheries Review* 31, no. 12:24–31.

Bell, Frederick W., and Canterbery, E. Ray. 1976. *Aquaculture for developing countries.* Cambridge, Mass.: Ballinger Publishing Co.

Bell, Frederick W., et al. 1975. A world model of living marine resources. In *Quantitive models of commodity markets,* ed. Walter C. Laby. Cambridge, Mass.: Ballinger Publishing Co.

Brennan, Michael J. 1960. *Preface to econometrics.* Cincinnati: South-western Publishing Co.

Burk, Marguerite C., and Ezekiel, Mordecai. 1967. Food and nutrition in developing economies. In *Agricultural development and economic growth,* ed. Herman M. Southworth and Bruce F. Johnston. Ithaca, N.Y.: Cornell University Press.

Cleary, Donald. 1969. *Demand and prices for shrimp.* National Marine Fisheries Service working paper no. 15.

Crutchfield, James, and Zellner, Arnold. 1962. *Economic aspects of the Pacific halibut fishery.* U.S., Department of Interior, Fishery Industrial Research.

Doll, John P. 1972. An econometric analysis of shrimp ex-vessel prices, 1950-1968. *American Journal of Agricultural Economics* 54:431-440.

Dow, Robert L.; Bell, Frederick W.; and Harriman, Donald H. 1975. *Bioeconomic relationships for the Maine lobster fishery with consideration of alternative management schemes.* National Oceanic and Atmospheric Administration technical report, National Marine Fisheries Service SSRF-683.

Farrell, Joseph F., and Lampe, Harlin. 1965. *The New England fishing industry: functional markets for finned food fish.* University of Rhode Island bulletin 380.

——. 1967. The revenue implications of changes in selected variables examined in the context of a model of the haddock market. In *Recent developments and research in fisheries economics,* ed. Frederick W. Bell and J. E. Hazleton. Dobbs Ferry, N.Y.: Oceana Publishing Co.

Food and Agriculture Organization. 1961. *Future development in the production and utilization of fish meal.* Vol. 2. Rome.

——. 1968. *International action to avert the impending protein crisis.* A report to the UN Committee on the Application of Science and Technology to Development.

——. 1973. *Yearbook of fishery statistics.* Vols. 36-37.

Fraser, Dean. 1971. *The people problem.* Bloomington: Indiana University Press.

Fullenbaum, Richard F. 1971. A general equilibrium demand model for living marine resources: an application of general equilibrium and common property resource theory to the U.S. seafood sector. Ph.D. dissertation, University of Maryland.

Gaede, Harold W., Jr., and Storey, David A. 1969. *An analysis of consumer purchases of seafood in the Springfield, Massachusetts, metropolitan area.* University of Massachusetts, Agricultural Experiment Station, bulletin no. 584.

Gates, J. M. 1974. Demand price, fish size and the price of fish. *Canadian Journal of Agricultural Economics* 22:1-12.

Gillespie, William C.; Hite, James C.; and Lytle, John S. 1969. *An econometric analysis of the U.S. shrimp industry.* Clemson University, Economics of Marine Resource no. 2.

Krebs, Edward H., and Storey, David A. 1969. *An analysis of consumer purchases of fresh haddock, flounder and cod.* University of Massachusetts bulletin no. 579.

Lovell, R. T. 1973. The value of fish for food in developing countries. Auburn University, Center for Aquaculture, mimeographed.

Meadows, Donella H.; Meadows, Dennis L.; Randers, Jorgen; and Behrens, William W., III. 1972. *The limits to growth.* New York: Universe Books.

Miller, Morton M., and Nash, Darrel A. 1971. *Regional and other related aspects of shellfish consumption.* National Marine Fisheries Service circular 361. Seattle, Wash.

Nash, Darrel A. 1967. Demand for fish and fish products with special reference to New England. In *Recent developments and research in fisheries economics,* ed. Frederick W. Bell and J. E. Hazleton. Dobbs Ferry, N.Y.: Oceana Publishing Co.

——. 1970. *A survey of fish purchases by socio-economic characteristics, annual report, Feb. 1969–Jan. 1970.* National Marine Fisheries Service working paper no. 50.

Nash, Darrel A., and Miller, Morton M. 1969a. *Industry analysis of Gulf area frozen processed shrimp and an estimation of its economic adaptability to radiation processing.* National Marine Fisheries Service working paper no. 16.

——. 1969b. *Industry analysis of West Coast flounder and sole products and an estimation of its economic adaptability to radiation processing.* National Marine Fisheries Service working paper no. 11.

National Industrial Conference Board. 1955. *Expenditure patterns of the American family.*

National Marine Fisheries Service. *Basic economic indicators (BEI)*

CFS no. 6131 *Shrimp* (June 1973)
CFS no. 6130 *Tuna* (June 1973)
CFS no. 5934 *Menhaden* (June 1973)
CFS no. 6273 *Oysters* (November 1974)
CFS no. 6127 *Scallops* (June 1973)
CFS no. 6273 *Clams* (June 1975)
CFS no. 6132 *Blue Crab* (September 1973)
CFS no. 6133 *King and Dungeness Crab* (August 1973)
CFS no. 6272 *American and Spring Lobsters* (August 1974)
CFS no. 6128 *Halibut* (June 1972)
CFS no. 6129 *Salmon* (August 1973)
CFS no. 6271 *Atlantic and Pacific Groundfish* (June 1974)

——. 1970. Joint master plan for the northern lobster fishery. Mimeographed.

——. 1976. *Shrimp statistics—December, 1975.* St. Petersburg, Fla.

Penn, Erwin S. 1974. *Price spreads and cost analyses for finfish and shellfish products at different market levels.* National Oceanic and Atmospheric Administration technical report, National Marine Fisheries Service SSRF-676.

——. 1975. Cost analyses of fish price margins, 1972–74, at different functional levels—for management, decisions on production, distribution and pricing policies. Unpublished draft.

Purcell, J. C., and Raunikar, Robert. 1968. *Analysis of demand for fish and shellfish.* University of Georgia research bulletin 51.

Robinson, M. A., and Crispoldi, A. 1971. *The demand for fish to 1980.* FAO fisheries circular no. 131.

Schary, Philip B., et al. 1971. Distribution of fresh and frozen salmon: analysis and simulation. National Marine Fisheries Service, field manuscript 94, unpublished contract report, vol. 1.

Scrimshaw, N. S.; Béhar, Moisés; Guzmán, Miguel A.; and Gordon, John E. 1969. Nutrition and infection field study in Guatemalan villages, 1959–1964. *Archives of Environmental Health* 18: 51–62.

Sukhatme, P. V. 1972. Protein strategy and agricultural development. *Indian Journal of Agricultural Economics* 28: no. 1, 1972.

Suttor, Richard E., and Aryan-Nejad, Parviz. 1969. *Demand for shellfish in the United States.* University of Maryland miscellaneous publication 695.

Sysoev, N. P. 1970. *Economics of the Soviet fishing industry.* Translated for the National Marine Fisheries Service.

Wilcox, Walter W.; Cochrane, Willard W.; and Herdt, Robert W. 1974. *Economics of American agriculture.* 3rd ed. Englewood Cliffs, N.J.: Prentice-Hall.

White Fish Authority, Fishery Economics Research Unit. 1971. *The market for fish in Britain.*

World Health Organization. 1965. *Protein requirements.* Technical report, series no. 301.

3. The Resource Crisis: The Supply of Fishery Products

The Nature and Extent of the Crisis

The exponential expansion of the world's population coupled with rising affluence in highly developed countries has placed increasing pressure on fishery resources, which have been harvested at a rate exceeding that of most other food products. Despite international agreements and conservation regulations imposed by various nations on their coastal fisheries, the world has witnessed widespread overexploitation and the depletion of some of its most valuable fishery resources.

Some specific illustrations would help highlight the nature of the resource crisis. In 1968, the FAO published a map of the North Atlantic Ocean; it shows the approximate date at which fishing on each stock reached a level at which a further increase in the amount of fishing would give no appreciable increase in the total yield. This is shown in figure 3.1. For example, herring in the North Sea was overfished in 1950, and redfish off the coast of Newfoundland was overfished by 1960. For our purposes here, overfishing merely means to fish to the detriment of a fishing stock or, ultimately, to depletion. A more precise definition will be developed later in this chapter. Figure 3.1 shows that as early as 1905 (haddock in the North Sea) overfishing was observed in the North Atlantic Ocean. Indeed, the North Atlantic was the first area of the world's oceans to witness substantial overfishing, a fact that is due to population increases, rising affluence, and the advent of the modern fishing technology that the industrialized countries have been able to generate.

Because of substantial fishing pressure on Atlantic fishery stocks, the International Commission for the Northwest Atlantic Fisheries (ICNAF) came into existence in 1950. The convention

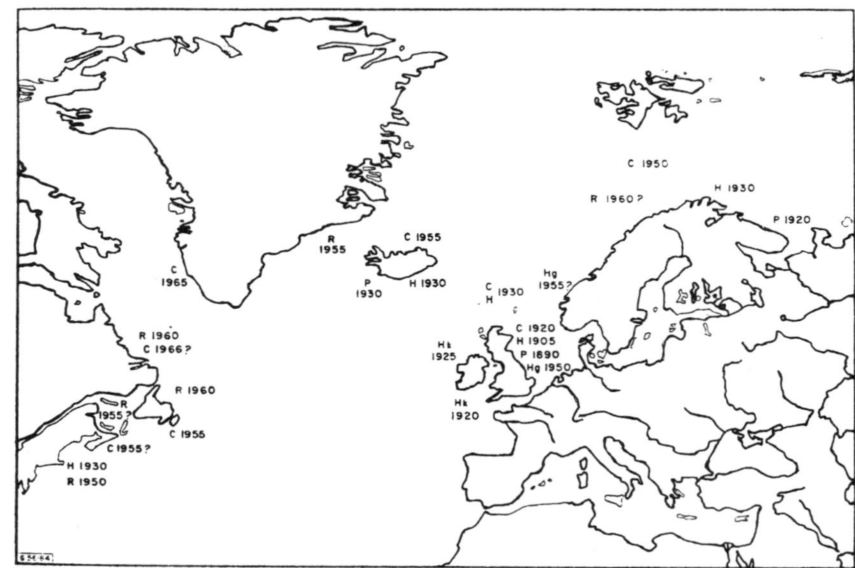

NOTE: The years are the approximate dates by which fishing on the stocks indicated reached a level beyond which further increases in fishing give no sustained increase in total catch.

C Cod H Haddock P Plaice R Redfish Hk Hake Hg Herring

SOURCE: Estimated from reports of working groups of the International Commission for the Northwest Atlantic Fisheries and the International Council for the Exploration of the Sea.

Figure 3.1: Spread of Overfishing in the North Atlantic
Source: FAO (1968)

area covers a large region north of the latitude of Cape Hatteras and west of the longitude at the tip of Greenland. The convention area is divided into numerous subareas roughly covering the location of certain, high-value stocks of demersal (bottom feeding, such as cod and haddock) species of fish. The growth in fishing effort has become so great that it is now possible to deplete a fish stock in one or two seasons.[1] An example highlights this point. In ICNAF subarea 5, the annual haddock catch had averaged 50,000 tons for many years. The catch jumped to 155,000 and 127,000 tons in 1965 and 1966, respectively. However, this combined catch by U.S. and Soviet fishermen was greater than its estimated maximum sustainable yield of 50,000 tons.[2] The haddock fishery fell off rapidly in the following years, reaching an exceedingly low catch of about 12,000 tons over the 1971–1974 period. It is clear that this serious overfishing contributed to the depletion of the fishery, which would have yielded 50,000 annually if catches had been restricted. Although ICNAF was created to *prevent* such resource crises, it unfortunately acted only after the fact. Finally, in 1973 a total *quota* of 6,000 tons was imposed on the haddock catch in subarea 5, and it has remained in existence through 1976, indicating little, if any, recovery of the haddock stock. Hence, the essence of the resource crisis is apparent—overfishing has reduced a 50,000-ton haddock fishery to a maximum of 6,000 tons, or a *loss* of 44,000 tons annually until, and if, the stock recovers.

Table 3.1 shows the maximum substainable yield (*MSY*) and the 1976 ICNAF quotas for yellowtail flounder, haddock, and redfish (i.e., ocean perch). The very existence of a quota is a fair indicator that the fishery stock is or will soon be overfished. For subareas 5 and 6 of the ICNAF convention area (i.e., generally within 200 miles of the United States), an ICNAF memorandum (November 14, 1972) stated that "The total effort deployed in sub-areas 5 and 6 reached the level which could produce the maximum yield by 1965 . . . and had exceeded it significantly in recent years." According to Christy (1973), "On this basis [memorandum of November 14, 1972, ICNAF], it can be inferred that the maximum [fishing] effort should be about 140,000 standard fishing days fished, rather than the 240,000 days fished in 1971." With respect to ICNAF subareas 5 and 6, there is no doubt that fishing effort is excessive and has greatly contributed to a decline in the yields from important commercial fishery stocks.

Table 3.1: Comparison of Estimated Maximum Sustainable Yield (MSY) and 1976 Quotas for Selected Species, ICNAF Convention Area

Species	(1) MSY[1]	(2) Quota[2]	(1)-(2)
1. Yellowtail Flounder (Sub-areas 5 and 6)	32,000	20,000	12,000
2. Haddock			
(a) Sub-area 5	50,000	6,000	44,000
(b) Sub-area 4X	18,000	15,000	3,000
(c) Sub-area 4VW	25,000	2,000	23,000
3. Redfish			
(a) Sub-area 5	30,000	17,000	13,000
(b) Subareas 4VWX	80,000	20,000	60,000

Sources
1. Bell et al.(1972)
2. ICNAF (1975)

Roger Payne's article "Among Wild Whales" (1968) graphically shows how extensive fishing effort depleted the blue whale, a marine mammal. In the early development of the United States and other countries, whales were hunted for oil used for lighting. Now, however, whales provide no unique food or intermediate product. Whale oil is used for oleomargarine, but it can be easily changed by chemistry to become anything from oleomargarine to lipstick or even transmission oil! Thus, many companies can use whale oil. Figure 3.2 illustrates the increasing world harvest of whales since the end of World War II. The increasing harvest has reduced the average size of the whale (twice as many whales were caught in 1966 as in 1933, although the same number of barrels of oil were produced). Whaling vessels have become larger with greater horsepower; however, the increasing pressure on the resource has resulted in declining productivity, or declining average production of whale oil per catcher boat per day's work (barrels). As Payne indicates, "Unbridled exploitation seems to have been the pattern taken by whaling throughout the 20th century. It is unlikely that whaling will ever stop in any region before it has brought stocks below a density at which whaling still pays" (1968, p. 130). As figure 3.2 clearly indicates, blue whale stocks have been almost completely depleted. Fin, sei, and sperm whales seem on their way to depletion and possible extinction. This pattern is not unlike that observed in the northwestern and northeastern Atlantic for food fish.

Another area of the world that is presently experiencing great fishing pressure is the northwestern Pacific and Bering Sea—including the Sea of Okhotsk and the Pacific waters of the Japanese islands. Japan, the USSR, South and North Korea, the People's Republic of China, and Taiwan are aggressive fishing nations that have fully exploited or overfished many species. In short, the demand pressures discussed in chapter 2 (population, affluence) have generated so much fishing effort that traditional species are apparently coming close to maximum sustainable yield. Some species are being significantly overfished.

The United States has recently reviewed the condition of its coastal species (NMFS 1975) with respect to overfishing or other factors (see chapter 5 on marine pollution). The following definitions were used:

Depleted: A fishery stock that has been so reduced through overfishing, or any other man-induced or natural cause, that a

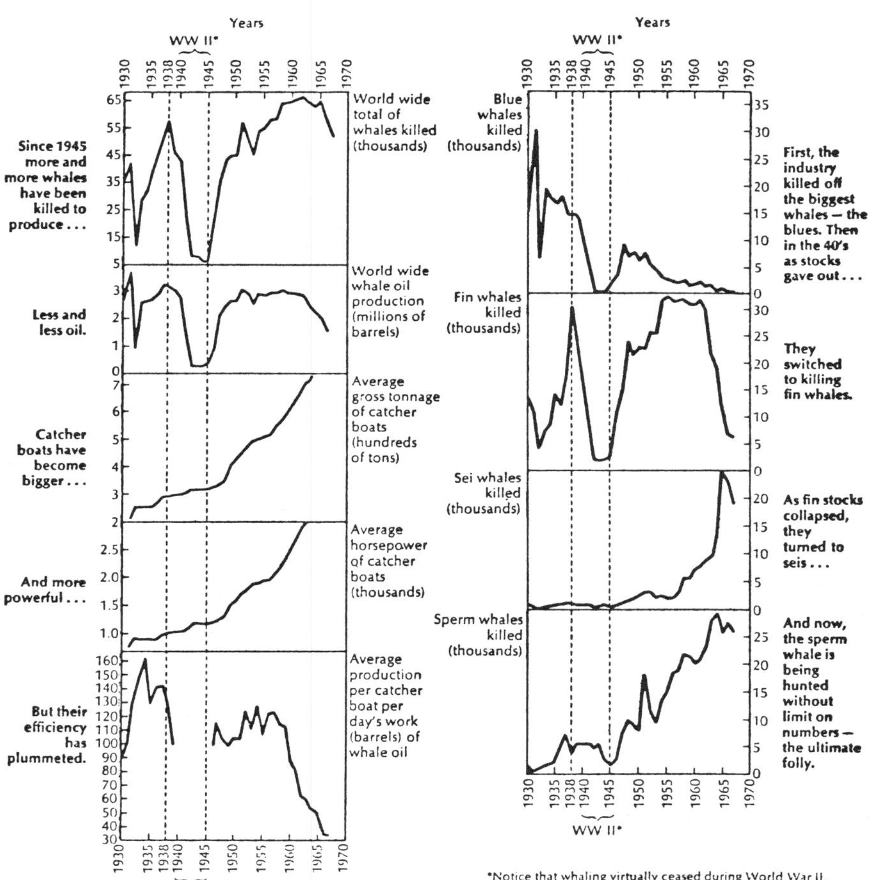

As wild herds of whales have been destroyed, finding the survivors has become more difficult and required more effort. As larger whales are killed off, smaller species are exploited to keep the industry alive. However, since there have never been species limits, large whales are always taken wherever and whenever encountered. Thus small whales are used to subsidize the extermination of large ones.

Figure 3.2: Modern Whaling

Source: Payne (1968)

substantial reduction in fishing effort must be achieved so that stock can replenish itself.

Imminent Danger: A fishery stock that has reached *MSY*, but no decrease in total catch has been observed, but the fleet operating in that area has sufficient fishing power to cause depletion.

Intensive Use: A fishery stock where *MSY* is being approached.

Table 3.2 lists the location and species identified as depleted, in imminent danger, and intensively used. On the U.S. Atlantic coast, haddock, herring, menhaden, yellowtail flounder, oysters, sea scallops, and shrimp have been important species from an economic standpoint. For example, yellowtail flounder and sea scallops are the principal species caught by the New Bedford, Massachussetts, fishing fleet. The Boston fleet *was* principally based on haddock. The Maine sardine industry has been seriously hurt by the depletion of the Atlantic herring. Schaaf and Huntsman have carefully documented the decline of the Atlantic menhaden: "At least since 1962 the fishery has been operating beyond the peak [maximum sustainable yield] of the catch-effort curve and has been on a decline typical of biologically overexploited populations" (1972, p. 293). In terms of landed weight, Atlantic menhaden reached a peak production of 712,100 metric tons in 1956, but by 1969 only 161,400 metric tons were harvested. As noted in chapter 2, menhaden is an important industrial fish used for animal feed.

Although the Gulf of Mexico shows no depleted stocks, menhaden and shrimp are in "imminent danger." Nichols and Griffin (1975) report that total fishing effort for shrimp in the Gulf of Mexico has increased by approximately 89 percent over the 1962–1973 period but that the catch has fluctuated greatly around 200 million round weight pounds. As shown in chapter 2, our example—demand for shrimp in the United States—showed an increasing per capita and aggregate consumption over the 1947–1971 period. This demand pressure was responsible for increasing fishing effort; however, no appreciable increase in the Gulf shrimp catch has been observed since 1962. U.S. demands, in part, have to be supplied by shrimp imports from other countries. Thus, the growth in the overall fish consumption in the United States has led to the depletion of many coastal and migratory species near the U.S. mainland. The subtle interaction between demand analysis and resource potential is well demonstrated by the Gulf shrimp fishery as well as by many North Atlantic species.

The management (see chapter 4) of trawl fisheries on a species-

Table 3.2: Selected U.S. Coastal and Migratory Fishery Stocks Classified As Depleted, Imminent Danger or Intensively Used, 1975

Area	Depleted	Imminent Danger	Intensive Use
A. Atlantic Coast (Maine-Florida)	Fluke[1] Haddock Menhaden Herring Yellowtail Flounder Oysters[1] Pandalid Shrimp[1] Sea Scallop	Grey Sole Mackerel	American Plaice Sea Bass Blue Fish Cod Ocean Perch Amer. Lobster Blue Crab Hard Clams Surf Clam Shrimp Spiny Lobster
B. Gulf of Mexico (Florida-Mexican Border)		Menhaden Shrimp	Grouper Snapper
C. Pacific Coast (Mexican to Canadian Border)	Mackerel Sardine Abalones[1] Halibut	Bonito Hake Rockfishes	Dungeness Crab Cod Flounder Sablefish Razor Clams Shrimp
E. Gulf of Alaska to Bering Sea	Halibut Pollock Yellowfin Sole Herring	Cod Sablefish	Tanner Crab King Crab Shrimp
F. Migratory and Anadromous	Bluefin Tuna Alewife Atlantic Salmon Pacific Salmon[1] Striped Bass[1]	Yellowfin Tuna	Albacore Tuna Amer. Shad

1. Not all stocks depleted

Sources: NMFS (1975)

by-species basis often results in damage to some of these or others. For example, the incidental catch of Pacific halibut (listed as depleted in table 3.2) by the pollack fishery in the eastern Bering Sea has caused an alarming depletion of the halibut.[3]

Our examples from around the world are not isolated cases, they are characteristic of a trend toward overfishing and depletion produced by the factors behind the demand for fishery products (see chapter 2). Certainly, the old myth that the sea possesses an inexhaustible supply of food cannot be taken seriously. Are all fishery stocks depleted, in imminent danger, or under intensive use? As of 1977, many areas of the world are not heavily fished. The species in these areas are just coming under exploitation or presently possess little commercial value. The latter species are called latent or underutilized (see chapter 8). We shall defer our discussion of the estimated potential of the sea to provide food, recreation, and industrial products until the end of this chapter, since this potential is quite controversial. More important, we must explore the area of population dynamics to understand more fully the use of *MSY*, or resource potential. However, let us take a brief look at who supplies the fish of the world, since these nations will suffer lost income and employment from the numerous resource crises occurring throughout the world.

Who Harvests the Fish?

In chapter 1, we looked at the aggregate physical harvest by country. In terms of metric tons, China, Japan, the USSR, the United States, Norway, and Peru were the six leading nations in 1973 (see table 1.3). In terms of ex vessel value, however, Japan, China, the USSR, and the United States were the leading countries; in 1973 they accounted for approximately 51 percent of the world's total income from the fisheries. We have consistently argued that a species breakdown will give us more information for analysis, since demand and supply for an individual fishery product on a world or regional basis can then be compared. Table 3.3 shows the four leading fish-harvesting nations by species for 1973. China is not included because of the lack of available detailed data, although it would certainly rank high in many fishery categories. First, it is interesting to note the concentration of fishery production. The four leading producers account for 70 percent or more of total world production in the groundfish, salmon, halibut, clams, scallops, and oysters species categories. Only in sardine and lobster categories do the four leading producers account for less than 50 percent of the world's harvest. The harvesting of lobsters is distributed fairly equally among eleven countries, and the sardine harvest is distributed among twelve countries.

Out of the twelve fishery categories, Japan is among the four leading producers in ten (shrimp and lobsters are the categories in which Japan does not rank in the top four). Japan is the world's leading producer of tuna, salmon, scallops, and fish meal. It is not only one of the leading fishing nations, but probably one of the most diversified in terms of species of fish caught. Not only has it exploited its coastal fisheries, but it has also developed large distant-water fleets, expecially for tuna. Over the 1962–1971 period, the Japanese increased their tuna vessel tonnage by 34 percent, the number of tuna boats by 31 percent, and horsepower by 47 percent—which greatly increased the pressure on the world's tuna stocks. According to Huang and Lee (1974), South Korea

Table 3.3: Rank of Four Leading Countries in the Harvesting of Selected Fish Products, 1973 (Round or Live Weight)

Species Category	Country	Total Harvest (mil. lbs.)	Percent of Total
1. Groundfish[1]	U.S.S.R.	7,834	30
	Japan	7,071	27
	Norway	1,861	7
	U.K.	1,549	6 (70)[2]
	World Total	26,050	100
2. Tuna	Japan	1,368	36
	U.S.A.	342	9
	S. Korea	309	8
	Spain	128	3 (57)
	World Total	3,763	100
3. Salmon	Japan	371	36
	Canada	245	23
	U.S.A.	213	20
	U.S.S.R.	195	19 (98)
	World Total	1,044	100
4. Halibut	U.S.S.R.	83	31
	Canada	40	15
	Norway	38	14
	Japan	33	12 (72)
	World Total	265	100
5. Sardines/Herring[3]	U.S.S.R.	1,849	16
	Japan	1,058	9
	S. Africa	1,047	9
	Denmark	844	7 (41)
	World Total	11,450	100
6. Shrimp	India	458	19
	U.S.A.	372	15
	Thailand	244	10
	Mexico	160	7 (51)
	World Total	2,441	100
7. Lobsters	Chile	56	13
	U.S.A.	40	9
	Canada	36	8
	France	29	7 (37)
	World Total	424	100
8. Crabs	U.S.A.	235	29
	Japan	172	21
	U.S.S.R.	41	5
	S. Korea	32	4 (59)
	World Total	805	100
9. Clams	U.S.A.	617	45
	Japan	491	36
	Malaysia	88	6
	S. Korea	77	5 (92)
	World Total	1,374	100
10. Scallops	Japan	136	28
	Canada	90	19
	U.S.A.	75	16
	France	56	12 (75)
	World Total	480	100
11. Oysters	U.S.A.	743	41
	Japan	504	28
	S. Korea	206	11
	France	159	9 (89)
	World Total	1,813	100
12. Fish Meal (Industrial)	Japan	8,581	20
	Peru	5,483	13
	U.S.S.R.	5,387	13
	U.S.A.	3,930	9 (55)
	World Total	42,659	100

1. Includes flounders, soles, cods, hakes and haddock.
2. Number in parentheses is cumulative percentage of world total for the four leading harvesters.
3. See footnote 3, Table 2.2 for a better description of this category.

Source: FAO Year Book of Fishery Statistics, 1973

and Taiwan, in competition with the Japanese for northwest Pacific tuna, increased their tuna boat tonnage by over nine times over the same period.

A country such as Japan faces the resource crisis because it depends on some of the most heavily exploited species, such as tuna, salmon, and groundfish. As Joseph has indicated:

> In summary, the information suggests that most of the principal market species of tuna are nearly or fully exploited. Yellowfin, albacore, and bigeye tuna appear to be nearly fully exploited, and increased effort on these species will result in at best small increased catches and even decreased catches. Northern bluefin tuna in the Pacific Ocean are probably fully exploited, and in the Atlantic Ocean are possibly overexploited. Albacore tuna appear to be fully exploited in all three oceans and increased production is not likely. Southern bluefin tuna has been heavily exploited recently and catches have declined by about 30% (1973, p. 2476).

In terms of resource potential, this is not an optimistic appraisal of future growth in the Japanese distant-water tuna fleet. Figure 3.3 is taken from Joseph (1973) and shows the *decline* in tuna catch per metric ton of fishing capacity for the world's oceans. Although tuna catch leveled off after 1969, fishing capacity continued to increase. As Joseph remarks about the trends shown in

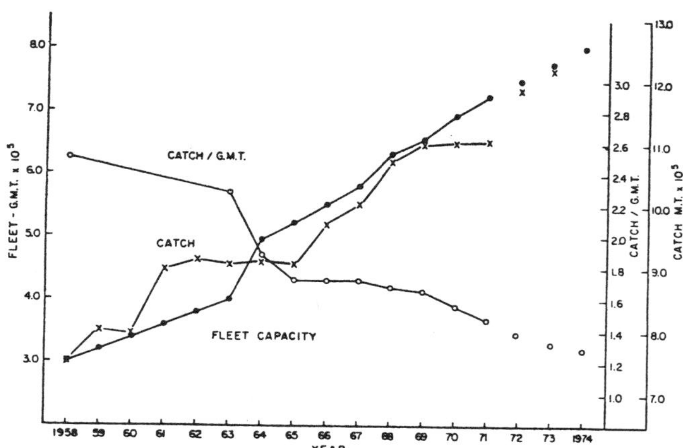

Figure 3.3: Recent World Trends in Fleet Capacity, Catch and Catch Per Gross Metric Ton of Capacity, for the Principal Market Species of Tuna, 1958-71, with Extrapolations Through 1974

Source: Joseph (1973)

figure 3.3, "Total fleet growth has been extrapolated (unconnected dots). If catches of tuna remain as in 1970 and 1971, then with the projected fleet expansion the catch per gross metric ton should drop substantially (unconnected open circles). By 1974 . . . the catch per gross metric ton would dip to 1.30 tons, about half that prior to 1960" (1973, p. 2475). The other leading tuna-harvesting nations, such as the United States, South Korea, and Spain will undoubtedly be adversely affected by this exploitation, which follows a pattern similar to that we have noted for other species.

The USSR is one of the four leading fishing nations in six of the twelve classifications and is the leading producer of groundfish, halibut, and sardines. All of these species are now under heavy exploitation. Sysoev states that "in the postwar period investment in the [Soviet] fleet grew continuously. The average yearly investment was 43.6 million rubles in 1946–1950 and . . . 410 million rubles in 1966–68" (1970, p. 224). Sysoev calls for international regional fishery agreements (see chapter 4) aimed at ensuring the rational utilization of the raw material resources in the seas and oceans. However, the USSR has shown little inclination to abide by this principle. It has engaged in massive fishing efforts in the Gulf of Alaska and northwest Atlantic without much regard for the state of the resource or the impact that incidental catches have on other fisheries.

The United States, despite a stagnation in the physical catch (chapter 1), is still a world leader in the harvesting of crabs, clams, and oysters. Of the twelve species categories in table 3.3, the United States is among the top four producers in nine classifications. Behind Japan and the USSR (and China, if statistics were available), the United States still remains a world fishing power. In both 1955 and 1967, the United States was among the top three producers of groundfish, but it slipped to tenth place in the 1973 statistics. The same general pattern has been true of halibut production. Except for sardines, herringlike species have not been pursued by U.S. fishermen because of the lack of domestic demand. Other nations, such as the USSR and the Scandinavian countries, have pursued herring and herringlike species off the U.S. and Canadian coasts in the northwest Atlantic. The adverse effect of declining groundfish resources on the New England fishing industry is one of the principal factors behind the U.S. enactment of a 200-mile fishing jurisdiction (see chapter 4).

As might be expected, few developing countries are large producers of major fishery products, as shown in table 3.3. The notable exceptions are India and Chile, which are the world leaders in the production of shrimp and lobsters, respectively. India exports about 20 percent of its production of shrimp, but Chile has no significant exports of lobsters. Spain (tuna), Thailand (shrimp), Mexico (shrimp) South Korea (tuna, crabs, clams, oysters), Peru (fish meal), Malaysia (clams), and South Africa (sardines) have emerged as major harvesters of fish. Most developing nations are able to achieve prominence in fishery production through the utilization of their labor-intensive coastal fisheries. The resource crisis has diverted the fishing effort of developed nations to the coastal areas of developing countries. The United States fishes off the coast of Brazil for spiny lobsters and shrimp.[4] To put it quite plainly, the crunch at sea for increasingly valuable fishery resources is rapidly changing the traditional law of the sea, which was predicated upon open access for all nations to the catching of *common property* fishery resources. This new world order is likely to be explosive, since it may adversely affect many of the traditional fishing powers. The economics of the fisheries will be extremely important in predicting who the future fishing powers will be.

Of the major world producers of marine fishery products, only the United States must significantly augment its domestic production with foreign imports to satisfy increasing demand. Although it has large fishery stocks of its own, the United States is not self-sufficient in fishery resources; its demand for fishery products is too high! The U.S. National Marine Fisheries Service states that "Most of the world's marine fisheries resources live in temperate and sub-arctic shelf areas of the oceans. Of the total of such resources, almost a fifth is found in waters within 200 miles of the U.S. coasts" (1975, p. 5). The dependence on foreign imports is one of the features that distinguish the United States as a major fish harvester from other world fishing powers. The growth of population and general affluence within the United States and domestic and foreign overfishing of coastal resources have produced this distinction. The lack of proper management of the U.S. coastal resources (chapter 4) and environmental deterioration (chapter 5) have greatly contributed to the problem.

Throughout this discussion, we have tried to give an overview of the fishery resource crisis and look at the major producers that must deal with this crisis at sea. In order to understand individual fisheries, however, the reader should be familiar with many technical aspects of the resource. The foundation of living marine resource analysis is called *population dynamics.*

Theory of Population Dynamics

Except for aquaculture, man still "hunts" for fish. However, we can predict the behavior of such wild animal populations. The study of the growth and general behavior of the wild animal population that man exploits for food or industrial purposes is called population dynamics. One model of population dynamics is sometimes called the *logistic,* or *Schaefer* (1968), explanation of the relation between fishing effort and catch. This model is probably the most simplified; nevertheless, it has been useful to fishery scientists in understanding catch-effort relationships.

Schaefer Model

More than twenty years ago, Milner Schaefer developed a framework to explain the relation between fishing effort and the resulting catch. Using available information on catch and effort, he (1954) developed a mathematical model that described the effect of fishing on the catch of yellowfin tuna in the eastern tropical Pacific. This model, termed the logistic, estimates the proportion of the stock (biomass) removed by a single unit of fishing, the intrinsic ability of the stock to increase, and the maximum size the stock can theoretically attain. These estimates can then be used to predict the *maximum* average yield (i.e., catch) the stock can support on a *sustained* basis as well as the average yield under any sustained fishing intensity.

First, let us hypothesize that *without* intervention by man, a wild stock of a specific species will grow in a logistic manner:

$$N_t = \frac{N^*}{1 + be^{-at}} \quad . \tag{3.1}$$

Equation (3.1) is shown in figure 3.4. The assumption behind this explanation of the growth in a fishery stock, or population N, is that at low population levels the stock will grow very rapidly, since space, food supply, oxygen, and other requirements are in ample supply. Furthermore, natural predators may be less successful in feeding on a small stock because of its dispersion over a wide area of ocean. However, as the stock grows over time t, it will eventually become subject to limiting factors, such as food supply, competition with other species, and predation. In theory, the stock will approach asymptotically a limit on maximum size N^*. One can debate the exact shape of the curve, but empirical studies have shown that wild stocks of animals experience some limit to their growth because of limiting environmental factors.[5]

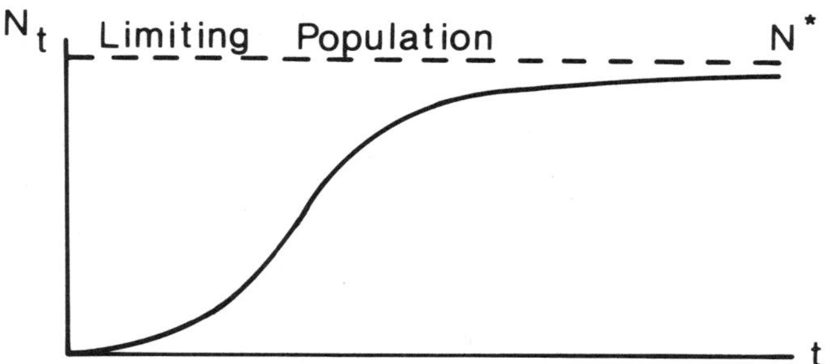

Figure 3.4: Logistic Growth Function

If we differentiate equation (3.1) with respect to time, we have

$$\frac{dN_t}{dt} = aN_t(1 - \frac{N_t}{N^*}) \ . \tag{3.2}$$

The first derivative is a parabolic function, or

$$\frac{dN_t}{dt} = aN_t - \frac{aN_t^2}{N^*} \ . \tag{3.2'}$$

That is, the *absolute* increments to the stock increase to a maximum and then decline (see figure 3.5). The maximum dN_t/dt is $aN^*/2$. Multiplying equation (3.2) through by N^{-1}, we can obtain the *rate of change* in the stock as a function of stock size:

$$\frac{dN_t}{dtN_t} = a(1 - \frac{N_t}{N^*}) \ . \tag{3.3}$$

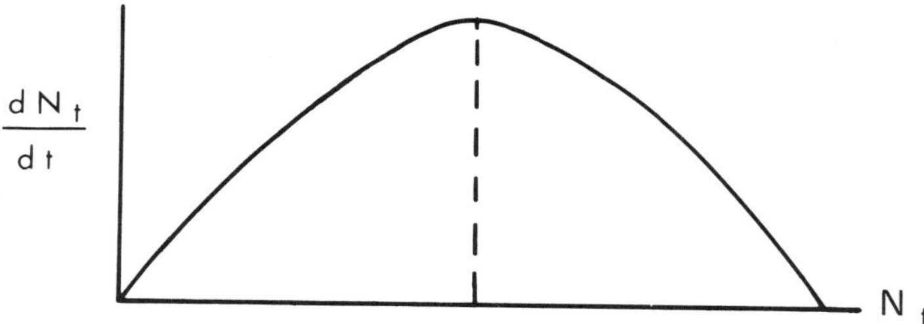

Figure 3.5: Relation of Increment to Stock

Up to this point, we have been discussing the growth in the fishery stock in the wild without intervention by man as a predator. We must stress that such wild stock growth is not as deterministic as the mathematics suggests. The assumptions behind logistic growth are that the environment is fixed or unchanging and that given the environment as a parameter, fish grow or respond in a systematic way. Alterations in seawater temperature, salinity, dissolved oxygen, currents, or ultraviolet radiation may change the environmental parameter on a daily basis. Little is known about the total impact of environmental changes on N_t. At best, the logistic growth function should be considered a crude approximation of reality, with possible wide fluctuations above and below this function from year to year. Longhurst et al. (1972) have commented on the great instability of ocean populations. Fishery biologists long ago discovered that the most difficult problem in monitoring man's effect on life in the ocean is how to separate man-made effects from those induced by natural climatic changes in the physical environment. For example, the tilefish was discovered on the continental shelf off the East Coast of the United States in 1879. A fishery developed rapidly, but by 1882 temporary flooding of the continental edge by abnormally cold, deep water depleted the stock. After this climatic aberration abated, the tilefish returned in great abundance by 1898. El Niño seriously reduced the catch of Peruvian anchoveta stock owing to an incursion of tropical surface water. Marine climatic data for the Pacific Ocean demonstrate that this must have occurred ten to twelve times during the present century. Despite these difficulties, population dynamics does help us formulate models for empirical testing similar to the demand models in chapter 2.

Man intervenes as a predator or, to be more dignified, as a hunter. Let us define *F(E)* as the rate of *loss* (i.e., a subtraction from the rate of growth of the wild stock) of the fish due to *fishing effort (E)*—inputs of capital labor, materials, and technology—or man-made mortality:[6]

$$\frac{dN_t}{dtN_t} = a(1 - \frac{N_t}{N^*}) - F(E) \quad (3.4)$$

rate of wild stock growth — rate of change in wild stock due to fishery effort

$F(E)$ is a general function expressing fishing effort. Let us specify the function more specifically:

$$F(E) = kE_t \; . \tag{3.5}$$

From man's standpoint, he is applying units of fishing effort in order to harvest fish from a growing resource. Let us assume a *steady-state* relation between the rate of growth in the stock and rate of loss due to fishing effort (i.e., the rate of growth of the stock is exactly equal to the rate of loss from fishing effort):

$$a(1 - \frac{N_t}{N^*}) - kE_t = 0 \; . \tag{3.6}$$

Equation (3.6) is a steady-state relation. In essence, it is an equilibrium relation. However, the reader should not be misled and assume that increases in E will not deplete the stock. In order to maintain equation (3.6), N_t must get smaller and smaller as E increases. This is because *large* amounts of fishing effort are counterbalanced by high rates of growth in stock *only* when N_t is relatively small. The reader should refer to figure 3.5.

Next, let us hypothesize a fishing *production function*[7]

$$Q_t = kE_tN_t \, , \tag{3.7}$$

where Q_t is the quantity of the stock harvested. That is, the catch depends on the *amount of fishing effort* and the *size of the stock.* kE_t is really the fraction of the stock removed by fishing effort. Let us respecify equation (3.7) by dividing each side by kE_t:

$$N_t = Q_t/kE_t \; . \tag{3.7'}$$

Substituting (3.7') into (3.6), we obtain

$$Q_t = kN^*E_t - (\frac{N^*k^2}{a})E_t^2 \; . \tag{3.8}$$

Letting $A = kN^*$ and $B = N^*k^2/a$, we have

$$Q_t = AE_t - BE_t^2 \; . \tag{3.9}$$

Hence, equation (3.9) expresses a parabolic relation between the catch (at time t) and fishing effort (at time t). This is known as a sustainable yield function, since it mathematically depicts the

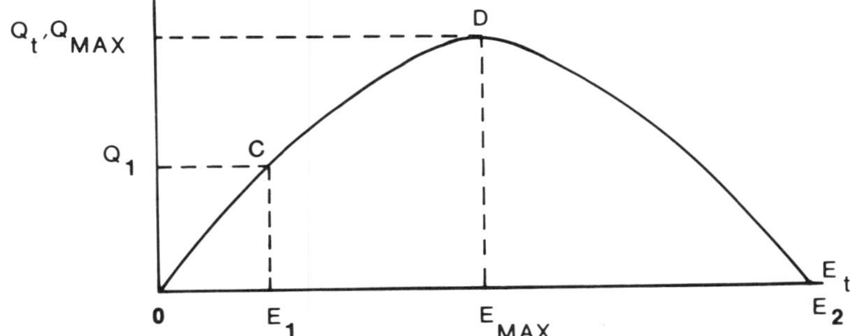

Figure 3.6: Sustainable Yield Function

renewable nature of the fishery resource. In the fishery literature it is also called the "surplus yield or production function." The word *surplus* refers to the incremental growth in the stock—equation (3.2)—which may be harvested or allowed to accumulate.

Consider point *C* in figure 3.6. If fishing effort is held constant at E_1, then Q_1 may be caught in year *t*. In contrast to an oil field or a copper mine, the inputs of capital and labor represented by E_1 each year and each subsequent year will produce the same output Q_1, assuming that all other factors remain constant (e.g., environmental variables such as El Niño). This is the essence of a renewable resource as opposed to a nonrenewable resource such as tin, iron, or coal. Constant inputs of the same capital (vessels) and labor (fishing effort) will yield a constant output into infinity for a fishery resource, but they will yield a dwindling output for a nonrenewable resource, *assuming no change in technology* that might increase the efficiency of capital and labor. The contrast between a renewable (living) fishery resource and nonrenewable resource is shown below schematically.

Fishery Resource

Economic Resources (E)

Capital
Labor
Materials

$$\rightarrow \quad \sum_{t=1}^{\infty} E_1 = \sum_{t=1}^{\infty} Q_1$$

Nonrenewable Resources (oil, uranium, copper)

Economic Resources (R)

Capital
Labor
Materials

$$\rightarrow \quad \sum_{t=1}^{\infty} R_1 = \sum_{t=1}^{T} Q_1$$

where T = exhaustion

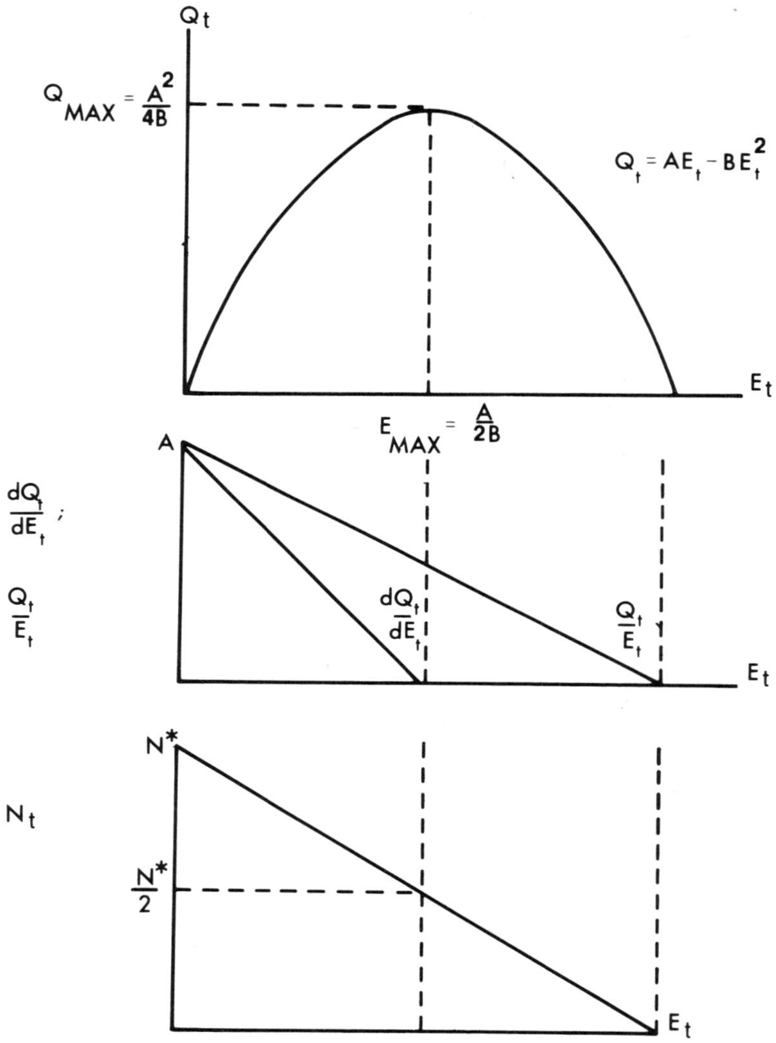

Figure 3.7: Relationships of Q_t, CPUE, dQ_t/dE_t and N_t to Fishing Effort (E_t)

Of particular interest, figure 3.7 illustrates that there is a *maximum sustainable production or yield (MSY)* at point *D.* In the earlier part of this chapter, we discussed *MSY* in general terms. Now the reader may see that it is the maximum perpetual yield from the fishery stock requiring E_{MAX} units of fishing effort. At *MSY*, the fishery is fully utilized. If A and B are known, MSY and E_{MAX} may be calculated by the following formulas:

$$MSY = Q_{MAX} = \frac{A^2}{4B} \tag{3.10}$$

$$E_{MAX} = \frac{A}{2B} \; . \tag{3.11}$$

Looking again at figure 3.6, we can see that $0E_{MAX}$ represents increasing production as fishing effort is increased. After E_{MAX}, further increases in fishing effort will result in falling catches. In other words, increases in vessels and fishermen result in declining production. From both the economic and biological point of view, $E_{MAX}E_2$ levels of fishing effort represent overexploitation or depletion of the fishery resource. In theory, extinction of the stock occurs at E_2 level of fishing effort.

Fishery population dynamicists generally look at a variable called Catch Per Unit of Effort ($CPUE$). This variable can be derived by dividing equation (3.9) by E_t, or

$$CPUE = \frac{Q}{E} = A - BE_t \; . \tag{3.12}$$

This is the *average* productivity of the sustainable yield function. Figure 3.7 shows the relation between N_t, Q_t, $(Q/E)_t$, dQ/dE, and fishing effort. At E_{MAX} (and consequently at Q_{MAX}), the maximum stock N^* is $N^*/2$. Thus, if a fishery were *not* exploited by man, one-half of the stock would be an estimate of MSY.

The GENPROD Model

In an attempt to provide more flexibility with respect to the growth function of the fishery stock, N_t, Pella and Tomlinson (1969) developed a generalized production (GENPROD) model or yield function, of which Schaefer's logistic growth function was but a special case. Using a somewhat different notation for their constant terms H and K, they hypothesized that the absolute growth, or more specifically incremental growth, of a fish population or stock could be described by the following:

$$\frac{dN_t}{dt} = HN_t^m - KN_t \qquad m < 1 \tag{3.13}$$

$$\frac{dN_t}{dt} = -HN_t^m + KN_t \qquad m > 1 \; . \tag{3.14}$$

Equations (3.13) and (3.14) are the first derivatives of a generalized growth function discussed by Richards (1959):

$$N_t^{1-m} = (N^*)^{1-m}[1 + be^{-at}] \ . \qquad (3.15)$$

If $m = 2$, equation (3.15) becomes the now familiar logistic growth function, or equation (3.1). Equation (3.14) assumes $m > 1$ and is similar in form to the first derivative of the logistic growth function:

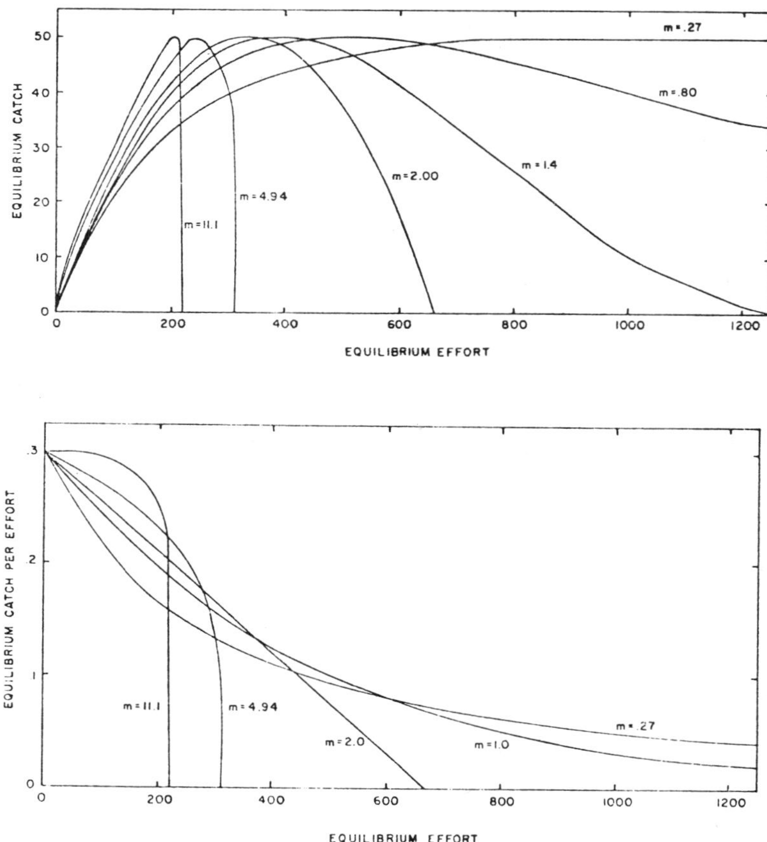

Figure 3.8: Relation between GENPROD "m" and Shape of Catch Effort Function

Source: Pella and Tomlinson (1969)

$$\frac{dN_t}{dt} = aN_t - \frac{aN_t^2}{N^*} \quad . \tag{3.16}$$

Rearranging terms and substituting *m* for 2, we have

$$\frac{dN_t}{dt} = -\frac{a}{N^*} N_t^m + aN_t \, , \tag{3.17}$$

where $(a/N^*) = H$ and $a = K$ using the Pella and Tomlinson notation. Using the *generalized* (GENPROD) derivative of the growth function and going through the same steps as in the development of the Schaefer model, we have the yield function

$$Q_t = \left[N^* k^{m-1} E_t^{m-1} - \frac{N^* k^m E^m}{a} \right]^{\frac{1}{m-1}} \quad . \tag{3.18}$$

One can easily see that when $m = 2$, we have equation (3.8), or the simple parabolic yield function. Figure 3.8 shows the relation between catch and fishing effort as we vary *m*. If the yield curve is skewed positively ($m < 2$), as many have suggested after they have looked at catch-effort data, fishing beyond the maximum of the yield curve (3.18) will result in a less pronounced decline in yield than would be predicted by the Schaefer model or by models with $m > 2$. As *m* decreases to zero, the yield curve is asymptotic to the maximum equilibrium catch. Obviously, an $m = 11.1$ or 0.27, as shown in figure 3.8, are rather unrealistic. Figure 3.8 also shows the catch per unit of effort curve (*CPUE*) for various hypothetical values of *m*.

Pella and Tomlinson present a computer program called GENPROD, which for a fixed *m* computes *H*, *K*, and P_0 (initial population size) from estimates of the optimum fishing effort (E_{MAX}), maximum catch per unit of effort, and a ratio of initial population size to maximum population size. The program requires observed catches, effort, and length of the time period covering each value of catch and effort. GENPROD undertakes an extensive computer searching routine to find which *m* minimizes observed values from predicted values.

In the Pella-Tomlinson paper, examples for stocks of two species of fish were worked out to explore the use of GENPROD. The first example came from Silliman and Gutsell's (1958) experiments on exploiting guppies. The second example was taken from

Bayliff's (1967) catch statistics on the yellowfin tuna in the eastern Pacific Ocean. The observed and predicted (by the use of GENPROD) catch statistics were extremely close. The empirical evidence for guppies suggests a yield curve skewed to the right (i.e., $m < 2$). The predicted and observed yellowfin tuna catch histories for $m = 1.4$, the best estimate of the skewed parameter, indicate a reasonably good prediction of fluctuation in catch. Pella and Tomlinson indicate that

> Clearly the fit to the yellowfin data is not of the quality of that for the guppies, but we did not expect it to be. There are sources of variation inherent in the yellowfin population dynamics which are absent or less influential in the guppy experiment. The guppies were kept in a relatively homogeneous environment where variation in reproduction, growth of individuals, or survival, induced by environmental perturbations should be of less importance than in the case of yellowfin tuna (1969, p. 443).

Schaefer (1970) also applied the GENPROD model to the world's largest fishery, the Peruvian anchoveta, over the 1960–1968 period. It was found that no modification of m from the logistic value of 2.0 resulted in significant improvement in the fit of the model to the data. Schaefer concluded that given the available data a linear relation between effort and catch per unit of effort prevailed.[8]

Dynamic Pool Models

Until recently, the most widely employed models in fisheries research have been dynamic pool models, often called Beverton-Holt models because of their detailed elaboration by Beverton and Holt (1957). Figure 3.9 may be useful in understanding the approach taken by Beverton and Holt.

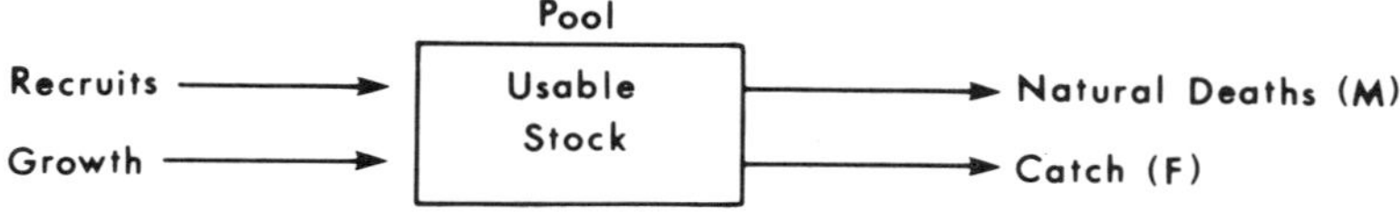

Figure 3.9: Dynamics of a Fishery Stock

The square in figure 3.9 represents the usable stock, or "pool," of fish *available to the fishermen.* The actual biomass as discussed above may be much greater than the usable stock. Why? In trawl fisheries, for example, the mesh size of nets may vary. Thus, the usable stock may be restricted to large fish; the smaller fish escape through the net, depending on mesh size (see chapter 5, on mesh size). *Recruits* are fish that have reached a certain age and size and have thereby become part of the usable stock. In figure 3.9, the arrow is drawn to indicate recruitment to the usable stock. The age at recruitment can vary from a few months in the case of shrimp to a couple of years for haddock and up to ten years for halibut. The Beverton-Holt model requires that the investigator must have some idea of the *number* of fish (recruits) entering the pool and their *average age*.

In the simplest model, the *natural* mortality rate is assumed to be constant, or what is called "density-independent," and the same at all ages. In the logistic model, it will be recalled, the additions to the stock declined after the stock had grown to a certain size or density. In the logistic formulation, natural mortality would thus be considered "density-dependent." The Beverton-Holt analysis begins by considering the history of a group of recruits through their life, after entry into the pool at a catchable age t_r. During any interval of time, an individual fish may be caught, die of natural causes, or survive to the beginning of the next period. In many fisheries, the fish do not begin to be caught immediately after recruitment. The survivors can be calculated as the difference between the number alive at the beginning of the period (t_r) and the deaths from *natural* mortality during the period. In figure 3.9, natural deaths of fish are represented by an arrow pointed away from the usable stock. The survivors at the time of availability for capture (t_c) are now subject to both natural mortality and fishing mortality. As the length of time increases after t_r, the year class will leave the "dynamic pool" as fish that have died from natural causes or fishing. Let us be more specific. Given R recruits and a constant rate of fishing mortality, the number of survivors (N_t) at age t is

$$N_t = [Re^{-M(t_c-t_r)}]e^{-(F+M)(t-t_c)} \tag{3.19}$$

where $[Re^{-M(t_c-t_r)}]$ = number of survivors to t_c

$e^{-(F+M)(t-t_c)}$ = mortality of survivors from t_c to t

t_r = age at recruitment to pool

t_c = age at minimum size of permitted capture

M = instantaneous rate of natural mortality

F = instantaneous rate of fishing mortality

and where the first term in (3.19) is the number of survivors up to possible capture and the second term is the combined *fishing* and *natural* mortality from t_c to t, an arbitrary time period. Let w_t equal the weight per fish at t. w_t can be calculated using the von Bertalanfy growth curve or from a table of weights at different times in the fish's life cycle. Now w_t can be multiplied by the number of recruits, N_t, or those still existing in the year class at age t to obtain the total weight of the year class. The rate of *catch* over a specific time interval such as a year may be obtained by multiplying the instantaneous rate of fishing mortality by the weight of the year class:

$$\frac{dQ}{dt} = FN_t w_t \quad . \tag{3.20}$$

Total catch Q from the year class during its *life* in the fishery is

$$Q = \int_{tc}^{tm} FN_t w_t dt \ , \tag{3.21}$$

where t_m = maximum age in the fishery.

According to Schaefer, "In the steady state, Beverton and Holt (1957), Richer (1948), and their predecessors have shown that the total catch each year is equal to the total harvest from a year-class during its life so that . . . [equation 3.20] represents the annual equilibrium yield" (1968, p. 233).[9] To avoid the problem of the relationship between stock N_t and recruitment R_t, the dynamics pool model expresses yield in terms of yield per recruit, or

$$\frac{Q}{R} = \int_{tc}^{tm} \frac{FN_t}{R} w_t dt \quad . \tag{3.22}$$

Figure 3.10 shows the relation between Q/R and F *assuming* N_t, R, and w_t are constant. The end result gives the yield per recruit as a function of two independent variables that may be controllable by management (see chapter 4)—the fishing mortality F and the age at first capture t_c (i.e., mesh size).[10]

The parameters (or constants) one must obtain are M, t_r, w_t and R. The last parameter is extremely difficult to estimate. In a number of fisheries, especially in temperate or subarctic waters—e.g., yellowtail flounder in the northwest Atlantic—the annual recruitment fluctuates very widely and apparently at random. According to Gulland, "Another, and in many ways, the

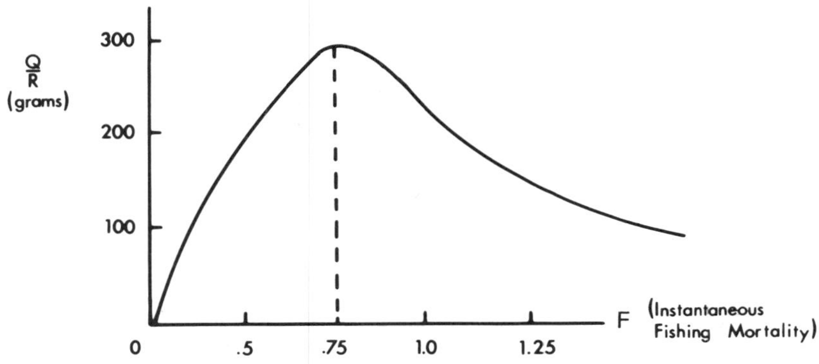

Figure 3.10: Yield Per Recruit and F

most important reason for expressing the results as yield per recruit is that the least satisfactory assumption made in the simple model is that the average recruitment is unaffected by the amount of fishing, or by the abundance of the parent stock" (1974, p. 90). With constant recruitment, the maximum Q/R or yield per recruit would maximize the catch on a sustainable basis. As shown in figure 3.10, fishing pressure (effort) in excess of F = 0.75, would be overfishing. As Cushing has pointed out, "Obviously, it is wrong that fishermen should have to exploit the stock at any point beyond or to the right of the maximum" (1968, pp. 96–97).

The reader can see that predicting recruitment is the main difficulty with the Beverton-Holt model. Ricker (1954) has proposed a dome-shaped curve to relate stock or population (i.e., total biomass) to recruitment, but Beverton and Holt (1957) have used a stock-recruitment relationship in which R increases with N_t throughout.[11]

Population Dynamics and Fishery Data

Measurement of Fishing Effort

To implement the population dynamics models discussed in the last section, proper fishing effort series must be developed. Fishing effort is a prime input for these models; it gives us indications of overfishing and depletion of fishery resources. There is a great diversity in the way fishing effort is expressed. It may be days fished, number of pots or traps set, or number of skates (a groundline 250 to 300 fathoms long from which short lines called gangions are attached at intervals; each gangion carries a hook). According to Gulland, "A simple form of operation is a day at sea of a single boat, so that a simple measure of effort is the number of days at sea" (1969, p. 47). The measured fishing effort per day for a given stock density (i.e., size of fish population in a specific geographical area) may increase: (1) because of an increase in "fishing power," or more fish are caught per day; (2) because more actual time fishing is spent per day; and (3) because the vessel operates in local areas of greater stock density.

The fishing power of a particular gear can be looked at in many ways: e.g., as (1) the extent (area or volume of water) over which the influence of the gear extends, or (2) the proportion of fish within this area that is in fact caught. For trawlers, for example, it is well known that the fishing power increases with both size (tonnage) and horsepower. Carlson (1973) found that fishing power for the New England groundfish fleet was increased with increases in *tonnage;* however, horsepower has a positive, but weak, influence. Both vessel capacity (i.e., tonnage) and horsepower were highly significant in explaining the fishing power of the Pacific yellowfin tuna seine vessel. The increase in tonnage and horsepower for modern whaling vessels was discussed at the beginning of this chapter. One of the most significant aspects of fishing power is technological change, which may well revolutionize the physical nature of the fishing vessel. Bell (1966) demonstrated that *stern ramp* trawlers had 29 percent more fishing power than *side* trawlers. Joseph (1970) indicated that tuna purse seiners possess much more fishing power than tuna bait boats, a fact that brought about a rapid conversion to purse seining technology beginning in 1962. As we have seen in this chapter, rapid technological change among world fishing fleets has, without management controls, depleted many stocks.

For any vessel with a *given* fishing power, the productive time spent fishing during a time period (i.e., usually a year) will determine its *total* fishing effort. "Productive" time includes: (1) time on the fishing grounds searching for fish and (2) time when gear is in operation. At one extreme is whaling, where the gear (harpoon) is only in operation for a short time and the most important factor is searching time. In contrast, groundfish trawlers spend little time searching and leave more time for the net to be on the bottom catching fish. In addition, the distribution of fishing may influence fishing power (measured in catch per day) depending on whether fish density is even or uneven over a given area. Furthermore, several kinds of gears (i.e., bait boats vs. seiners) may compose the fleet of vessels for which we must express or compute total fishing effort. All these factors influencing fishing effort are taken into account in computing what is called "standard" effort. Table 3.4 is an *illustrative* example of how standardized fishing effort is computed.

Table 3.4: Computation of Fishing Effort

Gear and/or Technology	(1) Estimated Fishing Power (one ship)[1]	(2) Number of Vessels Tons[2] (thousand)	(3) Standard Vessels Tons (thousand) (1x2)	(4) Productive Fishing Days (per ton)	(5) Standard Fishing Days (3x4)
Bait Boats	.25	100	25	100	2,500
Seiners	1.00 (standard)	50	50	200	10,000
		150	75		12,500

1. Fishing power is given as a reference point or standard from which to express all other technologies and or gear (i.e., a seine technology is arbitrarily set equal to unity). Fishing power also adjusted for differences in stock density.

2. Vessel tons are standardized or adjusted for differences in horsepower, age, crew size before entering in this table. Thus, fishing power differences are recognized between technologies in the table, but within technologies also.

Hence, over the period represented in table 3.4, the fishing fleet expended 12,500 (thousand) *standard* fishing days on the stock. Such detailed data are not often available for a fishery. If they are not available, researchers use the number of vessels, traps set, or even number of fishermen as a crude proxy for fishing effort. As one can see, a computation of fishing effort that adjusts for the factors discussed above involves large amounts of data and sophisticated statistical techniques. Dumont and Sundstrom (1961) have classified the commercial fishing gear used in the United States by function (such as encircling or impaling) as shown below:

Commercial Fishing Gear in the United States

Encircling or Encompassing: seines (haul, stop, purse), lampara, bag nets, trawls (beam, otter).

Entrapment: weirs, pound and trap nets, hoop nets, fyke nets, pots and traps, slat traps.

Entanglement: gill nets (anchor, drift, semidrift, runaround, stake, bar nets, riprap nets), trammel nets.

Lines: hand, troll, long or set with hooks, trot with baits, snag.

Scooping: dip nets, lift nets, reef nets, push nets, cast nets, Wheels.

Impaling or Wounding: harpoons, spears.

Shellfish: scrapes, dredges, tongs and oyster grabs, rakes, hoes and forks, shovels, picks, crowfoot bars.
Miscellaneous: frog grabs; brush traps; hooks (sponge, other); diving outfits; by hand.

Finally, the Beverton-Holt model uses instantaneous fishing mortality (F) rather than fishing effort, which is used in the Schaefer and GENPROD models. *Total* instantaneous mortality is equal to $F+M = Z$. Z may be calculated if we know the abundances, say N_0, N_1, of any group of fish at two known times; the fraction then surviving is $N_1 \div N_0 = S$ or $e^{-Zt} = S = N_1 \div N_0 = CPUE_1 \div CPUE_0$.[12]

Fishing effort, as computed from year to year, is often related *directly* to catch per unit of effort. The assumption here is that the fishery is in equilibrium from year to year (i.e., a steady state exists). Gulland (1969) has suggested that in a given year, the annual catch per unit of effort be related to fishing effort averaged over that year and over a certain number of previous years corresponding to the average number of years that a year class contributes to the fishery. For example, the yellowfin tuna and California sardines have year classes that contribute significantly to the fishery for about three years. Thus, the third year of *CPUE* would be related to the mean of the last three years of fishing effort. This is known as the *Gulland method.*

Empirical Example of Stock Assessment

Eastern Tropical Pacific Yellowfin Tuna. From 1965 to 1975, the capacity (measured in short tons) of the international fleet that fishes the Convention Yellowfin Regulatory Area, known as the CYRA, has increased from 46,743 to 169,300 short tons of carrying capacity (i.e., maximum weight of fish able to catch and store in hold), or a 262 percent increase. With the increase in fishing pressure, the catch of yellowfin tuna during the 1960s exceeded the capacity of the fish to replace themselves, and the catch and catch per unit of effort declined. Since 1966, the fishery has been under international regulation designed to restrict yellowfin takes to the corresponding *MSY,* in accordance with the convention establishing the Inter-American Tropical Tuna Commission (IATTC). One measure of fishing effort is the Class 4 baitboat (capacity 201–300 short tons) fishing day, since most of the yellowfin tuna caught in the surface fishery of the eastern Pacific before 1959 were taken by bait fishing. In Joseph (1970), the *logistic, GENPROD,* and *dynamic pool* models were applied to this fishery over the 1934–1968 period—as shown in figure 3.11. Effort in the fishery increased from about 10,000 standard days in

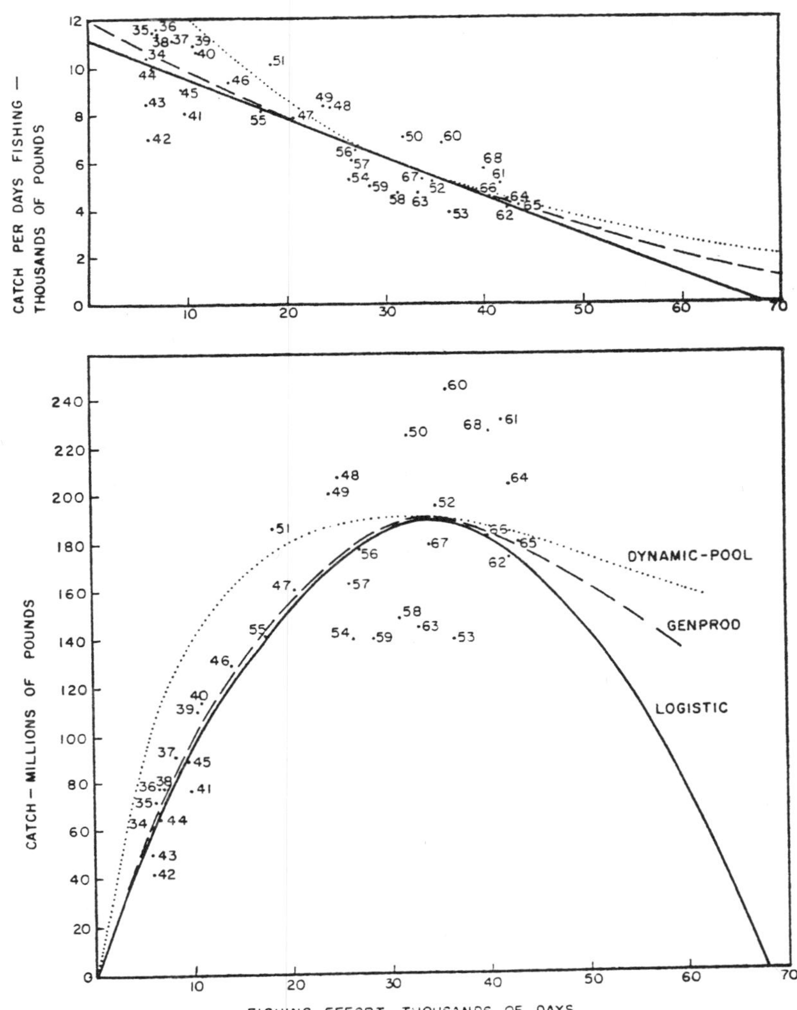

Figure 3.11: Relationship Between Catch Per Unit of Effort and Effort (upper panel) and Catch and Effort (lower panel) for Yellowfin Tuna in the Eastern Pacific Ocean, 1934-1968. The Effort is Expressed in Standard Baitboat-4 Units

Source: Joseph (1970)

the late 1930s and early 1940s to about 40,000 days during the mid-1960s. The *CPUE* (i.e., measure of abundance) dropped from 10,000 pounds per standard fishing day to about 4,000 pounds. *MSY* was estimated at 180 million pounds, taken with 32,000 standard fishing days. It was concluded that *m* in the GENPROD model was very close to 2.0, therefore indicating a logistic growth in the fish population. The IATTC (1971) estimated *MSY* at 218 million pounds attained in 21,000 standard fish days (SFD) of class III purse seine units of fishing effort for the years 1959–1970. This is shown in table 3.5. In 1969 and 1970, catches exceeded *MSY*, and effort was greater than E_{MAX} (21,000 SFD), thus indicating overfishing. In the early 1970s, it was discovered that purse seiners were catching yellowfin only in Zone 1 of the CYRA (i.e., inshore) during the early years of the fishery and that they were now expanding into Zone 2 (offshore), where the stock was partly independent of the stock in Zone 1. Therefore, the *MSY* was increased substantially. The *MSY* for the entire fishery was raised to 390 million pounds for 1975: this became the official quota. The population dynamics work on this fishery has been outstanding. The reader can easily see that since Schaefer's initial work in this fishery, the IATTC has made periodic adjustments to the *MSY* and, consequently, to its quota on fishing as new techniques and knowledge of the fishery have become available.

Peruvian Anchoveta. The commercial fishery for anchovy, called "anchoveta" in Peru, was once the largest single-species fishery in the world; in 1970, for example, it produced approximately 12.3 million metric tons. By 1973, the fishery had practically collapsed. It produced only 1.8 million metric tons, or 15 percent of its peak production in 1970. As already noted (see chapter 2), the anchoveta is used to produce fish meal and the disappearance of anchoveta off Peru has been reported to be the result of "El Niño," an oceanic phenomenon that is characterized by the failure of the Humboldt current to follow a seasonal decline in temperature. El Niño causes massive fish kills and drives the fish into deeper water, where capture is difficult. The warmer-than-normal surface conditions of the water have intensified since 1971; the effect has never been so prolonged or so devastating. Before El Niño, Schaefer (1970) published his famous article, "Men, Birds and Anchovies in the Peru Current—Dynamic Interaction." He pointed out that there are two distinctive predators of the anchoveta: men and the guano birds. Measuring effort in thousands of gross registered ton (GRT) trips, Schaefer applied

Table 3.5: Application of Schaefer Model To Selected Fisheries, Commercial and Recreational

Fishery	Equation	Estimated MSY or Q_{MAX}	Estimated E_{MAX}	Time Period	Latest Average Fishing Effort	Stock Analysis
A. Commercial						
(a) Pacific Yellowfin Tuna[1]	CPUE = 20.7 - .4925E	218×10^6 lbs	21×10^3 Standard Days	1959-70	23.6×10^3 Standard days	Overfished (1969 and 1970)
(b) Peruvian Anchoveta[2]	CPUE = 707 - .0146E	8.6×10^6 tons	24.2×10^6 in GRT Trips	1960-1 - 1967-8	20.8×10^6 GRT	Underfished (1967-68)
(c) American Lobster[3]	CPUE = 56.7998 - .000028E	28.8×10^6 lbs	1.0143 mil traps	1950-69	1.0618 mil traps	Fully utilized (1969)
(d) Gulf Mexico Menhaden[4]	CPUE = 2.1238 - .0026E	434,000 tons	407,000 Vessel Ton Weeks	1946-70	397,000 Vessel ton weeks	Fully Utilized (1970)
B. Recreational						
(e) California Yellowtail[5] (Party boat)	CPUE = 8.2681 - .000169E	10.1 mil lbs	2.43×10^6 Angler standard units	1947-54 1966-73	$.315 \times 10^6$ Angler standard units	Relatively Unexploited (1973)

1. IATTC (1970)
2. Schaefer (1970)
3. Bell and Fullenbaum (1973)
4. Chapoton (1971)
5. MacCall, *et al* (1974)

both the logistic and GENPROD models to this fishery over the 1960-1961–1967-1968 period. Two versions of each model were computed, one including only the fishing pressure exerted by man and the other using the combined fishing pressure of man and guano birds. The latter was a unique approach. Figure 3.12 shows the relation between catch per unit of effort (i.e., men as predators only) and effort. The computed *MSY* is 8.6 million metric tons taken with 24.204 million units of fishing effort (i.e., GRT trips). Figure 3.12 shows the relation between *CPUE* and fishing effort. The solid line is a result of relating *CPUE* to computed fishing effort each year, and the dashed line employs the Gulland method discussed above. As is evident, there is little difference in the results. The Peruvian Instituto del Mar was able to supply an estimate of the guano bird population and translated this predator impact on the anchoveta in terms of 1000 GRT trips. The logistic model was applied using men and guano birds as predators. The computed *MSY* reported by Schaefer was 10.29 million metric tons with 28.385 units of effort (i.e., GRT trips). Schaefer says, "The fishery could, in principle, take the entire maximum sustainable catch [10.29 million metric tons] if the birds were eliminated . . . . Presumably, if the bird population is maintained near its level of recent years, some 4.5 million adult birds, the annual anchovetta catch can be maintained indefinitely at 9.3 million metric tons, on the average, for the fishermen" (1970, p. 467). Finally, Schaefer could find no difference between the logistic and GENPROD models. That is, $m=2$ when the data were fit using the GENPROD model.[13] It is indeed interesting to note the complex web of biological and oceanographic (i.e., El Niño) circumstances that must be considered not only in explaining past behavior, but also in the more difficult job of predicting future changes in a fishery so important to the Peruvian economy.

Gulf of Mexico Menhaden. The major work on this fishery has been that of Chapoton (1971). Like the anchoveta, menhaden are used for fish meal. In 1970, the Gulf purse seine fishery landed 1.2 billion pounds, which constituted, by physical weight, 25 percent of all fishery products landed by U.S. fishermen. Chapoton calls Gulf menhaden the United States's largest fishery. The 1970 catch was worth \$26.2 million at dockside, which by economic standards is only about one quarter of the *value* of the Gulf shrimp fishery (0.23 billion pounds). Fishing power is directly related to the size of the vessel and its fish hold capacity. Chapoton expressed the fishing effort series in vessel ton-weeks. Applying

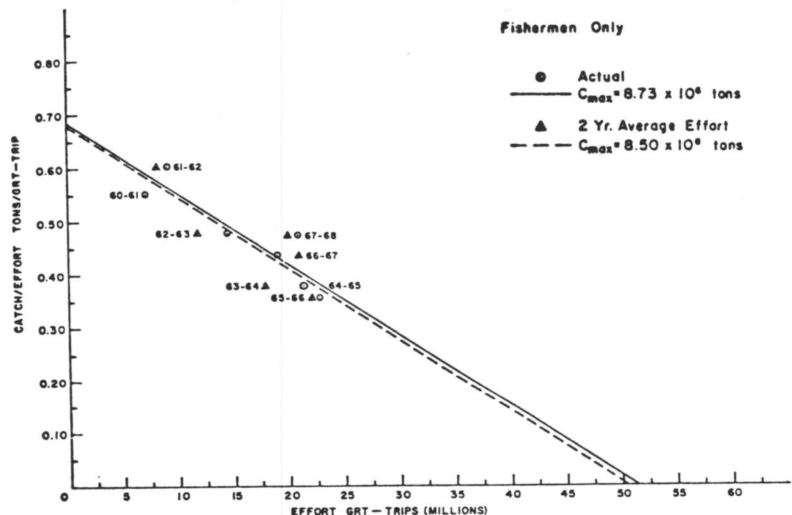

Figure 3.12: Relationship Between Effort by Commercial Fishermen and Anchoveta Abundance (Catch-Per-Effort), 1960-61 through 1967-68 with Estimates of Maximum Sustainable Yield (C_{max})

Source: Schaefer (1970)

the Schaefer model, Chapoton concluded that the *MSY* was 434,000 tons, taken by 408,000 vessel ton-weeks. The general conclusion is that the menhaden fishery is fully utilized, as shown in table 3.5, which also shows the extent of utilization for the American lobster and California yellowtail. Figure 3.13 shows the total and average yield relations for the Gulf menhaden.

The Bioeconomic Supply Curve

Construction and Implications

The *supply* of a product refers to the quantities (per unit of time, usually a year) that sellers are willing to place on the market at various prices—other things being equal. If wheat farmers were asked how much per year they would place on the market at different prices, their answers would provide the information for a supply schedule, or supply curve, for their product. Collectively, fishermen would be expected to harvest and market more fish at higher ex vessel prices, because, as we shall see, the supply curve slopes up and to the right. In the *short run,* the supply curve for most products such as wheat, corn, or beef also slopes up and to

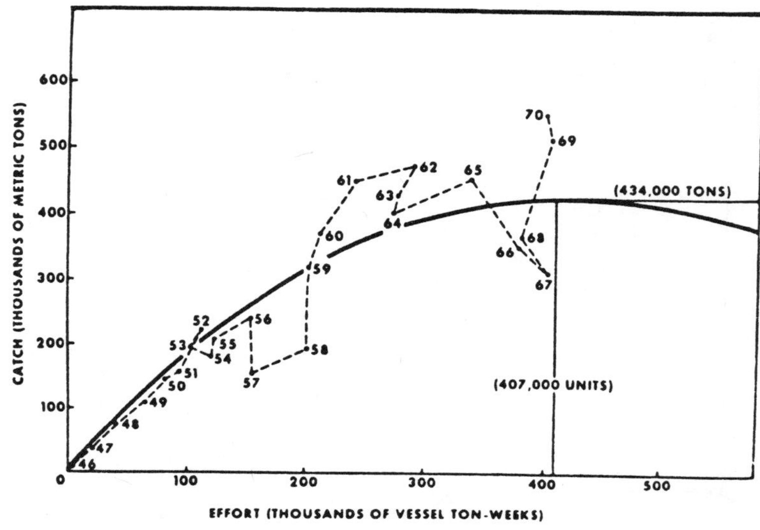

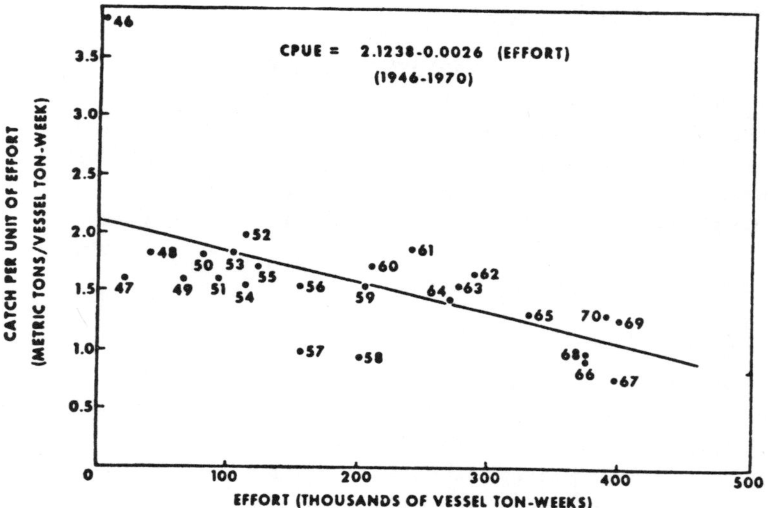

Figure 3.13: Equilibrium Relation (solid line) between Catch, Catch Per Unit of Effort and Fishing Effort in Gulf Menhaden Fishery, 1946-1970 (numbers refer to years)

Source: Chapoton (1971)

the right. What is the significance of the short run? All business firms must purchase what are called the *factors of production,* such as land, machinery, and labor, in order to produce, say, wheat, corn, or beef. They usually purchase land and machinery and hire labor according to their needs. If the price of wheat is expected to rise over a short period, say a year, the farmer will hire more labor or have his present workers put in more hours. Land and machinery are *fixed factors* of production in the short run, since the farmer may have to place his order for machines several months in advance and since it will take time to expand his investment in land. Labor is the *variable factor*—especially farm labor—and can be increased or decreased almost overnight. However, as more and more workers are hired to operate the machinery and work the land, the law of diminishing returns will set in; producing an additional bushel of wheat will incur a greater cost per unit. Why? As each worker is added, he is marginally less productive—machines designed for three operators will produce little more with four, and land intensively cultivated will yield less output per worker with more workers. This is the law of diminishing returns, which is especially prevalent in the short run. Thus, the supply or incremental cost of production will increase as output is expanded. And as prices rise, farmers or producers in general will supply more, since they can cover these rising incremental costs.

The supply curve that we shall consider here exists in the *long run,* where all factors of production are embodied in the one variable, "fishing effort."[14] The supply curve for fishery products is called a *bioeconomic curve* because it includes two factors: (1) the biological yield function and (2) the cost of fishing. Let us see how these factors interact. Using the Schaefer yield function as an example, we may specify catch as a function of fishing effort:

$$Q_t = AE_t - BE_t^2 \ . \tag{3.23}$$

Next, what is the total cost of fishing effort? An owner who manages a tuna vessel of a particular size incurs certain costs of fishing. Let us be more concrete. The vessel is 150 feet in length and has a hold capacity of 726 tons. In any given fishing year, the owner incurs the following expenses (hypothetical data):

1. Fuel	$ 100,000
2. Food	22,000

3. Crew Share	$ 500,000
4. Repair and Maintenance	75,000
5. Gear and Supplies	15,000
6. Taxes	16,000
7. Insurance	55,000
8. Interest	50,000
9. Depreciation	110,000
10. Return to Vessel and Management	150,000
	$1,093,000

Some of these expenses need no explanation, such as fuel, taxes, and insurance. However, those unfamiliar with the term *economic cost* might wonder about some of these items. More precisely, we are talking about "opportunity cost." An *opportunity,* or *alternative, cost* is the value of a resource (i.e., economic input such as labor or capital) in an alternative use. Opportunity cost is at the heart of the definition of cost because it shows that costs are "sacrifices," or alternatives foregone. How does it apply to fishing cost? The returns or profits to vessel and management must be included in cost because it is an excellent example of opportunity cost. Obviously, before purchasing the vessel, the owner had to consider his opportunities for using the money elsewhere. He could have entered the shoe-making business or, perhaps just as well, have invested the money in the stock market. All of these opportunities were alternatives to entering the fishing business. However, the vessel owner probably chose fishing because this is where his money and talents have their highest alternative use or opportunity cost. In effect, he gave up the shoe business or stock market (sacrifices or alternatives foregone) to become a fisherman. The $150,000 shown above is his opportunity cost. If *actual* profits should drop below this figure, the owner might consider leaving the fishing business and moving to some other alternative. For labor (fishermen) and capital (management), therefore, we must include an adequate return to induce them to stay in fishing. If returns to fishermen and vessels drop considerably below their opportunity cost, they will eventually leave the industry. We shall discuss this point in greater detail in chapter 9.

For this 150-foot vessel, biologists calculate that it will in the course of a fishing season (i.e., one year) exert 500 *standard* fishing days. This is not to say that the vessel will fish 500 days in one year (which is impossible), but that it has the size and technological capability relative to our "standard vessel" (see discussion on fishing effort above and table 3.4) of rendering 500 units of fishing effort (i.e., man-made mortality) to the stock. What, then, is the cost of one unit of fishing effort? To render 500 units of fishing effort to the stock will cost the vessel owner $1,093,000, or $2,186 per standard fishing day. Hence, fishing effort can be purchased on the open market at $2,186 per unit. The inputs to creating a unit of fishing effort are listed above. For the entire fishing fleet at time *t*, the total industry cost may be obtained by multiplying cost per unit of effort by total fishing effort exerted by the fleet, or

$$TC_t = CE_t , \qquad (3.24)$$

where TC_t = cost of operation for all vessels in the fleet
C = cost per unit of fishing effort, which is a constant
E_t = total fishing effort of the fleet.

Now let us consider the average or unit cost per pound of fish landed, or

$$AC_t = \frac{TC_t}{Q_t} . \qquad (3.25)$$

This can be illustrated using equations (3.23) and (3.24). Suppose E_1 units of fishing effort are presently expended in harvesting lobsters. To find the total industry cost (*TC*), we can use equation (3.24) and multiply C times E_1. C will depend on the cost of purchasing a unit of lobster fishing effort. Next, by inserting E_1 into equation (3.23), we can obtain the corresponding total catch, or Q. Hence, for E_1 level of fishing effort, we can, for example, calculate the average cost of harvesting one pound of lobsters. This process is a bit cumbersome, so let us use simple algebra and express AC_t as a function of the quantity of fish harvested by the fleet. By substituting equations (3.23) and (3.24) into (3.25), we have

$$AC_t = \frac{TC_t}{Q_t} = \frac{CE_t}{AE_t - BE_t^2} = \frac{C}{A - BE_t} . \qquad (3.26)$$

Solving (3.23) in terms of E_t we have the familiar quadratic

$$E_t = \frac{A \pm \sqrt{A^2 - 4BQ_t}}{2B} \quad . \tag{3.27}$$

Substituting equation (3.27) into (3.26), we have

$$AC_t = \frac{2C}{A \pm \sqrt{A^2 - 4BQ_t}} \quad . \tag{3.28}$$

By definition, equation (3.25) can be used to express total cost as a function of quantity harvested

$$TC_t = \frac{2Q_tC}{A \pm \sqrt{A^2 - 4BQ_t}} \quad . \tag{3.29}$$

Another extremely important cost concept is the notion of marginal or incremental cost (*MC*). Let us consider a simple numerical example where *Q, TC,* and *AC* are as defined as above and *MC* = marginal cost.

Q (pounds of fish)	*TC* ($)	*AC* ($)	*MC* ($)
0	0	0	0
1	5	5	5
2	12	6	7
3	24	8	12
4	48	12	24
5	125	25	100

Marginal cost is defined as

$$MC_t = \frac{\Delta TC}{\Delta Q} \quad . \tag{3.30}$$

Expressed verbally, marginal cost is the incremental cost associated with incremental changes in the quantity of fish harvested by the fleet. Our numerical example shows a number of interesting aspects of cost. To produce one pound of fish, the *TC* is $5. To produce four pounds of fish, it is $48. For any level of harvest, say three pounds of fish, the incremental cost of harvesting the third pound of fish is $12. That is, to increase production from two pounds of fish to three pounds, the total cost increases by \$12 (\$24–\$12); therefore, the *MC* of the third pound of fish is \$12. This is a very interesting concept: it tells us what additional cost must be incurred to produce an additional pound of fish.

If we differentiate equation (3.29) with respect to Q_t, we have the mathematical expression for marginal cost

$$\frac{\partial TC_t}{\partial Q_t} = MC_t = \frac{C}{\sqrt{A^2 - 4BQ_t}} \quad . \tag{3.31}$$

Let us look at AC_t and MC_t graphically. Figure 3.14 shows a plot of equations (3.28) and (3.31). AC_t rises from zero production and bends back at the maximum sustainable yield (*MSY*). Why does AC_t behave in this way? This is easy to explain. As fishing effort increases, according to equation (3.24), the total cost increases in the same proportion. That is, if fishing effort increases by 10 percent, the total cost of fishing will increase by 10 percent. The fundamental point is how the catch responds to the 10 percent increase in fishing effort. Equation (3.23) tells us that a 10 percent increase in fishing effort will result in a less than 10 percent increase in quantity harvested. This is true because the catch per unit of effort declines with increasing effort. Hence, total costs rise faster than the increase in the harvest; therefore, the average cost of production rises. Here we can see the interaction between biology and economics. Fishermen must pay for units of fishing effort at a constant cost per unit. However, the population dynamics of a fishery indicates that the yield, or catch per unit of effort, will in the case of the simplified Schaefer model decline as fishing effort expands. Thus, the average or unit cost of harvest increases with increasing fishing pressure.

Why does the average cost curve bend backward? Because, once we reach *MSY*, any additional fishing effort will reduce the catch. It is obvious that the average cost will still rise with not only a falling catch per unit of effort, but also with an absolutely falling harvest. Thus, the AC_t curve bends back, and the reader can identify the higher and higher per unit (average) cost with a reduced harvest. Consider points *c* and *b* in figure 3.14. To harvest Q_1, E_1 units of fishing effort must be used, which corresponds to point *b* on the yield function. The corresponding average cost is AC_1 at point *c*. If effort is increased to E_2, no change in catch will come about, *at least in theory*. However, E_2 corresponds to point *a* on the yield function and point *d* on the *AC* curve. It is obvious that

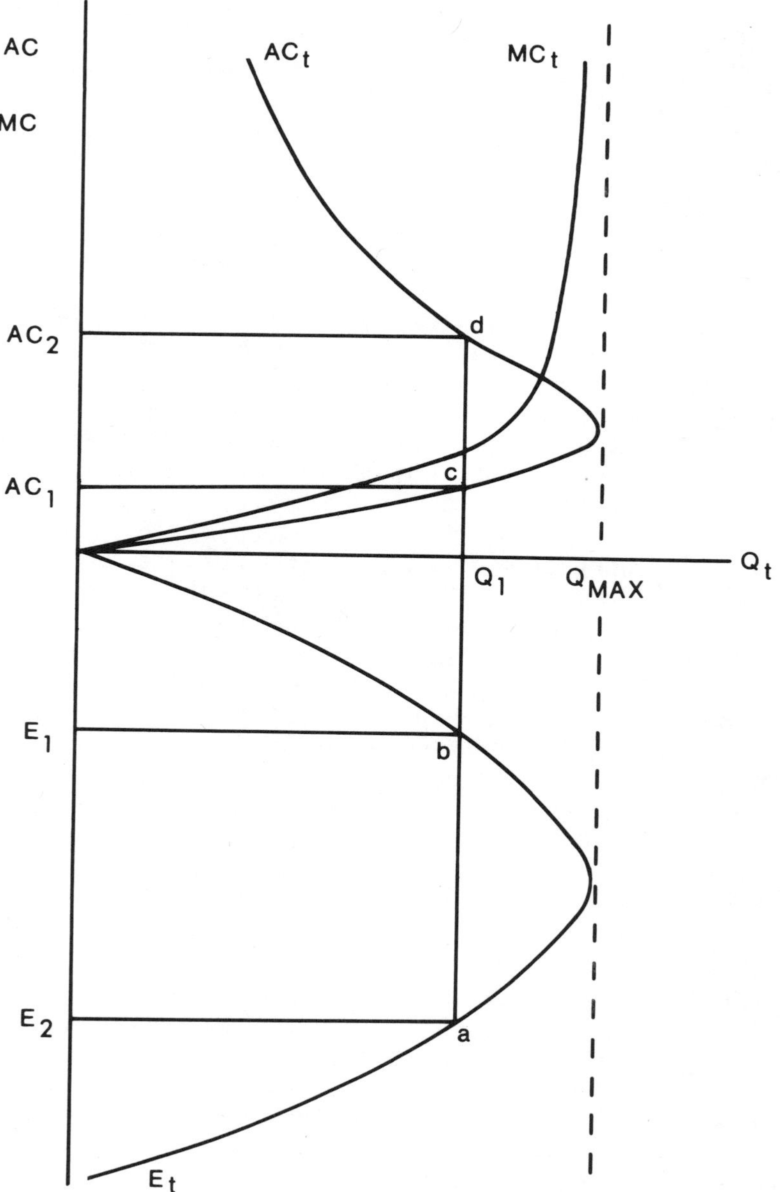

Figure 3.14: Relation between Average and Marginal Cost and Quantity of Fish Harvested and Fishing Effort

overfishing to E_2 is economically wasteful, since the same catch could have been obtained with less effort (E_1) and cost!

The marginal cost is, as in our numerical example above, the incremental cost of harvest and is asymptotic to Q_{MAX}. *MC* beyond Q_{MAX} would be negative and does not have any economic meaning, so we have drawn *MC* corresponding to *AC* up to Q_{MAX}.[15] The intersection of *AC* and *MC* has no meaning, because the *MC* curve only corresponds to production up to Q_{MAX} while the *AC* curve corresponds to all points on the yield function.

As we shall see in chapter 4, both the *AC* and *MC* curves are the *bioeconomic* supply curves, depending on how the fishery is managed.

Empirical Example of the Bioeconomic Supply Curve

In this section, we shall look at an *actual* bioeconomic supply curve. We will look at Gulf of Mexico menhaden in great detail. As indicated when we were constructing a bioeconomic supply curve, we are combining population dynamics with economic variables. For purposes of illustration, we shall use the Schaefer model, which has been applied to selected fisheries as shown in table 3.5. To compute the average and marginal cost per metric ton of Gulf of Mexico menhaden, equations (3.28) and (3.31) can be used. Three parameters are needed: *A, B,* and *C*. Table 3.5 contains estimates of the first two parameters (i.e., from the Schaefer population dynamics model):

$$A = 2.1238$$
$$B = -.0026$$

Remember, these are the parameters estimated by Chapoton (1971). The parameter *C* is the cost per unit of fishing effort. We have already discussed how this is obtained. At present, there are no detailed cost studies on the cost structure for the Gulf menhaden fleet. However, an approximation to *C* can be obtained if we assume that the total value of the catch is equal to the total cost of producing it, including adequate returns to vessels and fishermen. In 1970, the value of the catch was \$23.2 million dollars, and there were 397,000 vessel ton-weeks expended, or \$58.44 per vessel ton-week. Thus, *C* = \$58.44. Table 3.6 shows the average and marginal cost per metric ton using our biological and economic information.

Table 3.6: Estimation of the Average and Marginal Cost for the Gulf of Mexico Menhaden Fishery (1970 Cost Per Unit of Fishing Effort)

Landings (Q) (thous metric tons)		Average Cost (AC) Per Metric Ton[1] (Dollars)	Marginal Cost (MC) Per Metric Ton[2] (Dollars)
2.0		27.53	27.58
24.0		27.89	28.31
68.0	Fishery	28.69	29.96
124.0	Expansion	29.82	32.55
202.0		31.79	37.64
280.0		34.49	46.21
346.0		37.95	61.16
390.0		41.76	86.60
412.0		44.95	122.77
434.0	(MSY)	55.07	infinite
412.0		70.92	↑
390.0		80.65	
346.0		100.03	
280.0	Overfishing	136.02	
202.0		204.55	negative
124.0		355.15	
68.0		673.27	
24.0		1961.07	
2.0		23,853.06	↓

1. Equation (3.28) in text; 2. Equation (3.31) in text. Parameters: A = 2.124; B = -.0026; C = $58.44. Note: The negative sign for B should be ignored when using equations (3.28) and (3.31) since the sign is already included in their derivation. The plus sign in the quadratic (±) should be used in the expansion of the fishery; the negative sign should be used in the overfishing phase.

According to the yield curve for Gulf menhaden, we have increased landings from 2,000 to 434,000 metric tons, or *MSY*. After this point, landings decrease as overfishing occurs. As the fishery expands, the average cost per metric ton increases from $27.53 (2,000 metric tons) to $55.07 at *MSY* (434,000 metric tons), or 200 percent. This is directly attributable to the decline in the productivity of the fishery, or *CPUE*. As overfishing takes place, the catch declines absolutely, thereby accelerating the increase in average cost per metric ton as more and more fishing is expended to catch less fish. For example, if the Schaefer parabolic yield function exists throughout the entire range of fishing, the fleet could land 280,000 metric tons of menhaden at an estimated cost of $34.49 per metric ton compared to obtaining the same catch under conditions of overfishing at $136.02 per metric ton. Although the goal is to use vessels and fishermen in the most efficient manner, overfishing in this example has raised the average cost by $88.05 per metric ton. Neither the menhaden fishing industry nor the buyers of fish meal would benefit from this situation (i.e., overfishing). As shown in chapter 2, if fish meal prices increase, soybeans will be substituted for fish meal. The menhaden fleet will therefore contract, and most harvesters will suffer economic losses in the short run.

Why does overfishing take place? This is such an important subject that we have left it to an extended discussion in chapter 4. But there are indeed reasons for this madness! Marginal cost throughout the expansion of the fishery is above average cost and increases rapidly as *MSY* is approached. In reality, marginal cost approaches *MSY* asymptotically. When overfishing occurs, marginal cost (as discussed above) is negative and has no functional economic significance. In 1970, 397,200 vessel ton-weeks were expended, and the *predicted* catch was 434,000 metric tons using the catch-effort equation for menhaden in table 3.5. The *actual* catch was 546,000 metric tons (see figure 3.13), which illustrates our earlier point about the wide fluctuations in catch around any mathematical yield function. In 1970, the ex vessel value, or price per metric ton, was $43.19. According to the *predictive model* as demonstrated in table 3.6, the menhaden fleet's average *cost* of production would equal this market price at just under 412,000 metric tons. In chapter 5, we shall develop the concept of *market equilibrium* in great detail. Put simply, the market price for menhaden is determined by the world supply and demand for fish meal. At an actual price of $43.19 per metric ton, we would predict that the industry would produce somewhat *less* than 412,000 metric tons, since the predicted average cost of production is $44.95 per metric ton. The actual catch of 546,000 metric tons probably increased profits and wages above their opportunity cost *for 1970.* This usually has the effect of enticing more fishing effort into the fishery. As might be expected, 470,800 vessel ton-weeks were expended in 1971. Figure 3.15 shows the results of table 3.6 in graphic form.

The Fishery Productivity of the World's Oceans: The Final Word?

Potential from the Sea: Biological Productivity Analysis

In the nineteenth-century view, the high-seas fisheries were inexhaustible. After all, a little more than seven-tenths of the surface of the earth is covered by ocean. Although most of these waters are in areas where temperature and light conditions are

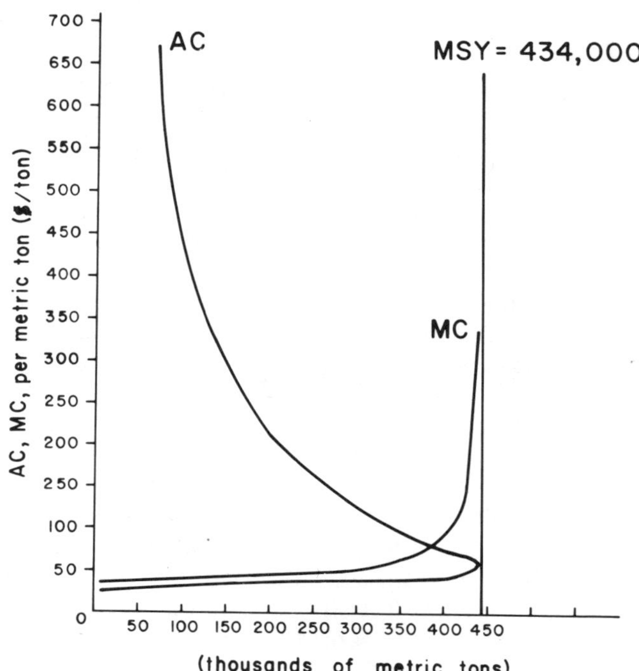

Figure 3.15: Average and Marginal Cost Curve for Gulf of Mexico Menhaden (1970)
Source: Table 3.6

favorable for plant growth, only 9 percent of the world's protein—adjusted for quality—comes from the sea (see chapter 2, table 2.1). Not long ago, Ryther (1969) calculated that "90 percent of the ocean and nearly three-fourths of the earth's surface . . . is essentially a biological desert." At the heart of the problem regarding the potential productivity of the sea is the "food chain."

As on land, animal life in the sea depends on plants, which either grow on the ocean bottom, such as seaweed and kelp, or float on the surface, such as microscopic phytoplankton. The growth of plant life is dependent on the rate of photosynthesis. In living aquatic plants sunlight produces the formation of carbohydrates from water and carbon dioxide. Thus, "primary production" involves the conversion of carbon (from carbon dioxide in the water) into organic substances. Marine plants must also have mineral nutrients, such as phosphates and nitrates. Under normal circumstances, these nutrients are the limiting factor to plant production in any marine ecological system (see chapter 5 for the implications of too many nutrients-fertilizers). The primary organic material produced is usually measured in terms of the element carbon; one gram of carbon is equivalent to ten grams wet weight of organic substances. By measuring photosynthesis rates at a series of depths at stations scattered over the oceans, estimates of organic production have been made. For example, Koblenz-Mishke (1965) recently estimated 13 billion metric tons of carbon fixed per year, or about 130 billion metric tons of primary organic matter. Earlier, Schaefer (1965) estimated 19 billion metric tons of carbon fixed per year. This translates into 190 billion metric tons of primary organic production (i.e., plant life).

It would seem that the ocean is just one giant vegetable garden, a panacea for the food crisis. Much of this vegetable garden, such as planktons, is available in extremely small concentrations per cubic meter of water. The process of filtering or centrifuging plankton is uneconomical. More important, as chapter 2 on consumer demand points out, affluent nations (and perhaps some less affluent nations) prefer beef, fish, sausage, or rice to plankton, which is salty and has a poor flavor. Remember, people would rather consume fish meal in the form of Colonel Sanders's chicken! Thus, there is little demand for the plant life in the ocean, and even if demand could be stimulated, the cost of harvesting would be prohibitive.

Let us climb the food chain a little higher. Herbivorous animals such as oysters, clams, mussels, and abalone "graze" in our gigantic vegetable garden in the sea. The Peruvian anchoveta depends considerably on plant plankton in its diet. About one-fifth (by weight) of the animals being harvested from the sea depend mainly on plant material.[16]

We can view the food chain as a pyramid having various floors called *trophic* levels. Figure 3.16 shows an aquatic food pyramid with the vegetable garden on the ground floor (trophic level 1). On the second floor are animals that consume plants directly, such as those mentioned above as well as zooplankton. There are over 200 species of typical plankton animals, the most numerous being crustaceans and coelenterates (e.g., hydra, jellyfish, and mollusks). Unfortunately, when an organism serves as food for another organism, there is a loss of matter and energy in the process.

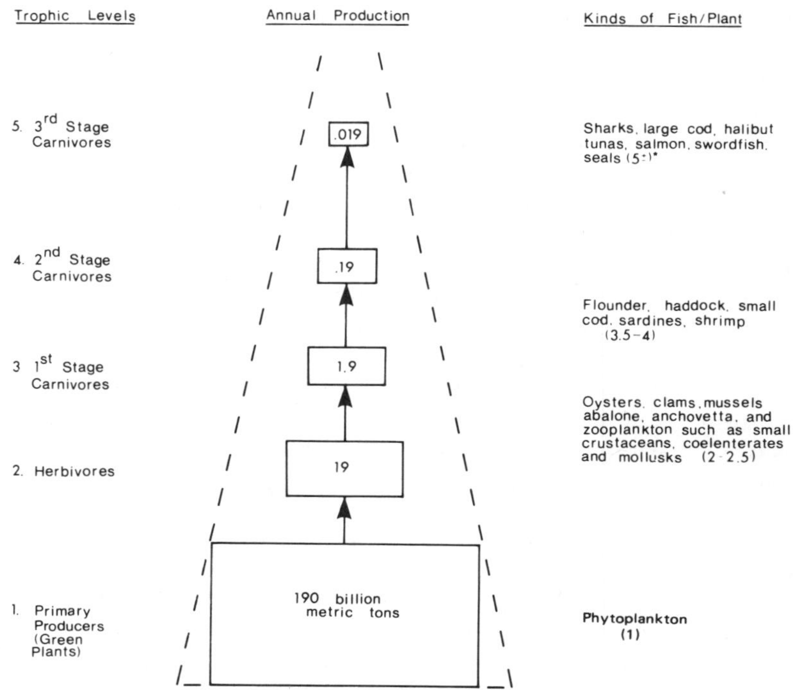

Figure 3.16: An Illustrative Aquatic Food Chain or Pyramid (Transfer Efficiency=10%)

Source: Schaefer (1965) for numerical examples

*Figures in parentheses indicate trophic level intervals where fish are found

It is generally regarded that yield decreases tenfold for every step up the food chain. This is called the *transfer efficiency,* or the percentage of the prey's annual production that is incorporated into the body tissue of the consumer species. For example, ten tons of phytoplankton would produce approximately one ton of oysters or clams. Note the reduction in organic weight as we ascend the food chain. Fish at trophic level 3 feed on zooplankton as well as smaller fish at their trophic level. Schaefer calls these species, such as sardines, groundfish, and shrimp, first-stage carnivores. It is impossible to classify some fish at an *exact* trophic level, since these levels are not discrete, but continuous. For example, small cod exist between the 3.5–4 trophic interval, and large cod at trophic level 5. The third-stage carnivores (trophic level 5), such as swordfish and tuna, produce only 19 *million* metric tons.[17] Let us illustrate the rapid decline in production as we ascend the food chain. One ton of swordfish would require the consumption of ten tons of flounder, which would require the consumption of a hundred tons of zooplankton, which would require the consumption of a thousand tons of phytoplankton. It is obvious that the lower man fishes on the aquatic food chain, the greater the potential food production. But, as pointed out in chapter 2, the demand preference for different kinds of fish determines where we fish on the food chain.

Where is all this leading us? Many biologists have used the food chain analysis to predict or estimate the *total world potential production of fish and shellfish from the sea.* We shall ignore the "apples and oranges" problem discussed in chapter 1. Schaefer (1965) believes that half of the production from trophic levels 3 and 4 might be available to man, with the balance going to predator–natural mortality. This would amount to over a billion metric tons. Since the actual fishery harvest can be only a small part of the product, Schaefer concludes that the possible marine catch should be about 200 million metric tons.

Ryther (1969) and Alverson, Longhurst, and Gulland (1970) have debated the potential fish production from the world's oceans. Ryther divides the world's oceans into oceanic (open sea), coastal, and upwelling provinces. For example, upwelling areas are characterized by water from below rising to the surface, where plankton production is often intense and where fish gather. It is in these areas of the world that Ryther claims food chains are the shortest—such as off the coast of Peru. He concludes that the open sea (oceanic province) is a biological desert, despite the fact that it covers 90 percent of the ocean. Ryther calculates the following potential production.

	Surface Area of Ocean (Percent)	*Fish Production (Mil metric tons-round wt)*
Oceanic	90	1.6
Coastal	9.9	120.0
Upwelling	.1	120.0
	100.0	241.6

Thus, 10 percent of the ocean creates through the food chain 99 percent of potential fish production. *Do not confuse biological production with potential harvest.* Man must share the production with other top-level carnivores, including the guano birds. Ryther concludes by saying, "it seems unlikely that the potential sustained yield of fish to man is appreciably greater than 100 million tons" (1969, p. 76).

The transition from estimated production through the food chain model to potential harvest is extremely ad hoc; therefore, the estimates are subject to great variability depending on assumptions employed.[18] Alverson, Longhurst, and Gulland (1970) take Ryther to task on a number of technicalities, such as his assumption about trophic levels removed from primary production. They indicate that his estimate greatly understates harvest potential. However, Ryther's rebuttal seems convincing: his estimate falls within the general range of other such estimates shown in table 3.7. Of course, Alverson, Longhurst, and Gulland claim Ryther may be right for the wrong reasons (e.g., high trophic levels for ocean and coastal provinces; low trophic level for upwelling areas).

Moiseev (1969) has made a detailed study of the fishery productivity of various areas of the world's oceans expressed in kg/km^2. He bases his production estimate upon *actual harvest,* which may well reflect the extent of primary food (plants) production. According to Moiseev, "there is no doubt that the primary production of a given region can generally be regarded as an index of the level of fish productivity" (1969, p. 270).

As we have argued in chapter 1, actual or potential total fish production is not very meaningful for economic analysis or even biological analysis when we are interested in the population dynamics of one species stock.

Estimated Utilization of World Fishery Potential by Species Categories

Groundfish. This category contains cod, haddock, flounder, sole, ocean perch, pollack, and other groundfish. These fish are generally classified as demersal (groundfish). For our analysis throughout this section, we shall draw heavily upon Gulland (1971) for estimates of resource potential.[19] Table 3.8 shows the 1973 world landings relative to *MSY,* or resource potential. The *world MSY* for each fishery category was obtained by summing the regional estimated *MSYs.* The reader should recall that in the simplified Schaefer model the ratio of catch to *MSY* is ambiguous if there is no information on which side of the parabolic yield function the fishery is operating. That is, if we are only catching 50 percent of *MSY,* there is no way of telling from this statistic

Table 3.7: Estimate of Total Ocean Yields of Aquatic Animals

Author	Potential Harvest (million metric tons)	Year	Method*
Chapman	1000	1966	FC
Pike and Spilhaus	200	1965	FC
Schaefer	200	1965	FC
Ryther	100	1969	FC
Ricker	150	1968	FC
Moiseev	80-100	1964	FC, Ext
Cushing	100	1966	FC
Bogorov	100	1965	FC
FAO[1]	120	1969	FC, Ext

*FC = food chain method; Ext = extrapolation of trends

1. FAO (1969)

Source: Moiseev (1969, p. 202); except FAO.

Table 3.8: The Relation of World Landings to Maximum Sustainable Yield (MSY) for Selected Established Ocean Fisheries (Thousands of Metric Tons--Round Weight)

Species	MSY (Estimated Potential)	1973 Landings	Percent Landings of MSY
Selected Food Finfish (Marine)			
(a) Groundfish	11,734	11,841	101[1]
(b) Tunas (includes bonito)	2,393	1,711	72[2]
(c) Salmon	484	475	98[3]
(d) Halibut	57	121[4]	N/A
(e) Sardines and Herring Like Food Fish	5,970	4,344	73[5]
(f) Mackerels	2,744	3,005	110[6]
Selected Food Shellfish (Marine)			
(g) Shrimp	1,492	1,109	74[7]
(h) Lobsters	193	193	100[8]
(i) Crabs	672	366	54[9]
(j) Clams	1,439	624	43[10]
(k) Scallops	1,491	218	15[11]
(l) Oysters	N/A	824	N/A
Selected Marine Food Fish and Shellfish		24,832	
Fish Meal Species	N/A	19,390	N/A
Other Marine Fish and Shellfish	N/A	11,047	N/A
All Marine Fish and Shellfish	120,000[12]	55,269	46
Fresh Water Fish and Shellfish	N/A	9,143	N/A
Aquatic Plants	N/A	1,088	N/A
Total World Harvest (Animals and Plants)	N/A	65,500	N/A

1. Fully utilized; 2. Fully utilized, except bonito; 3. Fully utilized; 4. 1973 landings include "greenland halibut" which is excluded from the MSY estimate which contains only halibut; 5. Overexploited, where further increases in fishing effort would decrease world catches; 6. Fully utilized; 7. Potential for increased production; 8. Fully utilized and overexploited in many areas; 9. Increased production possible; 10. Increased production possible; 11. Increased production poss sible; 12. FAO (1969).

Source: Gulland (1971), Fullenbaum (1970) and FAO (1973).

alone whether the resource is overexploited or underexploited. The world total may also seriously obscure regional overfishing: thus one must be very cautious in interpreting table 3.8. However, it is fairly safe to conclude that the world potential production of groundfish has been reached and that overfishing on many stocks is widespread. According to current estimates of groundfish potential, it would seem that no significant increases in production can be achieved on a sustainable basis.

Tuna. Earlier in this chapter, it was indicated (Joseph 1973) that for most commonly fished species of tuna, the world had reached its maximum sustainable potential of 1.893 million metric tons. The 1973 world landings include *bonito,* which is classified as a tuna. The world *MSY* for bonito is 500,000 metric tons, and the stock is relatively underutilized. With increasing world demand for tuna, however, expectations are that bonito will be used to supplement traditional species. Thus, any potential increase in world "tuna" production will have to come from the bonito resource or *frigate mackerel,* which is sometimes considered a close substitute for tuna.[20]

Salmon. World salmon production has not increased over the last thirty years. Unless aquaculture (chapter 7) or artificial means are extensively employed, there is relatively little potential for increasing salmon production on a world basis. The resource is fully utilized.

Halibut. Atlantic and Pacific halibut have been fully utilized for years. Because of intensive resource pressure over fifty years ago, the Pacific halibut is regulated by the International Pacific Halibut Commission. A close substitute for halibut is Greenland halibut. The landings figure in table 3.8 contains Greenland halibut obtained from the North Atlantic and Pacific. This writer is not aware of resource potential estimates for Greenland halibut; however, it is presently under the quota set by ICNAF. The best "guesstimate" is that the Greenland halibut is being fished near its sustainable potential.

Sardines and Herringlike Food Fish. This is one of the most difficult categories, since for many species there is a fine line between that used for direct consumption and that used for industrial (i.e., fish meal) demands (see footnote 3, Table 2.2). Table 3.8 indicates that 1973 world landings of sardines and herring were 73 percent of *MSY.* Pacific (i.e., off the coast of California) sardines underwent a decline in the late 1940s, a decline that reached catastrophic proportions in the early 1970s. The catch of sea herring in the northwest Pacific increased dramatically during the 1960s. The tremendous increase in fishing effort has reduced stock size and affected recruitment. Anthony's (1972) thorough study of the Atlantic herring in the Gulf of Maine concluded that catch quotas were necessary as a means of preventing further declines in stock size. Quotas have been placed on herring by ICNAF for areas of the northwest Atlantic. Thus, there is every indication that sardines and herringlike food fish are overexploited and that further increases in fishing effort will decrease catches.

Mackerel. There has been less study of the mackerel stocks; on a world basis, however, they are fully utilized. This is not to say that certain small regions may have potential for small increases.

Shrimp. Reports from the coastal areas of the United States indicate that the overwhelming majority of shrimp stocks are fully utilized. However, shrimp imports into the United States remain strong and are increasing in response to the demand factors discussed in chapter 2. Thus, shrimp production can be expanded *on a world basis.* Roughly three-quarters of the potential shrimp production is presently harvested. However, it is expected that this incremental production will go mainly to affluent nations and will not be a factor in the food-deficient countries of the world. In the Indo-Pacific area, however, much shrimp is produced through aquaculture (see chapter 7) and is consumed locally. For example, India consumed 80 percent of its shrimp production and exported the balance. Some of this production was derived from small family farms that used some of their produce for direct consumption.

Lobsters. It is quite safe to say that the world lobster production is at maximum sustainable yield. Recent indications are that overexploitation is now taking place, especially in the case of the American lobster, which is found off the Atlantic coast of the United States and Canada.

Crabs. As a general category, the world crab resource is only partly utilized with respect to its *MSY.* For example, the blue crab resource in the Gulf of Mexico still remains vastly underutilized. However, the global figures do obscure the overexploitation of king crab in Alaska, which has forced authorities to impose quotas. Unlike lobsters, crabs are mainly marketed as frozen or canned meat, except in areas in direct proximity to the resource (i.e., Maryland and Chesapeake Bay). Because the world resource is relatively underutilized, a large potential harvest is possible.

Clams. This resource has great potential for increase. Like oysters, clams can be aquacultured. The *MSY* from the wild stock alone is 1.432 million metric tons, but world landings from the wild stock and from aquaculture are only 624,000 metric tons.

Scallops. From the figures given in table 3.8, it may seem that the world scallop resource is vastly underutilized. The *MSY* of 1.491 million metric tons includes the recent discovery of 740,000 metric tons of calico scallops off the Atlantic coast of the United States. This estimate has been questioned. Mechanical shucking of calico scallops is still an obstacle to their development. The world totals also obscure substantial overfishing of northwest Atlantic sea scallops. However, it would generally appear that substantial world increases in production are possible from scallop resources.

Oysters. Because of the widespread aquaculturing of oysters, *MSY* is not generally estimated. Oyster production can be increased substantially given adequate, pollution-free coastal zones. Chapter 5 will discuss the role of water pollution, which has reduced the potential oyster harvest.

Of the current marine catch, over 35 percent goes directly into fish meal or industrial use. The major marine food fish species of the world account for 45 percent of the total marine harvest. The category *"other marine fish and shellfish"* accounts for 20 percent of the marine harvest and covers a wide diversity of fish and shellfish used for food and industrial purposes (e.g., cat food). It is quite apparent that among the marine food finfish the greatest pressure is on precisely these resources. Except for lobsters, there is increased potential for further production among the marine food shellfish. The FAO (1969) estimate of a marine animal potential production of 120 million metric tons obscures the fact of differential demand pressures for various species. As discussed above, the potential from the sea has been derived from food chain analysis. We shall explore the apparent potential for increased food production from latent resources in chapter 8.

Summary

In this chapter, we have surveyed some fundamental aspects of fishery resource analysis along with the derivation of bioeconomic supply or cost functions. The overexploitation or depletion of many of the world's most valuable fisheries is widespread and on the increase despite crude attempts at management. All this has occurred because of increasing world population and, in many countries, because of rising affluence coupled with the common property nature of fishery resources.

Relatively few countries exploit or harvest the fishery resource on a world basis. In 1973, for example, the four leading countries accounted for 70 percent or more of total world production in the groundfish, salmon, halibut, clams, scallops, and oyster species categories. Out of twelve major fishery categories, Japan was among the four leading producers in ten. Obviously, a country such as Japan must deal with the resource crisis, since it depends on some of the most heavily exploited species, such as tuna, salmon, and groundfish. The USSR was one of the leading four producers in six of the twelve classifications of fish and was the leading producer of groundfish, halibut, and sardines in 1973. Despite a stagnation in aggregate catch, the United States was still a world leader (as of 1973) in the harvesting of crabs, clams, and oysters. Of the twelve species categories, the United States was among the top four producers in nine classifications.

In trying to understand how fishery resources behave under exploitation, we have discussed a number of theoretical models to explain the relation between catch and fishing effort. The Schaefer model is probably the simplest to understand; it postulates a simple parabolic relation between catch and fishing effort. This model has been successful in analyzing eastern Pacific yellowfin tuna and Peruvian anchoveta fisheries. GENPROD, which generalizes the Schaefer model, gives the researcher more flexibility in testing catch-effort relations. An alternative model, which requires more data, is the Beverton-Holt formulation, which relates catch per recruit to fishing mortality. This model has been widely used to analyze the population dynamics of many fisheries throughout the world.

One of the most important features of this chapter is the derivation of a bioeconomic supply or cost function. In all the models, both the average and marginal costs of production increase as the output of the fishery expands—because the fishery resource is subject to diminishing returns as fishing effort is increased. The bioeconomic supply curve will be used in chapter 4 along with the demand curve developed in chapter 2 to analyze the behavior of market conditions in a fishery. This analysis will lead to a *bioeconomic model of a fishery.*

Finally, we have reviewed the overall potential for food production from the sea—which is based upon the food chain or pyramid. Given the current fishing effort at various trophic levels, it is estimated that the potential marine fishery production is approximately 120 million metric tons. Although current catches are running at only one-half of this potential, it is quite apparent that the most popular food finfish are either fully exploited or overexploited. The industrial catch used for fish meal is also in particular trouble owing to recurrent changes in the oceanographic environment. It is quite clear that demand factors determine harvesting trends, which are producing a crisis at sea. However, latent or *un*utilized species may still be very important in the future as a source of protein from the world's oceans.

Notes

1. Fishing effort is measured in terms of ships, men, other resources, and technology used in catching fish. The measurement of fishing effort will be discussed later in this chapter.

2. Maximum sustainable yield, or *MSY*, is the maximum catch that can be taken from a fishery stock on an annual basis into perpetuity without endangering the stock. A fishery resource is called renewable because its biological growth provides a steady stream of catches over a long time period.

3. Incidental catch is fish harvested as a consequence of trawling for another species. Some fish that are caught incidentally have little or no economic value and are discarded.

4. The Bahamas has prohibited foreign fishing off its continental shelf; Mexico has negotiated a phasing out of U.S. shrimp fishing off its coast.

5. In his original work, Schaefer (1954) postulated a somewhat different form of the *S*-shaped logistic, or

$$N_t = \frac{N^*}{1 + be^{-a^*N^*t}}$$

The first derivative is

$$dN/dt = a^*N_t(N^* - N_t) \ .$$

Using Schaefer's notation

$$\frac{dP}{dt} = k_1 P(L - P) \ ,$$

where $P = N_t;\ k_1 = a^*,$ and $L = N^*$.

Schaefer called this the Verhulst-Pearl logistic. For the mathematically minded reader, this will be helpful in comparing our exposition with some of the biology literature. For a discussion of logistic equations, see Blumberg (1968).

6. The techniques of the measurement of fishing effort are in the next section of this chapter.

7. In economics, this is a special form of the Cobb-Douglas production function (1948), where all exponents equal one. Although this formulation is used by biologists, it does have the undesirable characteristic that catch can be increased without bound by increasing effort. Carlson (1969) has modified equation 3.7 based on probability theory. This modification has

been empirically tested in Bell et al. (1973).

8. Fox (1970) modified Schaefer's technique by assuming a Gompertz growth function for the biomass, which resulted in better fits to catch-effort data for some species. Fox's model does not predict that catch per unit of effort goes to zero with high effort but that it approaches zero exponentially. Fox's catch-effort curve is similar to GENPROD when $m<2$ (i.e., skewed to the right).

9. If there are x years in the year class, there are also x age groups in any one year in the fishery. This is because, in all groups in one year, the same items are being summed as in the age groups within a single year class.

10. Biologists have developed a diagram called an "isopleth" to illustrate the relation between Q/R, F, t_c.

11. Ricker's equation for stock-recruitment is of the following form: $R = Ne^{a-bN}$. Beverton-Holt propose: $R = (a + b/N)^{-1}$.

12. M, or natural mortality, may be obtained by relating fishing effort to Z. The intercept will be an estimate of M (zero fishing effort)

F may be calculated as $F = Z-M$.

13. For an interesting attack upon Schaefer's standard population dynamics procedure, see Segura (1973), who argues that Schaefer has failed to adjust his fishing effort for technological change (i.e., introduction of the power block, echo sounders, steel vessels, and increasing skill of the fishermen) and has misspecified his model by not including separate environmental variables in his equation (i.e., bird predation, water temperature in Trujillo, Peru). Segura finds that optimal effort is approximately 16.2 million ton trips, or only 68 percent of the level of effort used in Peru in 1968-1969. Thus, he concludes that the anchoveta is overfished.

14. The short-run supply curve for a fishery is almost invariant to price, except where prices are so low that the revenue will not cover variable costs such as diesel fuel, supplies, and bait. This will be discussed in some detail in chapter 9.

15. Since $MC = \Delta TC/\Delta Q$, any increase in fishing effort beyond that necessary to harvest Q_{MAX} would result in ΔTC being positive while ΔQ would be negative. As we shall see later, negative MC has no economic function in market determinations of prices.

16. It is interesting to note that on land practically all major sources of animal protein get most of their nourishment from vegetable sources.

17. Third-stage carnivores are sometimes called nektons, implying large or relatively large animals possessing the capacity for active movement in the aquatic environment over considerable distance.

18. Most writers have assumed a utilization rate (defined as the percentage of ocean production available to man for potential harvest) of about 50 percent.

19. Much of the tabulation of resource potential by fishery category in

table 3.8 was taken from Gulland's compilations and analyses of others' work by Fullenbaum (1970).

20. See chapter 8 for a discussion of the central Pacific skipjack tuna and its possible potential in augmenting tuna supplies. It has not been included in table 3.8 under *MSY*.

References

Alverson, D. L.; Longhurst, A. R.; and Gulland, J. A. 1970. How much food from the sea? *Science* 168: 503–505.

Anthony, Vaughn C. 1972. Population dynamics of the Atlantic Herring in the Gulf of Maine. Ph.D. dissertation, University of Washington.

Bayliff, W. H. 1967. *Procedure for estimating the parameters of the Schaefer yield model for yellowfin tuna.* Inter-American Tropical Tuna Commission interim report 3.

Bell, Frederick W. 1966. *The economics of the New England fishing industry: the role of technological change and government aid.* Federal Reserve Bank research report no. 31. Boston.

Bell, Frederick W., and Fullenbaum, Richard F. 1973. The American lobster fishery: economic analysis of alternative management strategies. *Marine Fisheries Review,* paper 994.

Bell, Frederick W.; Carlson, Ernest W.; Hirschhorn, George; and Schaaf, William E. 1972. *Extent of capitalization in U.S. fisheries.* National Marine Fisheries Service working paper 129.

Bell, Frederick W.; Carlson, Ernest W.; and Waugh, Frederick V. 1973. Production from the sea. In *Ocean fishery management: discussion and research,* ed. A. A. Sokoloski. National Oceanic and Atmospheric Administration technical report, National Marine Fisheries Service CIRC-371.

Beverton, R.J.H., and Holt, S. J. 1957. *On the dynamics of exploited fish populations.* London: Her Majesty's Stationery Office.

Blumberg, A. A. 1968. Logistic growth rate functions. *Journal of Theoretical Biology* 21:42–44.

Carlson, Ernest W. 1969. *Bio-economic model of a fishery.* National Marine Fisheries Service, Division of Economic Resources, working paper no. 12.

——. 1973. Cross section production functions for North Atlantic groundfish and tropical tuna seine fisheries. In *Ocean fishery management: discussion and research,* ed. A. A. Sokoloski. National Oceanic and Atmospheric Administration technical report, National Marine Fisheries Service CIRC-371.

Chapoton, Robert B. 1971. The future of the Gulf menhaden, the United States' largest fishery. *Proceedings of the Gulf and Caribbean Fisheries Institute.* 24th annual session.

Christy, Francis T., Jr. 1973. Northwest Atlantic fisheries arrangements: a test of species approach. *Ocean Development and International Law Journal* 1:65–91.

Cushing, D. H. 1968. *Fisheries biology: a study in population dynamics.* Madison: University of Wisconsin Press.

Dumont, William H., and Sundstrom, G. T. 1961. *Commercial fishing gear of the United States.* U.S. Fish and Wildlife Service, Fish and Wildlife Circular 109.

Food and Agriculture Organization. 1968. *The state of world fisheries.* World Food Problems no. 7.

——. 1969. *Report of the FAO Committee on Fisheries.* 4th session.

——. 1973. *Yearbook of fishery statistics,* Vol. 36.

Fox, William W., Jr. 1970. An exponential surplus-yield model for optimizing exploited fish populations. *Transactions of the American Fisheries Society* 99:80–88.

Fullenbaum, Richard. 1970. A survey of maximum sustainable yield estimates on a world basis for selected fisheries. NMFS working paper no. 43.

Gates, John M., and Norton, Virgil J. 1974. *The benefits of fishery regulation: a case study of the New England yellowtail flounder fishery.* University of Rhode Island Marine Technology Report no. 21.

Gulland, J. A. 1969. *Manual of methods for fish stock assessment.* Rome: Food and Agriculture Organization.

——. 1971. *The fish resources of the ocean.* London: Fishing News (Books), Ltd.

——. 1974. *The management of marine fisheries.* Bristol: Scientechnica, Ltd.

Huang, David S., and Lee, Chae Woong. 1974. Externalities: the case of a Japanese fishery. Southern Methodist University, unpublished manuscript.

Inter-American Tropical Tuna Commission. 1970. *Annual report.*

International Commission for the Northwest Atlantic Fisheries. 1972. *Memorandum by the U.S. Commisioners on regulation of fishing effort.* November 14, 1972.

——. 1975. *1976 total allowable catches and national allocations for the convention area and statistical areas 0 and 6.* Circular Letter 75/58.

Joseph, James. 1970. Management of tropical tunas in the eastern Pacific Ocean. *Transactions of the American Fisheries Society* 99:629–648.

——. 1973. Scientific management of the world stocks of tuna, billfishes, and related species. *Journal of the Fisheries Research Board of Canada* 30:2471–2482.

Kasahara, Hiroshi, and Burke, William. 1973. *North Pacific fisheries management.* RFF Program of International Studies of Fishery Arrangements, Resources for the Future.

Koblenz-Mishke, O. I. 1965. The magnitude of the primary production of the Pacific Ocean. *Okeanologiia* 5:363–371.

Longhurst, Alan, et al. 1972. The instability of ocean populations. *New Scientist* 54:1–3.

MacCall, Alec D. 1974. Stock assessment, fishery evaluation, and fishery management of Southern California recreational and commercial fisheries. National Marine Fisheries Service, Southwest Fisheries Center, administrative report no. LJ-74-24. Unpublished.

Moiseev, P. A. 1969. *The living resources of the ocean.* Translated for the National Marine Fisheries Service.

National Marine Fisheries Service. 1975. A listing of marine fishery resources which are depleted, in imminent danger of depletion, or under intensive use. Unpublished paper.

——. 1976. National plan for marine fisheries. Draft manuscript.

Nichols, John P., and Griffin, Wade L. 1975. Trends in catch-effort relationships with economic implications: Gulf of Mexico shrimp fishery. *Marine Fisheries Review,* paper 1119.

Payne, Roger. 1968. Among wild whales. *The New York Zoological Society Newsletter.*

Pella, Jerome J., and Tomlinson, Patrick K. 1969. A generalized stock production model. *Inter-American Tropical Tuna Commission Bulletin* 13: 421–458.

Richards, F. J. 1959. A flexible growth function for empirical use. *Journal of Experimental Botany* 10:290–300.

Ricker, W. E. 1954. Stock and recruitment. *Journal of the Fisheries Research Board of Canada* 11:559-623.

Ryther, John H. 1969. Photosynthesis and fish production in the sea. *Science* 166:72-76.

Schaaf, W. E., and Huntsman, G. R. 1972. Effects of fishing on the Atlantic menhaden stock: 1955-1969. *Transactions of the American Fisheries Society* 101:290-297.

Schaefer, Milner B. 1954. Some aspects of the dynamics of populations important to the management of commercial marine fisheries. *Inter-American Tropical Tuna Bulletin* 1:27-56.

——. 1965. The potential harvest of the sea. *Transactions of the American Fisheries Society* 94:123-128.

——. 1968. Methods of estimating effects of fishing on fish populations. *Transactions of the American Fisheries Society* 97:231-241.

——. 1970. Men, birds and anchovies in the Peru Current—dynamic interactions. *Transactions of the American Fisheries Society* 99:461-467.

Segura, Edilberto L. 1973. Optimal fishing effort in the Peruvian anchoveta fishery. In *Ocean Fishery Management: discussion and research,* ed. A. A. Sokoloski. National Oceanic and Atmospheric Administration technical report, National Marine Fisheries Service CIRC-371.

Silliman, R. P., and Gutsell, J. S. 1958. *Experimental exploitation of fish populations.* U.S. Fish and Wildlife Service Fish Bulletin 58.

Sysoev, N. P. 1970. *Economics of the Soviet fishing industry.* Translated for the National Marine Fisheries Service.

4. *Why Manage the Fisheries?*

Common Property Resources and Technological Externalities

A basic characteristic of all wild stock fisheries is that they are a common property natural resource. Like many common property resources, such as water and air, they can be used *without cost* by economic enterprises. That is, no single user has to pay for the right to use the resource, nor does he have exclusive rights to the resource or the right to prevent others from sharing in its exploitation. One of the most fundamental characteristics of a fishery stock as a common property natural resource is that the amount of fishing effort (i.e., inputs of capital and labor) applied is not subject to the same restraints that govern the use of *privately owned resources,* such as farmlands, coal mines, or forest land. The commercial users of fishery resources are in competition with one another to get a larger share of the resource for themselves.[1] There is no limit on number of fishermen who can exploit a fishery so long as there is profit to be gained. As we saw in chapter 3, as the number of individuals entering the fishery increase (i.e., an increase in fishing effort), the marginal and average catch per unit of effort falls.

At this juncture the reader should separate two facts in his mind. First, the common property nature of the fishery resource has nothing to do with the behavior of the marginal and average catch per unit of fishing effort. This is due to the underlying population dynamics interacting with the fishing effort. The private or sole owner of a fishery resource would still be faced with these same relations; that is, he would face dwindling productivity of the resource as exploitation increased.

Second, the fact that the resource is available to all and not just one owner introduces a different set of circumstances. Each economic enterprise (vessel) is competing against all other enterprises for a "slice of the action." A curious thing happens: by allowing the resource to be common property, a vessel owner's productivity is directly influenced by how many firms are exploiting the resource. Consider this comparison. The output per man on a privately owned wheat farm is a function of the inputs of capital and technology that the farm owner selects. For purposes of simplicity, let us leave out the influence of weather. That is, the productivity—measured as output per man—is certainly independent of all other farms. Let us say the number of farms doubles or even triples. This would have absolutely no impact on the productivity of the wheat farm;[2] each farmer is the private or sole owner of his land. In contrast, the fisherman is not the sole owner of a fishery resource.

If the number of fishing firms doubles or even triples, will this have any impact on the individual fishing enterprise? The answer is a resounding *YES.* Why? Because the entrance of more firms increases the fishing effort on the common property resource. Everyone in the fishery is thereby influenced by what economists call a *technological externality*—technological because each fishing firm's productivity or catch per unit of effort declines, and external because the individual firm has no control over this aspect of its productivity. For example, the American lobster fishery off the New England states averaged forty-five pounds of lobsters per trap annually over the 1950–1952 period. In a fifteen-year period, the fishing effort (measured by the number of traps) nearly doubled. After all, the fishery is common property to everyone. As a consequence, annual yield per trap over the 1964–1966 period had dropped by nearly nine pounds, to approximately thirty-five pounds per trap.[3] Consider the farmer once again. If his land were common property, people might come onto his land and plant crops on land he was allowing to remain fallow. This would certainly influence the farmer's productivity. With fisheries, the average vessel operator's productivity is tied directly to the numbers exploiting the resource.

Thus, we can conclude that the common property nature of the fishery resource allows everyone to gain access to its use and that this produces a technological externality to all economic units exploiting the resource. Each operator has no vested interest in the resource (i.e., cannot gain ownership) and will operate in his own self-interest. As we shall see, the fisheries are an *exception* to the great law of private enterprise postulated by the father of economics, Adam Smith, in his famous *Wealth of Nations* (1776). According to Smith, the interest of the community or society is simply the sum of the self-interest of the individuals who compose it. Each man, if left alone, will seek to maximize his own wealth and is led by an "invisible hand" to promote social ends. Therefore, all men, if left unimpeded to compete with one another, will maximize aggregate wealth. Placed in a somewhat different framework, Smith's doctrine led to the conclusion that competition among entrepreneurs would lead to optimum use of society's scarce resources. Smith assumed that all the factors of production (including natural resources) were privately owned—not common property.

As Smith visualized competition, unimpeded fishermen, too, would maximize social welfare in their quest for individual profits—*but not with a common property fishery resource.* A rebuttal to Smith's theory is given by Harden in his "The Tragedy of the Commons" (1968). This will be illustrated in the next section, where we integrate the demand for fish (chapter 2) with the supply of fish discussed in chapter 3.

Market Equilibrium: Individual vs. Social Optimization

We are finally ready to bring together the demand for fishery products discussed in chapter 2 with the supply curve of fishery products developed in chapter 3. Let us assume that consumer income and the price of near substitutes are held constant so that we have only the relation between prices and quantities demanded. Let us further assume that the market for the fishery resource being harvested is local in nature. The New England yellowtail flounder fleet, for example, harvests fresh flounder, which is sold to buyers (i.e., middlemen) who market the product to restaurants, primarily in the large cities of New England and New York. Thus, the product has a limited local demand, and increases or decreases in supply will alter prices and hence the quantity demanded. Thus, where the supply is such that the fishery in question completely dominates the market, the demand curve will slope downward. But this is not always the case. In fact, the opposite may be more characteristic of the market for fishery products. Hence, the demand curve discussed in chapter 2 could be a market demand curve for the world. Conceptually, we would sum the market demand curve across all countries for each species. For example, tuna, frozen groundfish, fish meal, and shrimp have an international market, a market supplied, in many cases, by small fishery stocks of the individual species scattered throughout the world. Changes in the production from one stock of species may be too small to influence world prices appreciably. A small shrimping ground off Australia, for example, will not appreciably influence world prices regardless of its level of production. Therefore, world prices for many species of fish are determined by the aggregation of world supply and demand and become a *given* to each fishery stock and the associated fishing fleet.

Figures 4.1 and 4.2 illustrate the two cases. In figure 4.1 a given fishery resource (i.e., geographically specific) typically supplies the entire market. The market is usually localized and is usually determined by the demand for fresh fish or a unique kind of fish that is not eaten elsewhere in the world. In this case, the changes in quantity marketed will influence the price; therefore, the demand curve facing this fishery slopes downward. If the harvest level is increased from Q_1 to Q_2, price will decline from p_1 to p_2. Figure 4.2 is probably typical for those fishery products that are (1) demanded by many countries in the world, (2) have relatively *small* stocks of that species geographically scattered throughout the world, and (3) can be preserved for reasonable lengths of time. Frozen groundfish and shrimp are good examples of this latter case. Suppose panel (a) in figure 4.2 represents the

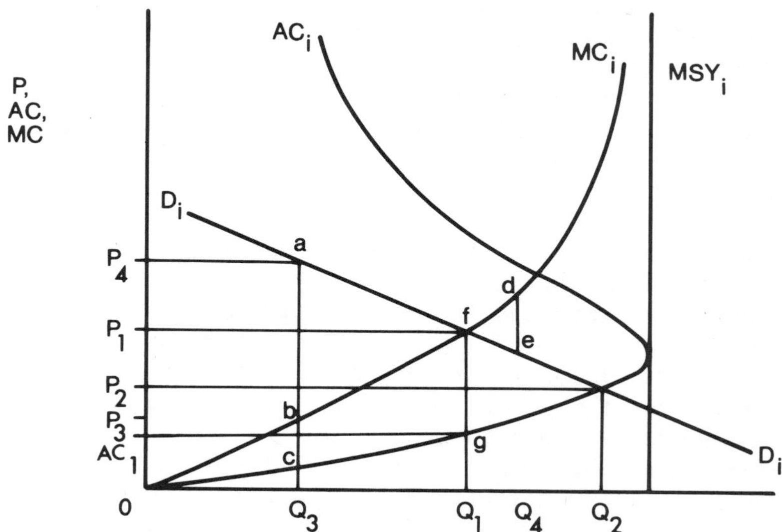

Figure 4.1: Supply from Fishery Resource Dominates the Market (i'th fishery)

world market for frozen shrimp. Since many countries compete on the demand side for shrimp, we have summed the demand curves for all countries running from one to *n*. Similarly, we have summed the *AC* curves (designated *S*) for all fishery stocks running from one to *v*. Up to this point, we have not talked about market equilibrium.

Consider figure 4.1. If Q_3 is produced, consumers will be willing to pay p_4, or point *a* on the demand curve. Fisherman would calculate the cost of supplying Q_3. Their average and marginal costs are shown as points *c* and *b* on the respective curves. The marginal cost curve is usually a good guide in pricing a product, since as production (harvest) increases, the entrepreneur will want to know the increment to revenue vs. the increment to cost. In this case, the price, or p_4, is well above MC_i at point *b*. Why not expand to the point where the incremental revenue, or in this case price, is equal to the incremental cost? Good suggestion! But what about further expansion, say to Q_4? At Q_4, the MC_i is at point *d* while the price is at *e*. Therefore, the incremental cost well exceeds the price, and the industry would experience incremental losses. Now we finally have it! The *market equilibrium* is where $p_1 = MC_i$, or point *f* at quantity Q_1. Wrong? The problem is that this equilibrium is unstable as far as the industry is concerned. Look at it from this point of view. At point *f*, what are total revenue and total cost? Total revenue is $OP_1 fQ_1$, and total cost is OAC_1gQ_1. Total cost is much less than total revenue. It should be remembered

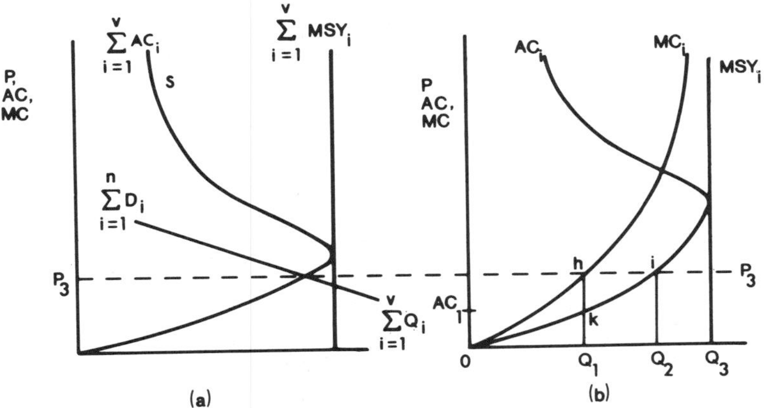

Figure 4.2: Supply from an Individual Fishery Resource too Small to Influence the Market

that total cost already contains a normal return on invested capital and wages for fishermen, a return that is comparable to those that could be obtained in competing industries for capital and labor. Hence, point *f* reveals that fishermen would be earning an economic profit defined as *a return well above normal.* Remember this important point! Where marginal cost equals price, excessive returns accrue to those engaged in fishing. Fishermen who previously drove Chevrolets are now driving Cadillacs, buying large homes, and smoking expensive cigars. Let us go back to Adam Smith. It would seem that each fisherman acting individually is maximizing his own income. True! What about maximizing society's welfare? Also true. Why? At point *f*, society through individual action has allocated so many units of effort where the marginal cost of producing that last fish (or pound of fish) is equal to what consumers are willing to pay. From society's standpoint, points *d* and *e*, for example, are clearly undesirable at harvest level Q_4, since more fishing effort will have to enter the industry and since the marginal cost of producing the last fish exceeds the price consumers are willing to pay. This is often a difficult concept for readers to understand. What we are really saying is that there are more vessels and fishermen in the fishery than are necessary, since the incremental cost at point *d* is above the incremental revenue (price) people are willing to pay. What about the individual fisherman? He may well be unconcerned about the industry or society, but he is indeed concerned about his own situation. Point *e* on the demand curve is above the average cost of production; total revenue would be greater than total cost, and a somewhat smaller economic profit would occur than at point *f* (i.e., maximum economic profits). Here is the crux of the problem. From society's point of view, Q_1 should be produced—where the capital, labor, and materials used in fishing are just enough for the marginal cost of producing a unit of fish to be equal to what people are willing to pay. The *additional* vessels and fishermen needed to produce Q_4 are misallocated to this fishery. They should be operating in another fishery or, as we shall see, producing in other sectors of the economy where $MC = P$. Because the fishery resource is common property, we have what economists call a divergence between private and social interests. Adam Smith's theory does not apply to the fisheries.

Let us turn the situation around. Consider figure 4.2, panel (b). This is, for example, a small lobster fishery off Australia. However, it is not common property; it has but one owner. At price p_3, how many pounds of lobsters will be supplied to the world market?

The answer is Q_1, or the sole owner will produce lobsters as long as the price is above the marginal cost of production. Point *h* is in the sole owner's self-interest, since it maximizes his economic profits. Point *i* is out of the question, since it dissipates all the sole owner's economic profits. If the common property characteristic of the resource is eliminated, society's and the individual's objectives coincide, at least in terms of the numbers of vessels and fishermen that should be producing lobsters. But is not the sole owner earning an abnormal return? Yes, he certainly is! Some economists would call this excessive return *economic rent.* Some would say that the sole owner is not really entitled to economic rents, since he would still produce Q_1 if the government owned the resource and rented it to him; he would be willing to pay an amount equal to the economic rents for the right to use the resource. The government would charge *hk* per unit of fish produced. Hence, the tenant would produce Q_1, where his total revenue would be OP_3hQ_1 and his operating cost (including a normal return) would be OAC_1kQ_1 plus the rent paid to the government of AC_1P_3hk.

But did we forget about all those fishermen driving Cadillacs at point *f* in figure 4.1? We should not forget about them because sooner or later other people will begin to notice these abnormal returns and will, acting in their own self-interest, enter the fishery—after all, they want a Cadillac, too. But the problem is more complex. The entrance of the new fishermen will increase production and, as our demand curve indicates, will lower the price since more is produced. Price will fall from P_1 to P_2, where price is equal to average cost. This point represents market equilibrium for the fishing industry in the local market case. The same is true in figure 4.2, panel (b). Point *h* is unstable because of the abnormal profit, assuming the resource reverts from sole ownership to common property; thus, new fishermen are attracted. The fishery will expand to point *i* and harvest Q_2. At the final equilibrium, all economic profits (or rents) will be dissipated since $TR = TC$. However, individual self-interest coupled with the common property nature of the resource has led to a marginal cost well above the price.

Why Do We Overfish?

In figures 4.1 and 4.2, the demand curve is such that it intersects the *AC* curve (i.e., market equilibrium) at a harvest level *below* maximum sustainable yield. Chapter 3 presents several cases of overfishing (i.e., increasing effort beyond that necessary to obtain *MSY*). Why does this occur? You don't have to be an economist to see how wasteful overfishing is: more vessels are entering a fishery while production is constant or falling. Certainly, this is economic waste. But what produces this phenomenon? Once again, the common property nature of the resource coupled with rising demand for protein or food in general is the problem.

Consider figure 4.3. We have omitted the marginal cost curve, since market equilibrium will finally result in price being equal to average cost; therefore, the average cost curve is our supply function for fish. At *DD*, Q_1 will be produced, and E_1 units of fishing effort will be utilized. Even though no overfishing occurs, this position is still suboptimal from an economic standpoint, as discussed in the previous section (i.e., $MC>p$). Next, population and per capita income increase and shift the demand curve for this fishery upward and to the right to $D'D'$. Q_2 will be produced with E_2 units of fishing effort. We can see that expanding population and increasing affluence are placing increasing pressure on a renewable, but not inexhaustible, natural resource. This has been the historical experience of all fisheries of the world, except for those where there is no market (i.e., no demand curve). Population again expands (and per capita income as well), shifting the demand curve to $D''D''$.

Let us look at the adjustment mechanism in two steps. In the short run, E_2 represents the vessels and fishermen in a fishery that suddenly experiences a rising demand for its product. Since these fishermen are working at full capacity, all they can collectively supply to the market is Q_2. With the new demand curve $D''D''$, the marketing of Q_2 will result in price p_1 and point *a* on the demand curve. This will be produced at a relatively low average cost, or point *c* on the average cost curve. Economic profits appear (and more Cadillacs), and more fishermen enter the fishery. Let us assume the fishing effort expands to E_5. Overfishing has occurred! Q_2 will still be produced, but now by E_5 units of effort when it could have been produced by E_2 units of effort. From society's point of view, this is a clear waste of economic resources: production has not changed, but more fishing effort has been introduced into the fishery by the combination of rising demand and *no* barriers to entry. Since more units of effort have been added (and they cost money) with no change in production, the average cost of producing a pound of fish rises to point *b* on the average cost curve. This closes the gap between the price people are willing to pay, or point *a* on the demand curve, and the average cost of

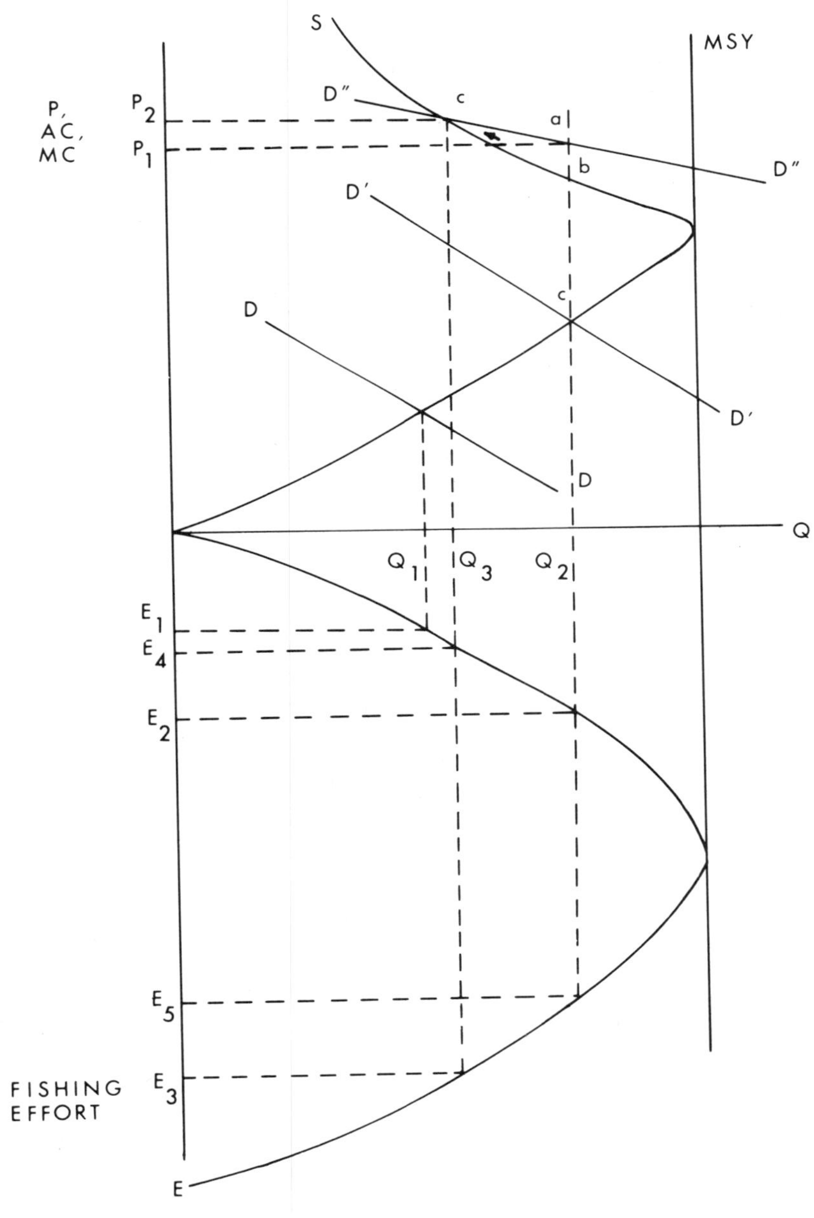

Figure 4.3: Economic Model of Overfishing

production (i.e., from *c* to *b*). The situation worsens. With the entry of all those redundant fishermen (redundant in the sense that E_2 units of fishing effort are all that is really required to harvest Q_2), we still have a spread, although somewhat smaller, between what people are willing to pay and average cost per unit (or pound) of fish, *a* - *b*. Hence, economic profit still exists as long as price is above average cost, which includes, as we have mentioned many times, a normal return to fishermen and vessels. The final equilibrium is disastrous. At point *c*, E_3 units of effort will produce Q_3 pounds of fish, and average cost will be equal to price. Note the little arrow showing the direction of adjustment until price is no longer above average cost. At *c* we have a real social calamity: private interests are fulfilled, but social disaster has occurred. Q_3 could have been caught with E_4 units of effort. Hence $E_3 - E_4$ represents complete waste: these fishermen and vessels could have been engaged in, let us say, another fishery where exploitation is not as intense. Or they could be working in other occupations, and the capital invested in vessels might have been used to finance the building of new houses, schools, or hospitals. This is "the tragedy of the commons." If a fishery is owned by no one and if demand for that fishery's products is growing, overfishing is eventually inevitable. Economists call this a *market failure.* Call it what you will—it is still a sad state of affairs. Our little tragedy may have one final act. This final act is presently going on in the northwest Atlantic with haddock, cod, and flounder and elsewhere in the world as discussed in chapter 3. As demand expands further, the catch will drop while the fishing effort increases. Yes, the curtain falls when the resource is extinct or close to it.

The tragedy of the commons has led to another interesting conclusion: the greater and greater the demand for a particular fishery product, the sooner less and less will be produced.[4] The normal economic mechanism usually works in the opposite direction. If the demand for toys, automobiles, mousetraps, or beer is increasing, prices will rise and induce more firms into the industry or provide more incentive for existing firms to expand production. (In some industries, such as automobiles, entry of new firms is difficult, especially in the United States.) "Well," you say, "if we made all the fishery resources private property, what would happen?" Would more and more be supplied in the face of rising demand? Consider a large number of sole owners of various lobster stocks throughout the world. If we go back to panel (b) in figure

4.2, the sole owner will equate his marginal cost to those rising prices; however, he would never fish to the point of overfishing.[5] He would be wasting money since Q_3, or *MSY*, is the largest production level on a sustainable basis he could supply. So his supply curve is approximately *MSY* at very high prices. The answer to the question of making all fishery resources *private property* is that increasing demand *would not result in overfishing* under most conceivable circumstances. However, since the fishery resource is limited to *MSY*—in contrast to other industries—increasing demand may result in no increase in supply. A recent study by Agnello and Donnelley (1975) of property rights and efficiency in the U.S. oyster industry concluded that "The empirical findings suggest that private property rights do in general make a significant difference in a state's average labor [fishermen] productivity in oyster harvesting. Common property rights are associated with low labor productivity resulting from . . . over exploitation. . . .Furthermore if all states had relied entirely on private property in oyster harvesting in 1969, R [common-private property variable] would have increased by over 70 percent implying an increase in oystermen's incomes of around $1,300 or almost fifty percent of 1969 average income." As for overfishing, we must qualify our remarks in light of alternative points of view. Fisheries are traditionally rural in nature, and unemployment rates are high. In addition, the national economy goes through periodic recessions; therefore, jobs are scarce. We have looked at overfishing as a waste of capital and labor; these resources are redundant, add nothing to, or actually subtract from production. Some critics have argued that for people in various fishing communities overfishing is preferable to unemployment. Furthermore, jobs are not available even on a national basis during recessions. So why not overfish? It is a fair question, and there is no absolute refutation to this argument. However, some responses can be made to anyone who espouses overfishing as a means to create jobs. First, he must be quite sure of his population dynamics, which by all standards are quite removed from a fine-tuned predictive model. Can the proponent guarantee that overfishing will not result in a precipitous decline in the resource, that mass unemployment may not result? This has happened with the California sardine, the blue whale, and many stocks in the northwest Atlantic. Just how far do we go? Second, overfishing can be handled by fishery management—rents or economic profits could be used to create local jobs if laws were written to give local governments management authority over the resource. The latter approach is theoretically possible, but hard to implement, since fishery resources do not obey legal, geographical boundaries. In point of fact, we do not know of a single case in which overfishing has in the long run benefited any community. The more progressive policy is to provide alternative employment locally by seeking to minimize the national rate of unemployment.

Finally, this discussion of overfishing would not be complete unless we consider what economists call society's time preference. In simple terms, if a society is on the verge of starvation, its members could not care less about overfishing. There is no tomorrow for such a society, so they may be persuaded to take sizable chunks of the entire fishery biomass for food. We say they have a short time preference: they would rather consume today than save the fish for harvest sometime in the future. There is no easy rebuttal to this argument—except to inquire what such a society will do when it faces tomorrow or what the fate of its future generations will be. It is hoped that the food problem will be mitigated by that time. Now let us consider fishery management practices. First, we shall consider those designed to enhance or improve the productivity of the resource by instituting laws and methods that regulate what one can do and what one cannot do *within a common property framework.*

Management Schemes to Enhance the Resource

Biological investigations of the entire life cycle of various fish have been going on for years. Their purpose is to understand more about the organism's behavior in both normal and abnormal environments. Much of this research is knowledge for knowledge's sake and is not directed at any specific objective. In this section we shall consider that segment of biological knowledge that can be used to develop rules and regulations to increase the productivity from the fishery resource.

Kinds of Gear

Certain kinds of gear may interact with the resource more productively than others. The best-known illustration is *mesh size.* One of the most important considerations in the use of fishing gear and impact on the fishery is the selection of mesh size for the trawl net. As the size of the mesh opening is increased, larger fish can escape through the mesh so that fish of the minimum age and size are taken. The act of varying mesh size must be considered in two respects. On the one hand, an increase in mesh size will affect recruitment (i.e., the number of fish that can actually be caught by the fishermen). The larger the mesh size, the larger and older the fish will be when they are actually recruited to the fishery population. At the same time, it is possible that if the fishing effort is held constant and the mesh size is increased through a range of values, the actual landings will drop, not because of a decrease in the population of fish, but because of the limitations on catching ability posed by the larger mesh gear. Hence, the yield curve will be affected by mesh size as well as by the biological interrelationships within the fishery. If the mesh size is small, very young fish will be taken in the fishing process. If these fish are left in the ecosystem longer, the weight added per fish as they mature will be greater than the loss of weight due to natural mortality. Thus, an increase in mesh size will tend to increase recruitment and permit a higher level of sustainable yield for some levels of fishing effort. After a certain point has been reached, however, a further increase in mesh size will result in a downward shift in the yield function owing to the limited ability of the gear to take fish (i.e., because the biomass or stock of fish—recruits—that the gear can catch is greatly reduced). This is discussed and illustrated (the Beverton-Holt model) in chapter 3.

Introduction of mesh size can be illustrated graphically as follows. For any given mesh size, a separate landings-effort function will obtain. Thus, curve 1 in figure 4.4 illustrates one mesh size; curve 2 represents a *larger* mesh size. If fishing effort is held

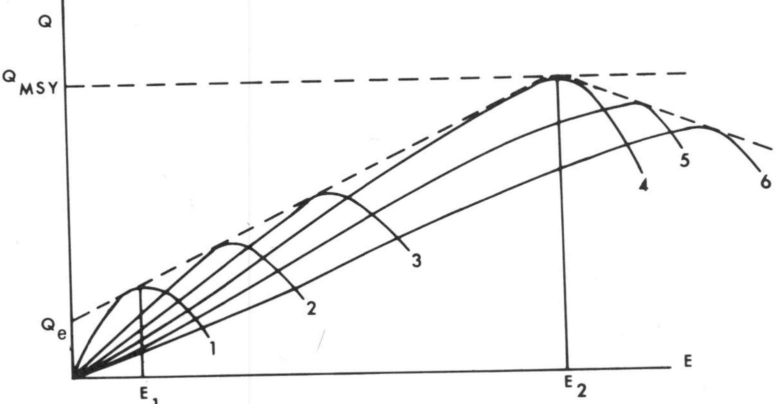

Figure 4.4: Impact of Mesh Sizes on Yield Functions: Eumetric Yield Curve

constant at E_1, the larger mesh size produces a smaller catch than the smaller mesh sizes. However, after the landings function has reached its peak and turned downward, landings continue to increase as effort is increased if we select larger and larger mesh sizes. This impact of mesh size has its limitations on sustainable yield as indicated in figure 4.4. With the selection of mesh size 4, the sustainable yield is maximized for the fishery. If mesh size increases to 5 or 6, all points on these landing functions will be lower for a given fishing effort. A continuous curve Q_e can be fit tangent to all such yield-mesh curves or passes through the maxima of all such landings curves. In biological population dynamics, such a curve has been termed the *eumetric yield* curve. Why not set all mesh sizes by legal regulation at 4? This would maximize the productivity of the resource. True enough. However, this may be a little shortsighted because we are considering only physical yield (i.e., pounds landed). Perhaps there is an economic qualification to this reasoning.

The economic significance depends on the effect mesh size has on the average size of fish landed and the tendency of the fish market (i.e., consumers) to discriminate in terms of price for different sizes of fish. This was discussed extensively in chapter 2. If the consumer prefers larger fish (e.g., shrimp) to small fish and if this is reflected in a substantial price differential, the yield function that will produce the maximum sustainable *physical* yield is not necessarily the yield function that will produce the maximum market value of fish landed. Consumer preference is especially marked for *yellowtail flounder, American lobster,* and *shrimp,* that is, in the difference in price people are willing to pay. The moral to this fish story is that economics and biology complement each other in deciding on management policies. Mesh size alterations are probably the most notable regulation variable under the gear classification. The reader should not confuse limiting certain types of gear in certain fisheries with controlling fishing effort. This will be discussed below.

Kinds of Catch

In order to enhance the fishery resource, regulations have been imposed on fish size (i.e., mesh size) and on other characteristics such as male or female. For example, the Maine American lobster fishery has set a legal limit on the size of lobster caught. The minimum size was set to allow smaller lobsters to reach maturity. Wilson (1973) states that "the most prominent alternative policy suggested by biologists is the minimum legal carapace requirement, specifically the current proposal to raise the legal minimum to 3 1/2″ by five 1/16″ annual increments" (p. 23). Biologists calculate three benefits.

1. The primary benefit is a predicted 18 percent increase in landed weight once a new equilibrium is reached.
2. The *possibility* that the policy will assure a greater number of mature females will lead to greater numbers of future recruits into the harvestable size class. This would seem to imply that the predicted increase in landed weight will be greater than 18 percent if any number of recruits is actually realized.
3. A third benefit is a hedge against the sudden collapse of the fishery.

Thus, the productivity of a fishery can be greatly influenced by laws regulating the kind of catch.

Closed Seasons and Areas

This is probably one of the most widely used tools to enhance the fishery stock and thereby increase the catch. *Area* closures are direct means of conservation for small fish that congregate on nursery grounds; here, therefore, they may for a time be more or less separated from concentrations of large fish. *Closed seasons* are generally designed to reduce the total fishing pressure on a stock, and *closed areas* are generally designed to protect *part* of the stock. There are closed seasons for salmon to protect them during critical periods in their migrations.

Hatchery Operations and Other Physical Schemes to Increase Production

Although we have argued that man has limited control over a wild stock of fish and that he essentially remains a hunter rather than a farmer, there are limited means, depending on the fishery in question, to augment the harvest.

At this time the largest hatchery program in the United States is for the Pacific salmon. The Columbia River has been used more and more extensively over the past thirty years, and valuable stocks of salmon have been depleted. To mitigate against added fishing pressure, federal funds have been used for hatcheries, fish ladders, and spawning channels in order to supplement the declining natural production of the Columbia River salmon and steelhead trout. According to Wahle, Vreeland, and Lander (1974), total benefits to both the sports and commercial fisheries were $8.58 million for the 1965 brood (hatchery baby salmon) and $9.11 million for the 1966 brood. Corresponding costs (operating and depreciated equipment) were estimated at $1.29 million for the 1965 brood and $1.23 million for the 1966 brood. For the twenty Columbia river hatcheries, the benefit-cost ratio was 6.6 for 1965 and 7.4 for 1966. Hatchery operations have been proposed for American lobsters, and oyster seeding is quite prevalent throughout the world.

Moiseev (1969) has pointed out several ways to increase the productivity of the ocean—such as transportation and acclimatization of food organisms from one ocean to another. For example, shad have been successfully introduced along the Pacific coast of the United States, and the humpback salmon has been introduced into the Barents Sea. As these examples demonstrate, such successful introductions may significantly increase fish productivity.

Management Schemes to Prevent Overfishing or Restore an Overfished Stock

Even though our resource analysis of chapter 3 called attention to widespread overfishing and depletion of fishery resources, various fisheries management techniques have to arrest the expansion in fishing effort in cases where overfishing threatened or where stocks were in serious jeopardy of extinction. One general theme has carried through these management problems: an inordinate emphasis upon the control of fishing effort by means that would still preserve the common property nature of the resources regardless of the economic impact of the selected management tool. This would certainly seem a strange statement, especially since the common property nature of the fishery resource is at the heart of the "fishery problem."

Historically, management schemes have rarely taken into consideration the direct economic impact or consequences they might have. They emerged without any regard for good economic analysis. Fishery management programs are the product of state laws or national treaties and involve legislators as well as physical scientists. Historically, two techniques that have been used extensively are (1) the control of the types of gear used in exploiting a fishery stock and (2) the use of a quota, in which fishing effort is indirectly controlled by placing limitations on catch per year. Both techniques preserve the basic common property nature of the resource—everyone still has equal access to the resource. Here is the rub! By not attacking the cause of the disease (i.e., common property), it has been easier to sell these management techniques politically. After all, they apparently discriminate against no one, infringe on no one's constitutional rights. We are not limiting entry to the fishery. Historically, economists have kept talking about *limited entry,* which were considered dirty words until just recently. Limited entry cannot be precisely defined, since it has so many forms. In general, limited entry is a fishery management tool that attempts in part or in total to deal with the common property problem by either (1) selecting those that may have access to the resource or (2) allowing individuals to qualify for access to the resource by use of economic criteria such as taxes, auctions, leases, or outright endowment. Let us consider one of the oldest management schemes first, one that has come to be called "regulated inefficiency."

Regulated Inefficiency: Restrictions on Gear

One of the most celebrated books in fishery economics is *The Pacific Salmon Fisheries: A Study of Irrational Conservation* (1969) by James A. Crutchfield and Giulio Pontecorvo. The authors meticulously documented the contention that the management of the U.S. Pacific salmon fisheries in Oregon, Washington, and Alaska—which was based upon gear regulations to limit fishing power and thereby effort—had resulted in economic disaster and had treated a symptom, not a cause. Let us consider the specifics.

Most of the regulatory methods the Pacific salmon fishery has employed generally affect the catching power of the fleet as a whole. Although area closure and fish size limitation have been extensively used in the Pacific salmon fishery, we shall concentrate upon the "almost endless series of limitations on fishing equipment and vessels; most of these limitations are deliberately designed to reduce technical efficiency of the individual operating units" (Crutchfield and Pontecorvo 1969, p. 45).

In Alaska no salmon fishing vessel over fifty feet could be registered. This is prima facie evidence that larger vessels are considered more efficient and better able to exert more fishing effort on the salmon stock. According to Crutchfield and Pontecorvo, "All the common types of nets used in salmon fishing are restricted in various ways that limit efficiency" (p. 46). For example, monofilament nylon, which is stronger and more durable than other material used in the manufacture of salmon nets, is banned from use in the salmon fishery. Fish traps were eliminated because of their efficiency in catching salmon. Salmon fishermen are forbidden *by law* to use electronic fish-finding gear, although their potential use in purse seining operations is widely accepted. And, of course spotter planes or helicopters, which are used in other fisheries, are outlawed in salmon fishing. For those just learning about the fisheries, it may strike them as indeed ironic that U.S. agriculture is revered throughout the world for its efficiency but that its counterpart in the food-producing area has chosen, in many fisheries, to impede technological change. Is the Pacific salmon fishery an isolated case, or is regulated inefficiency widespread in the United States?

Examples of actual or possible regulated inefficiency are numerous from state to state. The regulations on shrimp catching in Florida are nearly 300 pages long. Does anyone know the economic impact or cost of such legal passages as "It shall be unlawful for any person, firm or corporation to take shrimp from the waters of Escambia and Santa Rosa counties with a net or trawl that exceeds forty-five (45′) feet on a cork line length" or more dramatically, "The use of airboats by any person, firm or corporation in the taking or trawling for shrimp in the waters of Lee county, Florida, is hereby prohibited." Chapman (1968) has also commented on regulated inefficiency, citing such examples as the use of sailing vessels and rakes in oyster harvesting in Chesapeake Bay and the restriction on halibut fishing to long-line gear in the Gulf of Alaska. According to Chapman, "While the excuse used has ordinarily been conservation, the purpose has actually normally been social reform—to divide the proceeds of the fishery among more hands" (1968, p. 268). Table 4.1 gives a selected cross section of gear prohibited in major U.S. fisheries.

How can economic analysis help us through this "cat's cradle" of laws and regulations that produce regulated inefficiency? The central economic problems are highlighted by Crutchfield and Pontecorvo. Pacific salmon, despite millions of dollars spent on extensive biological research and attempts at artificial propagation (see hatchery operations above), is barely holding its own and in some areas is seriously depleted. These regulations have undermined the economic strength of the salmon industry by reducing its efficiency. The failure to come face to face with the common property situation combined with the rising demand for protein has led "to a new equilibrium, no more satisfactory than the previous one, with a net loss to the economy as a whole as more factors of production [fishermen and vessels] are trapped in the fishery" (Crutchfield and Pontecorvo 1969, p. 196).

Table 4.1: Some Selected Illustrations of Gear Restrictions For Various Fisheries

Fishery	Gear (Technology) Restrictions (Examples)
1. American Lobster (Maine)	Lobster pots and traps only
2. Pacific Halibut	Nets prohibited
3. Sea Scallops (New Jersey)	Dredging prohibited
4. King and Dungeness Crabs (California)	Crab pots only
5. Alaska Salmon	Single or multiple strand monafilament purse seine web prohibited
6. Gulf Shrimp (Louisiana)	Prohibit vessels rigged for double trawls or Bifoxi type vessel
7. Atlantic Clams (New Jersey)	Power dredges prohibited
8. Atlantic Menhaden (Delaware)	Trawl nets and drift nets prohibited
9. Atlantic Blue Crab (Maryland)	Pound nets and electronic devices prohibited
10. Atlantic Oysters (Virginia)	Patent tongs and dredges prohibited in certain areas

Source: Sokoloski (1970)

At best, regulated inefficiency is a short-run solution. It treats symptoms, not causes. The short-run benefits, if any, are outweighed by the intermediate and long-run deterioration of the fishery. The appeal to constitutional rights or spreading employment will hardly lead to an upgrading of economic conditions in many areas that depend on the fisheries. Over the 1927–1966 period, Alaska salmon landings declined by approximately 300 million pounds, average catch per fisherman fell by 50 percent, and the total number of fishermen increased threefold. In 1973 the Alaska legislature finally passed a Limited Entry Bill, which we shall discuss below. But the burning fact still remains that our social institutions, scholars, scientists, conservationists, and industry took over a century to see that regulated inefficiency was not working.

The Quota System: Control of Catch

The quota system is really quite simple in principle. Fishing effort is controlled indirectly by setting catch quotas, usually on an annual basis. Who determines the quota and on what basis? Let us consider the latter part of the question first. The size of the catch quota will depend largely upon the "development" of the fishery. That is, no one really wants a quota until the resource is either in danger of being overfished or is already seriously depleted. Man has become especially good at locking the stable door after the horse or, in this case, after the fish have been stolen. But fish can never be stolen—they are common property—however, they can certainly be seriously depleted. As history indicates, there are occasionally a few visionaries around to stimulate governments to act before overfishing takes place. In this case, the catch quota might be set at maximum sustainable yield if the fishery is following a steady-state yield function (see chapter 3). However, a seriously depleted fishery must be managed quite differently. Because fish intermingle in the ocean, it is especially difficult to control one fishery without at least considering the impact on other fisheries. The terms *directed* and *incidental* fishery stocks are useful in this respect. In a directed fishery, fleets fish at will, concentrating their entire efforts either on one stock or on several stocks. Sometimes, however, the stock is so depleted that it cannot support a directed fishery; but it may be *incidentally* exploited by fleets that are in pursuit of other stocks. Needless to say, the stock is in real trouble if quotas are set to restrict the catch to incidental catch. Obviously, stock estimates may reveal that the fishery is overfished but that it can still be a directed fishery. In this case, quotas will be set well below maximum sustainable yield to allow the stock to recover. If stock growth exceeds mortality, it may recover to a point where quotas can gradually be increased. The Beverton-Holt model (chapter 3) is especially useful in estimating recovery time. So what is wrong with a quota? To illustrate this problem, we shall consider three cases: (1) Pacific yellowfin tuna, (2) the Pacific halibut, and (3) the northwest Atlantic yellowtail flounder. Each of these fisheries is unique and illustrates certain problems associated with the quota management tool.

Pacific Yellowfin Tuna. The tuna fishery of the West Coast of the United States began at about the turn of the century and has since become one of our largest and most important fisheries. The very rapid rise in fishing for yellowfin and skipjack tuna in the eastern tropical Pacific caused the United States and several neighboring countries to the south to show concern about the conservation of these stocks of tuna. In 1950, a convention between Costa Rica and the United States established the Inter-American Tropical Tuna Commission (IATTC). Since then, other nations have joined the convention.[6] In a classic and visionary article published in 1957, Milner B. Schaefer calculated the maximum sustainable yield for the yellowfin tuna fishery: "It may be seen from the foregoing analysis that we may expect the maximum average equilibrium catch of yellowfin tuna to be obtained with about 35,000 standard days' fishing effort, and to be in the vicinity of 195 million pounds" (p. 260). In 1956, according to Schaefer, approximately 25,000 standard days were expended on the yellowfin stock; therefore, the fishery was not overfished, and the parties to the convention had a chance to prevent overfishing. Despite Schaefer's early work, overfishing of yellowfin tuna occurred in the early 1960s (Joseph 1973), which finally prompted the IATTC to implement catch quotas in 1966. The management program is based on a general catch quota, to be taken on a first come, first served basis. Obviously, this procedure does not attack the common property nature of the resource. The quota is established each year on the basis of the abundance of the stocks and applies only to the Commission Yellowfin Regulatory Area (CYRA).

What results has the quota system had? Because of the rising demand for tuna in the United States (see chapter 2) and elsewhere and because of federal subsidies (see chapter 10) to the tuna fleet, capacity has increased over three times since 1966. More and more vessels await the starting signal by the commission on January 1, and the open season has decreased from ten months to less than three. In 1971, *MSY* was estimated at 216 million pounds of yellowfin tuna. The quota system had numerous undesirable economic results:

1. The smaller vessels in the fleet (baitboats and small seiners) that were not flexible enough to fish for tuna in other areas of the world during the closed season experienced financial losses or a drop in annual income.
2. Because of the general overcapitalization of the fleet, which trebled in less than fifteen years, it was found that catches even for the flexible vessels that could fish in the Atlantic Ocean during the off-season were only 80 percent per vessel of previous levels. Therefore, vessel productivity declined.
3. Although the quota has managed to protect the resource, *it has been a complete failure* in an economic sense: the existing fleet is three to four times that necessary to harvest *MSY* and must operate in the Atlantic during the closed season. This greatly increases fuel and other costs.

Saila and Norton make the following observation about the IATTC quota system: "It would seem appropriate to dismiss overall quotas in that it is a technique that encourages economic waste. . . . The race to catch as much fish as possible before the quota is reached . . . should not be considered as a viable or acceptable long-term approach by the world community" (1974, p. 51).

Financial data now indicate that tuna vessels are obtaining a small, if not negative, return on investment and that fishermen's incomes have dropped considerably (see chapter 9). Some of this decline in earnings has been due to increased fuel cost and a leveling off of ex vessel tuna prices. From society's point of view, the quota system for yellowfin tuna has been an economic disaster.

Pacific Halibut. The Pacific halibut fishery became important to the United States and Canada around 1888. After a record catch of 69 million pounds in 1915, the catch dropped precipitously and remained at a low level well into the 1930s. In 1924, Canada and the United States concluded the original convention after both realized that the resource had been overfished.

Both Crutchfield and Pontecorvo and Crutchfield and Zellner (1963) have analyzed the results of regulation by quota for this fishery. According to Crutchfield and Pontecorvo:

> A rapid increase in real prices of halibut, together with a substantial recovery in the magnitude of the stocks—both numbers and average size of fish—brought a surge of new entrants to the industry, particularly in the immediate postwar period. In addition to the expected dissipation of economic rents in excess capacity, a variety of secondary effects stemming from the profit-maximizing reaction to the control program brought further increases in costs. These adjustments include: the necessary shortening of the fishing season; use of combination boats able to fish several types of gear in order to permit reasonably full utilization of the hull and engine; non-optimal geographical distribution of loading facilities; and some increase in marketing cost and loss in quality as a result of longer storage periods (1969, pp. 197-198).

The last points made by Crutchfield and Pontecorvo are especially important for fresh fish. As the season declines under the quota system, all the fish is landed in a short period of time. Despite a fall in price, the marketing channels may not be able to absorb such increases in so short a period. Much of the fish may have to be frozen, and as pointed out in chapter 2, this lowers the prices consumers are willing to pay. Hence, the quota system for fisheries, as in the case of halibut, has brought only excess capacity and higher costs. The failure of the International Pacific Halibut Commission (IPHC) to consider economic as well as conservation objectives has led to numerous inefficiencies.

Crutchfield and Zellner concluded that "The halibut fleet has shown little technological progress during the past 20 years. . . . As long as entry remains uncontrolled, it is not practical to test new fishing methods that might be more efficient" (1963, p. 102).

The present quota is set at 33 million pounds, and the season runs from 121 to 14 days depending on the subarea fished within the convention area.

Yellowtail Flounder. The yellowtail flounder fishery is one of the most valuable fin fisheries of the northeastern states. In 1972 landings were approximately 72 million pounds with a value of $10 million. Landings of yellowtail flounder have declined by almost 50 percent over the last ten years in the face of heavy foreign fishing pressure. The International Commission for the Northwest Atlantic Fisheries (ICNAF) has primary responsibility for regulation of this fishery. After the yellowtail flounder fishery had been seriously depleted, quotas were instituted in the early 1970s. The proposed 1976 quota for subareas fished by the United States is 44 million pounds to allow for recovery of the stock. Although the quota system has just been implemented for yellowtail flounder and related species such as cod and haddock, the ICNAF has failed to learn much from the experience of the IATTC or the IPHC. A recent study by Gates and Norton of the yellowtail flounder fishery indicates that "in fisheries, however, if there are no restrictions on entry, changes such as these [increased demand] tend to generate short-term profits that are quickly dissipated as more vessels are attracted to the fishery" (1974, p. 21). More pointedly, they say that "it is clear to us that an approach which sets only aggregate quotas on landings or effort will simply result in economic waste and is not an effective approach. It is ineffective from an economic viewpoint, because it does not provide any incentive for least-cost fleet size or least-cost factor combination" (p. 22). Yellowtail flounder is marketed fresh and, unlike tuna or salmon, cannot be easily stored without a loss in value.

These three examples clearly illustrate that international commissions have been unsuccessful in their use of the quota to achieve economic progress in the fisheries. They have turned their backs on the fundamental issue and have really compounded the "tragedy of the commons." The quota is the easy way out, and economic blame is seldom directed at a commission that allows everyone to fish. It certainly does not discriminate against anyone. In fact, commissions that adopt the quota system are often complimented for their conservation efforts. According to H. Scott Gordon "While it is true that a great deal (perhaps the greater part) of what has been done in the name of 'conservation policy' turns out, upon subjection to economic analysis, to be worthless, or worse, it is nevertheless also true that economic theory can offer a formulation of the conservation objective sufficiently clear and precise to permit the derivation of rational policies in the future" (1958, pp. 110-111).

The failures of fishery management are a direct result of a misunderstanding of the essential purpose of natural resource use: to enhance the economic well-being of man. But essential resource use should be governed by rational *economic policy,* not by quotas or regulated inefficiency. Economics and economic policy prescription are foreign to those from other disciplines who have guided fishery policy in the United States and elsewhere.

The following U.S. fisheries are now under the quota system (*in addition* to our illustrations above):

1. Atlantic cod
2. Atlantic haddock
3. Atlantic ocean perch
4. Atlantic red and silver hake
5. Atlantic mackerel
6. Atlantic herring
7. Alaska king crab

In 1974, this represented almost 18 percent of the *value* of total U.S. landings, or $162 million (ex vessel).

How much economic waste have the quota system and regulated inefficiency produced? Estimates are at best fragmentary. In the Pacific salmon fishery of the United States and Canada, Crutchfield and Pontecorvo (1969) have estimated that the same annual catch (and total revenue) could be taken with about $50 million less capital and labor than are currently employed each year. With respect to the Pacific halibut fishery, and the period during which the quota has been in force, Crutchfield and Zellner note that "in 1951, 820 American and Canadian vessels engaged regularly in the halibut fishery. This was more than double the number participating in 1932, yet the total catch was only 27 per cent higher" (1963, p. 44). Here are two examples of waste under regulated inefficiency and the quota system. Remember, at whatever level the momentary quota is set, there is still no explicit limitation on fishing effort. Devanney has noted that "this race [to take the quota] takes on a particularly obvious form in the Peruvian anchovy fishery where on the opening day of the legal fishing season the fleet steams offshore, fishes intensively until the total catch has reached the level thought wise by the authorities, whereupon the fishery is closed down and men and vessels sit idle for seven months or so" (1976, p. 30).

Direct Controls of Fishing Effort: Limited Entry 1

Recently, there have been attempts to abandon the quota system and regulated inefficiency in favor of *direct control over fishing effort itself:* i.e., limiting the number of fishermen, vessels, tonnage of vessels, units of gear, etc. Direct control has come in the form of Canadian and Alaskan limited entry legislation. South Africa and Japan have also instituted programs to limit effort directly through licensing. We call this Limited Entry 1, because it is a limited entry that discriminates or rations fishing effort on the basis of criteria established by law and not by the market mechanism or by what we shall call Limited Entry 3. Let us first consider the Alaskan limited entry program. The purpose of the Alaskan Act (HB126) is "to promote the conservation and the sustained yield management of Alaska's fishery resources and the economic health and stability of commercial fishing in Alaska by regulating and controlling entry into the commercial fisheries in the public interest and without unjust discrimination." This same act established the Alaska Limited Entry Commission, which is directed to (1) establish the maximum number of permits for each area and (2) establish qualifications for the issuance of entry permits.

Each entry permit is issued for a term of one year and is renewable annually. The permits are also transferable. This legislation attempts to come to grips with the problem of common property. Not everyone has the inalienable right to fish! But this is discriminatory and may border on a violation of one's constitutional rights.[7] Generally, it has been held that the state does have the right to discriminate if it is in the interest of the general welfare. Priority classification for applicants is based upon the (1) degree of economic dependence on the fishery and hardship resulting if not issued a permit and (2) extent of past participation.

The Alaskan limited entry program limits the number of participants (but not necessarily fishing effort, since participants may increase their effort per year) on what may seem to be humanitarian grounds, but it is still questionable from the point of view of economic efficiency. The main criticism is that other criteria might be used in awarding permits. Why not award permits to the most efficient vessel operations? The people now in the fishery and with permits may not be the most skilled or potentially skilled. Since entry is limited, those currently holding permits will eventually obtain economic rents or profits. Thus, the indirect effect of the program may be to use the fishery as a means of welfare, especially since the salmon fishery is so seasonal (about four months of the year) and since alternative opportunities are now limited in Alaska.

Christy (1976) reports that the value of the licenses issued in Alaska varies from $150 to $11,035, depending on the type of permit. In this case, the economic rents (profits) are appropriated by the fishermen who have obtained licenses in the *initial* allocation.

In the 1960s, Canada introduced legislation to limit fishing licenses in two fisheries—the Atlantic Coast American lobster and the Pacific salmon fishery. Let us consider the latter fishery as an example. The program divided salmon vessels into "A" and "B" categories based upon salmon production in either *one* of two base years, 1967 or 1968. The "A" category vessels, or high productivity vessels, could be *retired* and *replaced,* but the "B" vessels, or low productivity vessels, could not be reconstructed or enlarged and would be eliminated by attrition. The Canadian government was empowered with the right to charge fees for such licenses. The program began with a limit on the *number* (not size or kind) of vessels, but the economically minded fishermen replaced small vessels with big ones. The game continued as the administrators responded by placing a limit on total tonnage, so the fishermen smartly substituted more efficient gear and equipment (e.g., shifted from gill nets and trolls to purse seiners). As part of the license limitation program, the Canadian government instituted a "buy back" plan. It will purchase vessels and *thereby reduce fishing effort.* Although the license fees for salmon vessels have been increased and although vessels have been retired since the beginning of the program, all these measures have reduced the *number* of fishing vessels, but not the overall fishing power of the fleet. As Christy points out, "Limited access systems, based on input controls, provide an incentive for fishermen to exercise their ingenuity and the result is likely to be continued overcapitalization even though the number of inputs may be limited" (1976, p. 22). This is not to say that the Canadian plan will not ultimately be successful in reducing fishing effort (i.e., power). In contrast to the Alaskan program, some of the economic rents are being used through license fees to pay for the management program and to retire vessels through the "buy back" plan. For a discussion of this program, see Campbell (1973).

In South Africa, the coastal pelagic shoal resource was undergoing rather a large expansion in fishing effort and processing plants. Licenses were introduced for both fishing boats and factories. In 1949, the government refused to license more factories, but it issued licenses to more vessels. Pressured by fishermen, the South African government agreed to a freeze on the *number* and *hold capacity* of vessels fishing for pilchards and maasbankers (i.e., mostly industrial fish). As indicated by Gertenbach (1973), one result was larger boats with more powerful engines and equipment. This response is consistent with Christy's observations.

In Japan, almost all fisheries are limited in the number of vessels that may take part. As in Canada, where licensing controls only *numbers* of vessels, the Japanese vessels have grown larger, and their equipment better. There is no charge for licenses, and renewal is almost automatic. Therefore, the licenses acquire value, and investments and loans are made with the license as security. Asada (1973) has urged that it would be better to charge a fee and use the income to compensate those who are forced out. The Japanese have introduced a "tonnage supplement system" to outwit the fishermen who have increased their fishing power. That is, to replace a vessel with another of larger size, the equivalent extra tonnage must be retired by purchase of a license or some other means. Examples of the various systems to control fishing effort directly may be summarized as follows:

Table 4.2: Systems to Control Fishing Effort

Political Entity	License Fee Charged	Transferable	Value of License	Have Limited Fishing Effort
1. Alaska (mostly salmon)	No	Yes	\$150-\$11,035	No[1]
2. British Columbia (salmon)	Yes	No	Increased value of vessel	Partially[2]
3. So. Africa (pilchard)	No	No	Value of license increasing	No[3]
4. Japan (many fisheries)	No	No	License so valuable used as security	No[3]

1. Period of programs too short to make definite statement.
2. Some economic rents obtained from industry via license fees which inhibit fishing effort.
3. Fishermen have made substitution to increase fishing power (see discussion preceding page).

Fishermen or Boat Quotas: Limited Entry 2

Under this system, quotas are allocated to individual fishermen on the basis of past catch or other criteria (see Alaskan program above). The quota may or may not be transferable through lease or sale. Hence, if you cannot obtain a quota, you cannot fish. This limits fishing effort. If the quotas are transferable, those who wish to get out of the fishery will be induced to do so if they receive some return for their quota. According to Christy, "The chief advantage of this approach is that the fishermen would be free to take their shares however they wish, subject only to a modicum of conservation constraints. The incentive to reduce cost and maximize profits would lead them to adopt technological innovations at an orderly rate, rather than finding themselves faced by increasingly severe restrictions as will be true in most cases when inputs are controlled" (1976, p. 16). Devanney (1976) states that this scheme can work only if there is an "omniscient, omnipotent, and incorruptible administration." The surplus generated by a fisherman's quota is an open invitation to corruption in the form of kickbacks, post-government jobs, and the like. In contrast to Christy, Devanney states "Given the entrenched power of the present exploiters of the resource, they would undoubtedly be first in line for the quotas behind which protection, in the manner of all protected industries, they would become lazy and inefficient" (1976, p. 31). The chief difference between Christy and Devanney is that the former assumes transferability of quotas and the latter does not. With *transferability,* Christy and Devanney agree that technological advancement will be promoted. In addition to the initial allocation—which might be very arbitrary—and the potential corruption Devanney mentions, fishermen who consider themselves high producers and can always "outfish" others may reject it. Bell and Fullenbaum (1973) have suggested a *stock certificate* plan, which is identical to a fisherman quota with transferable rights. This is discussed below with some numerical illustrations.

Control of Fishing Effort by Economic Means: Limited Entry 3

In our review of management schemes, we have reached the conclusion that the essential problem is to manage the resource in such a way that it does not use excessive amounts of capital and labor. According to Crutchfield and Zellner (1962), there are five criteria for economic performance:

1. optimal output
2. efficiency (i.e., no excess capacity)
3. progressiveness (i.e., technological change)
4. proper distribution of income
5. stability of income, employment, and prices

Most economists agree that the Crutchfield-Zellner criteria are those associated with the efficient working of the private market system. For example, economic variables could be used to satisfy the criteria listed above and also prevent overfishing. We shall first consider the imposition of license fees on fishing vessels and gear as an economic measure to control fishing effort. Second, we shall compare this with a fisherman's quota. Third, competitive bidding for quotas (i.e., economic means) will be explored. Dow, Bell, and Harriman (1975) have simulated the impact of charging license fees for the Maine American inshore lobster fishery given *three* different objectives. All these objectives have two common characteristics: (1) to protect the resource from further over-exploitation, and (2) to allow maximum freedom for fishermen to function in a free enterprise fashion (i.e., achieve criteria for economic performance).

Freeze on Existing (1969) Fishing Effort by Placing a License Fee on Traps. Under this scheme, the regulatory authority would calculate a license fee on traps, a fee that would keep the level of fishing effort constant despite any increase in the demand for lobsters. The fee should not be levied on the individual vessel; this would not control the number of traps fished per vessel. The license fee would increase the cost of operations and thus make it uneconomical for vessels to enter the fishery even if ex vessel prices increase. In essence, it would siphon off the increased revenue (or profits) from an increase in ex vessel prices (assuming the latter increase faster than the cost of operations). For purposes of illustration, Dow, Bell, and Harriman (1975) assumed that the desired objective was to manage the inshore lobster fishery commencing in 1974. Given the trends in U.S. population, personal income, consumer prices, lobster imports, and other domestic production to the year 1974, it would be necessary to place an estimated annual license fee of $2.27 on each lobster trap fished in order to keep fishing effort at its 1969 level, as indicated in table 4.3. The regulatory authority would collect approximately $1.93 million in license fees, which could be used to finance resource research, enforcement, and surveillance.

Table 4.3: Projected Impact of Various Management Schemes Imposed on the Maine Inshore American Lobster

		Impact after the imposition of selected management strategies for 1974				
		(1)	(2)	(3)	(4)	(5)
Economic variables	Estimated values before imposition of management strategies (1969)	Freeze at 1969 level of fishing effort	Reduce fishing effort to E_{max}	Reduce fishing effort so MC=P	Issue stock certificate to vessel owner while freezing effort at 1969 level	Do nothing
1. Catch (mill. lbs.)	22.1	22.1	22.5	19.0	22.1	21.7
2. Value of catch (mill. $)	19.9	28.3	28.8	24.9	28.3	28.0
3. Vessels (full-time equiv.)	1,508	1,508	1,339	810	1,508	1,594
4. Traps	848,825	848,825	753,589	455,868	848,825	897,329
5. Ex-vessel price ($/lb.)	0.90	1.28	1.28	1.31	1.28	1.29
6. Total license fees collected (thou. $)	0	1,926	5,378	10,774	0	0
7. License fee per vessel ($)[2]	0	1,277	4,016	13,300	0	0
8. License fee per trap ($)	0	2.27	7.14	23.63	0	0
9. Return per vessel and fisherman ($)	6,365	8,400	8,400	8,400	11,966	8,400

[1]Projection of 1974-impact of selected management strategies. Assumes that F° = 48°; Y = $677.9 billion, (1969 prices); POP = 212.4 million; $Q_0 + I$ = 190.4 million pounds, and $\hat{\pi}$ = $15,292. All prices and dollar values projected for 1974 are expressed in 1972 dollars.
[2]The license fee per vessel was obtained by multiplying the license fee per trap by the average number of traps (562.8) fished per full-time vessel.

Source: Dow et al. (1975)

The license fee plan, however, has many disadvantages. First, a license fee on traps fished does not really get at the utilization rate. If a license fee were imposed on an individual trap, fishermen would probably fish each trap more intensively and thereby reduce their number of traps. At this point, we do not have any information on utilization rates, information that could be used to adjust the license fee upward if utilization increased. Second, enforcement and surveillance might be difficult along the coastline of Maine (i.e., licenses would be issued by the state of Maine to both residents and nonresidents, but unlicensed intruders from all over the coast would be a major problem). Third, and most important, the quantitative tools and projected figures needed to calculate a tax are crude at best and would have to be used each year for computation of the license fee.

Reduce the Existing Level of Fishing Effort to That Necessary to Harvest MSY by Placing a License Fee on Traps. With this scheme, the regulatory authority would calculate a license fee on traps that would reduce the level of effort to that necessary to harvest *MSY* (estimated at 753,589 traps), regardless of whether the demand for lobsters increased. Because we are actually reducing fishing effort as opposed to freezing it at the 1969 level, the estimated 1974 license fee per trap must be higher, or $7.14;

actual catch will not be significantly higher. The regulatory authority would receive approximately $5.38 million in license fee revenues. However, this plan has all the disadvantages of a general license fee plan.

Reduce the Existing Level of Fishing Effort so That the Marginal Cost of Landings Equals the Ex Vessel Price. The idea here is to obtain the greatest "net economic benefit," as has been suggested by such economists as Crutchfield and Pontecorvo (1969).[8] If a regulatory authority had tried this for the year 1974, it would have had a drastic impact on the fishery: the number of full-time equivalent vessels and traps would have been reduced by almost 50 percent. To accomplish this objective, an estimated 1974 license fee of $23.63 per trap would be needed. The regulatory authority would receive approximately $10.8 million in revenue.

From an economic point of view, it is argued that with this management strategy, the fishery will operate in the most efficient manner if fishermen and vessels can move easily to other fisheries or industries. However, this strategy may be particularly unwise in rural areas such as Maine, where labor mobility is low. A drastic cutback in the number of fishermen may increase social problems, and the social cost could greatly exceed any social benefits derived from such a management strategy. Therefore, this management strategy is difficult, if not impossible, to justify on economic grounds for many rural areas where the fishing industry is located. Let us now contrast the imposition of license fees with a fisherman's quota, or what Dow, Bell, and Harriman (1975) call the stock certificate plan.

Issue "Stock Certificates" to Each Vessel Owner Based on Average Catch Over the Last Five Years While Freezing the Existing Level (1969) of Fishing Effort. This scheme would recognize the historic rights of each fishing firm. As in private land grant procedure, the regulatory authority would simply grant each fisherman a "private" share of an existing resource or catch. The stock certificate would be evidence of private ownership. Individual fishermen would be free to catch up to their allotted share through the use of pots or other biologically permissible technology. Or, if they so desired, they could trade their stock certificates to others for cash. Remember the question of transferability vs. nontransferability discussed above.

Suppose the regulatory authority were to freeze the level of fishing effort at the 1969 level and distribute the catch through stock certificates to the existing fishermen. It should be pointed out that the regulatory authority fixes effort when it selects a catch. The selected catch could be either *MSY* or any other level of catch that the regulatory authority deems not injurious to the viability of the stock. The expansion in demand for lobsters by 1974 would generate excess profits for those individual fishermen who were initially endowed with the property right. By 1974, it is estimated that a full-time lobsterman would be earning $11,966 a year, of which $3,566 would be excess profits (i.e., above opportunity cost). To insure against increasingly excessive returns, fishermen holding stock certificates might be charged a fee to provide the regulatory authority with funding to conduct scientific investigations and enforcement.[9]

It should be noted that this plan is identical to the license tax scheme that freezes efforts at the 1969 level. In the latter case, however, excess profits are taken by the regulatory authority, but in this strategy fishermen are allowed to hold on to the abnormal profits the fishery generates. Since several fisheries are located in rural areas where earnings are traditionally low, this strategy might be justified on the basis that it will raise income levels and thereby help improve living standards to levels comparable to those prevailing in urban areas. This management strategy would, of course, be popular with those already in the fishery. However, new entrants would have to buy stock certificates from those initially in the fishery. This would pose certain questions of equity and legal precedent, questions that are beyond the scope of this book. In addition, the reader should remember Devanney's criticisms of this scheme.

Other Suggested Management Strategies, Including Competitive Bidding. Reeves (1969) has proposed a hike in license fees to "eliminate" marginal or part-time fishermen. He suggested that the $10 yearly fee in Maine be raised $10 a year over the next nine years to a limit of $100. In 1969, a little fewer than one-half of the lobster fishermen were part-time. As defined by Reeves, a part-time lobster fisherman is one who gains less than one-half of his annual income from lobstering.

The first step in most suggested limited entry schemes is usually to restrict the fishery to full-time utilization of capital and labor. Two problems occur with this policy. First, the part-time fishermen may represent the most efficient way of taking the catch. If so, the full-time fishermen may be eliminated by increased license fees. Second, license fees do not directly control fishing effort, since fishermen may fish more traps.

Rutherford, Wilder, and Frick, in their study of the Canadian inshore lobster fishery, endorsed the system suggested by Sinclair:

> An alternative management system is that suggested by Sinclair (1960) for the salmon fisheries of the Pacific coast. This would use the licensing of fishermen to limit entry into the fishery. In the first stage, lasting about 5 years, licenses would be reissued at a fee but no new entries would be licensed and it would be hoped that during the period there would take place a reduction in the labor and capital input, to take the maximum sustainable catch of salmon at a considerably lower cost. After the end of the first stage, licenses would be issued by the government under *competitive bidding* and only in sufficient numbers to approximate the most efficient scale of effort; the more competent fishermen would be able to offer the highest bids and it would be expected that the auction would recapture for the public purse a large portion of the rent from the fisheries that would otherwise accrue to the fishing enterprises under the more efficient production conditions in the fishery.
>
> An arbitrary reduction in the number of fishermen by restriction of licenses to a specified number would entail injustice and inequity as well as grave administrative problems in determining who should be allowed to continue fishing. The auctioning of licenses to exploit a public property resource is justifiable in a private enterprise system of production, particularly when the state is incurring heavy expense to administer and conserve the resource; the recovery by the state of some part of the net economic yield by means of a tax on fishermen (or on the catch) would recoup at least part of such public expenditures, or could be used to assist former fishermen (see strategies discussed above), for instance, by buying their redundant equipment. A tax on fishermen through the auctioning of licenses has, at least, the merit of using economic means instead of arbitrary regulations to achieve a desired economic objective—the limitation of fishing effort to increase the net economic yield from the fishery. Regulations have to be enforced, usually at considerable cost, but economic sanctions tend to be, if not impartial, at least impersonal and automatic in their operation (1967, pp. 99–100).

Actually, this latter management scheme is similar, in theory, to the taxing scheme, but it uses an auction rather than a direct tax. Devanney also supports competitive bidding, since it will "solve the corruption problem and will also greatly reduce the need for the regulatory agency to determine the exact composition of the optimal fleet" (1976, p. 32). The auction would be similar to a forest or oil auction, which is now standard practice in the U.S. government. The regulatory agency would be required to take the high bid, which would reduce the potential for corruption. The purchased quotas should be transferable to enable a maximum degree of economic freedom.

The management schemes discussed above are the major ones either now in use or proposed under the general category of limited entry. If we look at the Crutchfield-Zellner criteria for industry performance, it is fairly clear that limited entry schemes 2 and 3 are the most likely to accomplish these objectives (i.e., fisherman's quota, license fees, or competitive bidding). As we shall see below, any one of these schemes must be "hand-tailored" to the socio-economic conditions now existing in individual fisheries in order to be successful. However, let us turn to the international arena once again for a brief analysis of how the law of the sea and the present wave of extended fishery jurisdiction influence fishery management.

The Law of the Sea: Fable or Reality?

For over 300 years, one of the most basic principles of the freedom of the seas has been the freedom of fishing. That is, states have generally claimed and been accorded relatively narrow limits of jurisdiction, and fishermen have had free and open access to all stocks on the "high seas"—those waters outside the waters of coastal states. Is the freedom of the sea a fable or a reality? In part, it is fable: during the fifteenth and sixteenth centuries many claims of exclusive jurisdiction were made and enforced. The best known is the division of the newly discovered areas of the Atlantic, Pacific, and Indian oceans between Spain and Portugal. The Scandinavian countries have asserted similar rights over the Baltic sea. Thus, the question of law depends basically on the power governments have had to enforce proclamations. This has led to conflicts that are with us even today. Obviously, the law has not come only through force or unilateral declarations. International cooperation has also been reasonably effective in establishing a form of law through agreement.

A major problem of the law of the sea has been the width or extent of territorial waters. By the nineteenth century, after numerous disputes, most nations of the world had confirmed the three-mile territorial limit for coastal fisheries.[10] Was this a law? The answer lies in the enforcement, in other words, in the will and the power to proclaim this principle *the law of the sea.*

Until 1945, the territorial sea differed from country to country, but it was usually three miles wide. Over the years, there had been many international fishery agreements on conservation and fishing rights among various states. In 1925, for example, the United States and Mexico set forth fishing rights in the Gulf of Mexico. Until recently, the traditional rule of law—no single state or group of states had the right to exclude others from freely exploiting common property resources—created a workable arrangement for harvesting the fishery resources of the ocean.

In the history of the law of the sea, however, specific multilateral or bilateral international agreements for the conservation of fisheries are of recent origin. They have arisen because the traditional rule of freedom of fishing has been unable to conserve fish (see chapter 3) and settle international controversies. Since 1940, fishing agreements have been evolved quite actively, and many bilateral or regional arrangements seeking to control fisheries have been negotiated. From time to time, however, some countries have acted unilaterally and have altered some aspect of existing, conventional international practice or law—and fishing jurisdictions are a good example.

The most sensational unilateral action was the "Truman Proclamation" on coastal fisheries in certain areas of the high seas in 1945:

> In view of the pressing need for conservation and protection of fisheries resources, the Government of the United States of America regards it as proper to establish conservation zones in those areas of the high seas contiguous to the coast of the United States wherein fishing activities have or in the future may be developed and maintained on a substantial scale . . . and all fishing activities in such zones shall be subject to regulation and control. . . . The right of any State to establish conservation zones off its shores . . . is conceded. . . . The character as high seas of areas where such conservation zones are established and the right to their free and unimpeded navigation are in no way thus affected.

President Truman's proclamation was motivated by the incursion of Japanese fishermen into the Bristol Bay red salmon fishery in Alaska. Using this proclamation as a precedent, many countries extended their fishing jurisdiction. Chile, Peru, and Ecuador declared jurisdiction out to 200 miles. In 1952, South Korea proclaimed the so-called Rhee line, which in effect closed fisheries to Japan up to 250 miles from Korea. The Truman proclamation never actually became law; it served only as a basis for U.S. international negotiations to protect certain species of fish.

In 1958, a conference on the law of the sea was convened in Geneva—a very popular place for international conferences! Delegates from eighty-six countries went to Geneva to labor over the conventions some two months. The United States proposed a six-mile territorial sea and an adjoining six-mile exclusive fishing zone, with free access to countries with "historic rights" in the newly enclosed fisheries. Many countries preferred a "six plus six" rule without historic rights. Despite Law of the Sea conferences in 1960, 1967, 1974, 1975, and 1976, the question of fisheries jurisdiction proved to be the point on which agreement could not be reached. Thus, international law on fishery jurisdiction has become more fable than fact!

As a matter of policy, the United States once opposed the extension of fishery jurisdiction beyond twelve miles. The executive branch of the government had supported this policy, which is attributable to its strong naval interest, the need to import large amounts of energy and raw materials by water, and distant-water fishing interest, notably in tuna and shrimp. However, the United States has moved closer and closer to the widespread international desire for a 200-mile fishery management jurisdiction zone. This culminated in the *Fishery Conservation and Management Act of 1976,* which we shall discuss below under extended jurisdiction. The subject of the international fishery commissions and their effectiveness will be discussed in chapter 10.

Extended Fishery Jurisdiction: Folly or Realistic Solution?

On April 13, 1976, President Ford signed the Fishery Conservation and Management Act of 1976, or Public Law 94-265. In March 1977 the United States extended its fishery jurisdiction over coastal fisheries out to 200 nautical miles from its shores. The failure of the Law of the Sea conferences *could have been predicted* on the basis of foreboding world economic trends. As discussed in chapter 2, the pressure of world population on natural resources, especially food resources, has given rise to the specter of Malthusian stagnation. Increasingly, further disputes over such common property resources as the fisheries will undoubtedly occur. The Club of Rome (Meadows et al. 1972) forecasts a collapse in food production per capita early in the twenty-first century. As will be explained later in greater detail, unilateral extension of fishing jurisdiction for food, foreign exchange, and even recreation is becoming more and more common throughout the world. Open conflict has even broken out between Iceland and Britain over fishery jurisdiction. As noted above, *quotas* and *regulated inefficiency* have been the most common management schemes. But they have had very limited success in preventing the fishery resource from being depleted. What is done to "save the fish" actually depletes valuable reserves of capital and labor.

Hence, the purpose of extended jurisdiction management should be to promote the highest efficiency in the use of vessels and fishermen, as is our national purpose in every sector of our economy. If this is done by making the common property resource quasi-private property, the fish will be conserved. This is a hard lesson for some to learn, but it is one of the basic themes of this book.

Thus, the roots of the current fishery mess are basically economic. That is, the world is short on food and long on people. Add to this a common property resource and the failure of the world community to come to any agreement on how to make it private property and it is not difficult to understand why states are taking unilateral action.

The Developing Countries and Extended Jurisdiction

The principle of extended jurisdiction offers opportunity for advancement in many coastal developing countries. Let us consider some common characteristics of lesser-developed countries:

- capital is short, but labor is abundant
- constant or declining per capita food production
- foreign exchange short
- full or quasi dictatorships

An underdeveloped country that recognizes the potential value of its coastal fisheries either in foreign exchange or direct harvest may profitably extend its jurisdiction over these resources. Because deep-sea fishing is rather capital-intensive, it is quite likely that these countries will impose license fees on developed countries that wish to fish within their zones. As Saila and Norton maintain "The nature of world demand and international trade for tuna products dictate that tuna will be primarily consumed in higher income countries. Thus, the distribution of the benefits from tuna resources among developing states will not include substantial consumption of tuna. Rather, the developing countries must look principally to sharing in the income and employment, at the harvesting, processing and international trade levels or by extracting revenues from users" (1974, p. 49). If the underdeveloped countries relied principally upon extracting revenues from the users they would (1) not be required to invest in fairly capital-intensive technologies to harvest protein; (2) could place strings upon the resource harvesters, thereby forcing, for example, tuna processors to build processing facilities in their countries (which might be in the short-run interest of the processor since labor costs are lower; however, the usual threat of expropriation is great and the payback period would have to be extremely high on such shore facilities); (3) derive foreign exchange not only through the license fees but also through the value added in processing; and (4) possibly protect their resource better, since the dictator will be the sole owner—there would be no common property problems to deal with. Two problems would remain: (1) enforcement and (2) proper scientific investigation to manage their fisheries.

It is indeed ironic, if not tragic, that countries such as the United States should have to learn from less-developed countries how in effect to reduce overcapitalization by charging for the use of the resource. This general model has already been used in the exploitation of tuna and is very similar to Limited Entry 3 discussed above. In 1973, Ecuador set up a license system that would allow foreign fishermen to catch tuna off Ecuador but that would also provide Ecuador with much needed foreign exchange. The Ecuadorian license fee in 1973 was $20 per ton of hold capacity for either one round trip or forty-five days. Since a tuna vessel's hold capacity can vary from 300 to 2000 tons, the cost of a license could be anywhere from $6,000 to $40,000 per voyage. Because of U.S. unwillingness to purchase these licenses in what we once regarded as international waters, 156 U.S. vessels were seized and fined approximately $3.9 million by Ecuador from 1962 to 1972. Of course, the U.S. tuna fleet cannot ignore these charges, especially when the United States now has a 200-mile fishing zone and can charge foreign vessels for fishing within its jurisdiction. The following South American countries now require licenses:

Brazil: (mainly lobsters) $1,215 per vessel annually
Mexico: (mainly shrimp) $18 per capacity ton

Ecuador: (mainly tuna) $700 registration fee; $60 per net registered ton for fifty days or one full load, whichever comes first

Peru: $500 registration fee annually; $20 per net registered ton for 100 days

During the period when the United States did not recognize the 200-mile jurisdiction, what was actually happening? During this period, tuna vessels could *well* afford to pay such fees. In the late 1960s economic rents were so excessive that payback periods for tuna vessels were *four* years, or well over 25 percent annual return on investment (after taxes). The problem was compounded by U.S. subsidies to the tuna industry under the Fishing Fleet Improvement Act. The federal government extended over $15 million in construction cost grants and nearly $6 million in mortgage insurance coverage to the tuna industry from 1966 to 1970.

Table 4.4 shows the less-developed coastal countries along with their population growth, the time it takes for population to double, and present fishery jurisdiction. The countries on this list have (1) a per capita gross domestic product (at 1970 constant market prices) that is less than or equal to one-fourth that of the United States ($1,200); (2) a population equal to or greater than 4 million in 1970; or (3) appear on the UN list of underdeveloped countries. In all, there are sixty-eight underdeveloped countries that have some coastline. Obviously some coastlines are very short and would not constitute a great jurisdictional claim. Although there is now no significant move among the African and Asian underdeveloped countries to extended fishery jurisdiction beyond twelve miles, the trend is quite clear among South American countries. The crushing population realities of the less-developed areas may soon see the emergence of noncooperation and the charging of fishery fees for coastal and migratory species. It would not be surprising to see 300-mile or 400-mile jurisdictions in the near future. The FAO projects that in every less-developed country, except in the Near East, food production, if past trends are allowed to continue, will grow more slowly than population. This would be a widespread phenomenon and would not reflect low production rates in only a few big countries.

Using Gulland's *The Fish Resources of the Ocean* (1971) and our list of coastal less-developed countries, we have made some rough calculations of the "potential yield" for various classifications of fish off the coasts of the world's LDCs. In many cases,

Table 4.4 Population Growth Rates for Coastal Underdeveloped Countries

	% annual pop. growth	years to double pop.	Fishing Juris-diction (miles)
AFRICA (22)			
Algeria	3.3	21	12
Angola	2.1	33	12
Cameroon	2.0	35	18
Egypt	2.1	33	12
Ethiopia	2.1	33	12
Ghana	2.9	24	30
Guinea	2.3	30	130
Ivory Coast	2.4	29	12
Kenya	3.0	23	12
Morocco	3.4	21	70
Mozambique	2.1	33	12
Nigeria	2.6	27	30
South Africa	2.4	29	12
Sudan	3.1	23	12
Tanzania	2.6	27	65
Tunisia	2.2	32	50
Upper Volta	2.0	35	12
Congo	2.1	33	15
Dahomey	2.6	27	12
Gabon	.8	87	100
Libyan Arab Republic	3.1	23	12
Sierra Leone	2.3	30	200
SOUTH AND CENTRAL AMERICA (18)			
Argentina	1.5	47	200
Brazil	2.8	25	200
Chile	1.7	41	200
Columbia	3.4	21	12
Ecuador	3.4	21	200
Peru	3.1	23	200
Venezuela	3.4	21	12
Costa Rica	2.7	26	100
El Salvador	3.2	22	200
Honduras	3.2	22	12
Nicaragua	2.9	24	200
Panama	2.8	25	200
Uruguay	1.4	50	200
Cuba	1.9	37	12
Dominican Republic	3.4	21	12
Guatemala	2.6	27	12
Haiti	2.4	29	15
Mexico	3.3	21	12
WORLD	2.0	35	

	% annual pop. growth	years to double pop.	Fishing Juris-diction (miles)
ASIA AND MIDDLE EAST (21)			
Bangladesh	NA	NA	12
Burma	2.3	30	12
China (Mainland)	1.7	41	12
Hong Kong	2.4	29	12
India	2.5	28	12
Indonesia	2.9	24	12
Iran	2.8	25	12
Khmer	3.0	23	12
Korea (Dem. Peoples Rep.)	2.8	25	12
Korea Republic	2.0	35	Archipelago
Malaysia	2.7	26	12
Pakistan	3.3	21	50
Philippines	3.3	21	Archipelago
Syrian Arab Republic	3.3	21	12
Thailand	3.3	21	12
Turkey	2.5	28	12
Viet Nam	NA	NA	50
Yemen (Arab Rep.)	2.8	25	12
Cyprus	.9	77	12
Israel	2.4	29	12
Lebanon	NA	NA	12
EUROPE (7)			
Iceland	1.2	58	100
Greece	.8	87	12
Poland	.9	77	12
Portugal	1.0	70	12
Romania	1.0	70	12
Spain	1.1	63	12
Yugoslavia	.9	77	12

Source: World Population Data Sheet-Population Reference Bureau, Inc. (1973). Magnuson Fisheries Management and Conservation Act, Senate Report No. 94-416 (1975).

these resources may extend beyond 200 miles, but as indicated above there is nothing sacrosanct about 200 miles. According to our very rough calculations, approximately 64 percent of the world's fishery resource potential is off the coastal areas of the less-developed countries. This is shown in table 4.5 by continent.

Table 4.5 Potential Yield of Various Marine Animals Off the Coast of Lesser Developed Countries

					Potential Yield (thousands of tons)	Percent
1. Lesser Developed Coastal Countries					65,193	64
	Demersal	Pelagic	Crustacean	Total		
Europe	2,600	2,480	48	5,128		
Africa	2,135	4,500	228	6,863		
Asia	19,020	5,342	660	24,022		
South America	4,715	15,635	203	20,553		
Central America	2,150	5,150	327	7,627		
Total LDC	30,620	33,107	1,466	65,193		
2. Developed Coastal Countries					35,507	36
3. Estimated World Potential					102,700	100

Source: Calculated from Gullard (1971) and list of LDC's in Table 4.4.

Of course, the economic value of these resources may be quite another story. Adding tons of different kinds of pelagic fish (i.e., tuna and anchoveta) is, as we have stressed before, like adding apples and oranges—it is extremely ambiguous. As of 1973, the estimated value per metric ton of fish landing by continent or region was:[11]

	U.S. Dollars Per Ton
Africa	125
South America	132
Europe	257
Asia	304
USSR	307
North and Central America	331
Oceania	766
World weighted	275

Africa and South America have the lowest value per metric ton. These countries have limited coastal fisheries that are characteristically very labor-intensive and have low per unit economic value. Gulland's estimated potential for LDCs in terms of aggregate tons contains many species of relatively high value. The wave of extended jurisdiction obvious in South America will undoubtedly spread to Africa and Asia, where LDCs are quite prevalent. From an economic standpoint, this may well be a desirable trend. Why?

1. LDCs will gain in direct food production or foreign exchange or both.
2. Excessive overcapitalization and the corresponding depletion of world fishery resources may be prevented if LDCs charge foreign fishing nations for deep-sea fishing.
3. The needless dissipation of economic rents may be prevented, but more importantly *common property resources* in the LDC areas (because of strong dictatorial governments) will finally be owned by someone. This may help preserve the resources and prevent needless inefficiencies and may thus mitigate world increases in the price of protein from the sea.

Saila and Norton (1974) have some reservations about each coastal state acting independently. They believe that each state could attempt to attract more and more effort (and thus revenue) into its jurisdiction by lowering the license rate. But the rate might be then too low to control catch. Some states might be more interested in generating revenue than in conserving the stocks. What Saila and Norton are really saying is that the LDCs (or more generally, coastal states) may have a low positive time preference or place more value on revenue now than later. This would be especially true of less-developed areas that have the choice of allowing thousands to starve or of feeding them. If this is indeed the choice, one may have to face the tradeoff between starvation and the depletion of a fishery resource. In this case, it may come down solely to the preference of the LDC dictator, who may act either to preserve the resource for future generations or to deplete the resource.

The Soviet Reaction to Extended Jurisdiction: War or Peace?

The worldwide proliferation of extended fishery jurisdiction will directly affect but a few nations that fish off foreign shores. This may provoke confrontations such as the "lobster war" between France and Brazil over lobster resources on the continental shelf, the "tuna war" between the United States and several Latin American countries, and more recently the "cod war" between Iceland and Great Britain. Will the world be going to war over fish?

Moiseev, the Soviet author, has complained about the United States' twelve-mile fishing zone: "Thus, even the United States, which has well-developed fisheries, promulgation of the 12 mile fishing zone cannot be regarded as a truly protective measure favoring American fishermen whose activities in this zone are relatively insignificant. The measure is patently discriminatory toward fishermen of other nations who exploited this zone for aquatic items almost completely ignored by American fishermen" (1971, p. 288). In 1974, the United States derived approximately $146 million (ex vessel) from the 3–12 mile coastal zone, representing approximately 15 percent of the value of the catch. If there are species within the U.S. 200-mile fishing zone that are not exploited by U.S. fishermen, the Soviets can always obtain a permit from the secretary of commerce to fish such species (see the detailed discussion of PL 94-265 below).

In 1973, the Soviets landed 8.6 million metric tons of fish, of which 1.36 million metric tons came from the northwest Atlantic (ICNAF area) and only 0.35 million metric tons from subareas 5 and 6, which would generally coincide with the United States'

200-mile fishing jurisdiction. In the northeast Pacific (coastal Alaska, Washington, and Oregon for the United States), FAO statistics indicated a Soviet catch of 0.38 million metric tons (0.662 million metric tons on average over 1969–1973).

Hence, the USSR in 1973 derived 4.0 percent of its total catch within 200 miles of the New England coast and a maximum of 4.6 percent of its catch in U.S. coastal waters in the northeast Pacific.[12] Sysoev (1970) reported that in 1968 the Soviet fishing industry employed 11.3 percent of all workers in the Soviet food industry nationally and contributed 7.2 percent to gross food output. When we see the importance of the fishing component to the USSR food industry and the rather severe food shortages in the Soviet Union, we are inclined to regard the Soviets with more concern in fishing areas from which they derive the bulk of their catch (FAO 1973):

	Catch (million tons)
Northwest Pacific	2.2
Northwest Atlantic[1]	1.0
Northeast Atlantic	1.6
Central Eastern Atlantic	1.0
	5.8

1. Excluding U.S. 200-mile coastal zone.

Hence, it is not really the United States, but other countries, that could come into confrontation with the Soviets, especially Western Europe, Japan, South Korea, and many Latin American countries, if they extend their fishing jurisdiction to 200 miles.

The Japanese and Their Dilemma with Extended Jurisdiction

In 1973, Japan caught 10.7 million metric tons of fish with a dockside value of $2.5 billion. This ranked Japan first in landings and second in value—second only to China (People's Republic of China and the Republic of China or "Taiwan"). According to FAO statistics, nearly 84 percent of the Japanese catch comes from the northwest Pacific. Extended jurisdiction by North Korea, the PRC, South Korea, and the Soviet Union may influence Japan's access to many fishery resources.

Japan now harvests little off the New England coast or in the northwest Atlantic in general; however, some have feared repercussions in the northeast Pacific, where Japan derives about 9 percent of its catch. At present, the Japanese agree to *abstain* from fishing North American halibut, salmon, and herring stocks under the supervision of the International North Pacific Fisheries Commission. A provisional *abstention line* was established, east of which the Japanese agreed not to fish. Their present catch off the United States (Alaska, Washington, and Oregon) is primarily pollack and other groundfish. Senator Gravel of Alaska has argued that if the U.S. fishery jurisdiction is extended unilaterally, the Japanese will regard their abstention agreement as null and void. The primary fear is that the 200-mile jurisdiction would not cover *North American* salmon on the high seas. As a result, Japanese fishing effort would be diverted from pollack to the more lucrative salmon, and there would be an adverse impact on employment in the Alaskan fishing and processing industries. Of course, the secretary of commerce can easily grant Japan a permit to continue its normal fishing pattern (see provisions of PL 94-265 below).

It is difficult to predict what Japan will do on the international scene if extended jurisdiction becomes worldwide. Japan has about the highest per capita consumption of fish in the world (seventy pounds per person); therefore, it will not be pleased if its primary source of protein is threatened. However, Japan does not have the capability to violate other nations' jurisdictional claims by force. With respect to the United States, extended jurisdiction should not induce Japan to attack our exposed salmon resource. Since Japan is dependent on the United States for export markets, it certainly would be ill advised for it to violate an international agreement that *does not have to be influenced* by U.S. extended jurisdiction. However, Japan may be increasingly faced with a world trend of extended jurisdiction that may reduce its catches and increase its domestic prices.

The United States and Extended Jurisdiction: A Blessing or a Curse?

A Summary of the Fishery Management and Conservation Act of 1976 (PL 94-265). Under this act, the outer boundary of the fishery conservation zone extends seaward 200 nautical miles from the coastline of the United States; its inner boundary is a line "coterminous with the seaward boundary of each of the coastal States." This definition reserves to the states the authority to continue regulation of fishing within their boundaries, but only if such measures do not conflict with fishery management plans for the protection of species within the entire conservation zone.

The legislation creates U.S. authority to manage fishing of all species within the 200-mile zone, all anadromous species throughout their migratory range, and all continental shelf fishery resources beyond the zone, but it does not include authority over the fishing of "highly migratory species" such as tuna, which in the course of their life cycle spawn and migrate over great distances in the waters of the ocean.

The national fishery management program established by the act defines general standards to guide the preparation and implementation of specific management plans by regional councils and the secretary of commerce. These standards recognize the principle of optimum yield for each fishery, which is defined as that amount "which will provide the greatest overall benefit to the Nation, with particular reference to food production and recreation opportunities."

Fishery management plans are to be prepared by the regional councils or by the secretary of commerce, if necessary, on an interim basis. Measures the councils may recommend for implementation of fishery management plans *include the requirement of a permit and fees for any vessel fishing within the conservation zone;* limitations or prohibitions upon fishing in designated areas and descriptions of authorized fishing gear; specific limits on the catch of fish; restrictions or prohibitions upon the types and quantities of gear and equipment fishing vessels may carry; the establishment of a system of limited access; and other measures as necessary and appropriate. Of great importance is section 304(d), which states, "The Secretary shall by regulation establish the level of any fees which are authorized to be charged . . . (1) *Such levels should not exceed the administrative cost incurred by the Secretary in issuing such permit*" (emphasis added).

The act provides for a careful scrutiny and review of proposed international fishery agreements and individual applications of foreign countries for permits on behalf of their nationals within the conservation zone. The secretary of state has major negotiating authority, but he is required to cooperate with the secretary of commerce in determining the total allowable level of foreign fishing as well as in allocating this level among the foreign countries competing for limited resources within the U.S. conservation zone. A foreign nation entering into a fishery agreement with the United States must submit to U.S. authority, which may include boarding and inspecting any vessel permitted to fish within the conservation zone. U.S. authorities may make arrests and seizures whenever there is reasonable cause for belief that a violation of the provisions of the fishery agreement and vessel permit has occurred.

The act also contains a provision of reciprocity: a foreign country seeking access for its fishermen to U.S. conservation zone waters must extend substantially the same fishing privileges to U.S. fishing vessels.

Evaluation of Impact of PL 94-265. Some advocates of the U.S. extended fishery jurisdiction have argued that since the resource will come under the exclusive jurisdiction of the United States, the United States will have an incentive to protect and preserve this fishery resource and to work toward increased stock productivity. The U.S. catch of New England groundfish (cod, flounder, haddock, perch, and pollack) from the offshore region (ICNAF areas 5 and 6) is about 173 million pounds (1974). The foreign catch of these same species within a possible U.S. 200-mile zone is about 39 million pounds. If such catches were allocated to the U.S. groundfish industry under extended jurisdiction, there would be a 25 percent expansion in available resources. As mentioned in chapter 2, a number of these resources are "depleted," "under intensive use," or in "imminent danger" of depletion. A quick review of recent sustainable yield estimates for these groundfish species indicates that about a 100 percent increase in today's groundfish catch could be attained *if depleted stocks are restored.* It is possible that the United States alone could restore these stocks more readily than the ICNAF could. In theory, the impact of PL 94-265 on the New England groundfish industry might be to increase available catch from 100 to 200 million pounds. Given the high price elasticities for groundfish, this would increase ex vessel sales. Table 4.6 shows the present U.S. catch from 0-3, 3-12, and 12-200 miles, including what is caught in international waters off foreign shores. In 1976, only 11 percent by weight and 24 percent by value of the total U.S. catch was caught between 12-200 miles (i.e., 188 miles of extended jurisdiction over the present 12-mile national fishing zone). In contrast, more than 12 percent by weight and 15 percent by value is caught in international waters, usually off foreign shores by U.S. distant-water fleets (i.e., mostly tuna, spiny lobsters, and shrimp). The shrimp and tuna industries strongly opposed U.S. extended jurisdiction, since it would set a precedent for other countries. Table 4.7 will give the reader some idea of the foreign fishing vessels operating off the U.S. coast (but not necessarily *within* 200 miles) during 1975. As is evident, these foreign vessels exploited resources primarily in the northwest Atlantic (U.S. Atlantic coast) and northeast Pacific. It is not surprising that the political pressure

Table 4.6 Commercial Landings of Fish and Shellfish by U.S. Fishing Craft: By Species, By Distance Caught off U.S. Shores, and Caught in International Waters, 1976[1]

	Species	0 to 3 miles		3 to 12 miles		12 to 200 miles		International Waters	
		Thousand pounds	Thousand dollars	Thousand pounds	Thousand dollars	Thousand pounds	Thousand dollars	Thousand pounds	Thousand dollars
1.	Anchovies	12010	2001	224714	3370	20349	304	-	-
2.	Atlantic cod	4242	984	11235	2689	39563	10432	805	245
3.	Atl. & Gulf Flounder	21136	7889	18986	6919	66261	26900	206	94
4.	Pacific Flounder	7615	1246	23902	4311	23322	4159	3254	489
5.	Haddock	147	56	1274	534	9097	3974	2243	987
6.	Halibut	2608	2446	5096	4826	12901	12137	9	9
7.	Pacific Sea Herring	40392	6297	-	-	-	-	-	-
8.	Atlantic Menhaden	758292	21982	42364	1214	1042	16	-	-
9.	Gulf Menhaden	1068494	37860	169285	6107	-	-	-	-
10.	Pacific Salmon	276281	156476	27257	31682	5642	8254	62	84
11.	Red Snapper	124	145	632	589	7505	6666	955	624
12.	Tuna	757	455	3601	1943	33041	15096	622453	182308
13.	Clams	34339	40249	12506	4642	34156	17818	-	-
14.	Crabs	179907	60175	140714	61289	24189	15491	-	-
15.	Am Lobster	23109	36665	3184	5714	5448	10278	-	-
16.	Spiny Lobster	1810	2871	1975	2963	615	923	489	734
17.	Oysters	54391	53098	-	-	-	-	-	-
18.	Sea Scallops	806	1702	1265	1963	17769	31396	-	-
19.	Shrimp[2]	174235	116949	107530	54537	112911	146192	16662	28753
	Sub-total	2660695	549546	795519	195319	413811	309036	647138	214327
20.	Other	582557	88669	230492	24692	184374	31685	17984	4519
	Total	3243252	638215	1025948	220011	598185	340721	665122	218846

[1]Statistics on landings are shown in round weight for all items, except univalve and bivalve mollusks, such as clams, oysters, and scallops, which are shown in weight of meats excluding the shell.

[2]Includes shrimp landed at Gulf Coast and foreign ports. Data do not include production of artificially cultivated fish and shellfish.

Source: Fisheries Statistics of the United States, 1976 (NMFS)

Table 4.7: Foreign Fishing Vessels Operating Off U.S. Coasts During Fiscal Year 1975*

Fishing grounds	Stern trawlers [1]	Medium trawlers [2]	Other fishing vessels	Process and transport vessels	Support vessels [3]	Research vessels [4]	Total
Off Pacific coast:							
Off Alaska:							
Japan	534	751	1,098	208	23	2	2,616
Poland	6	0	0	0	0	0	6
Republic of China	5	0	0	0	0	0	5
Republic of Korea	17	61	37	6	0	0	121
Soviet Union	250	374	1	59	40	5	729
Other	0	0	0	0	0	0	0
Total	812	1,186	1,136	273	63	7	3,477
Off Pacific Northwest:							
German Democratic Republic	0	0	0	0	0	0	0
Japan	14	0	10	1	1	0	26
Republic of Korea	0	0	24	0	0	0	24
Poland	33	0	0	2	2	0	37
Soviet Union	215	3	11	38	14	12	293
Federal Republic of Germany	2	0	0	0	0	0	2
Total	264	3	45	41	17	12	382
Off California:							
German Democratic Republic	0	0	0	0	0	0	0
Japan	10	0	0	0	1	0	11
Poland	53	0	0	4	1	0	58
Republic of Korea	0	0	1	0	0	0	1
Soviet Union	339	1	3	24	18	11	396
Federal Republic of Germany	6	0	0	0	0	4	10
Total	408	1	4	28	20	15	476
In the Gulf of Mexico:							
Cuba	0	49	40	0	4	0	93
Japan	0	0	26	0	0	0	26
Mexico	0	6	2	0	0	0	8
Soviet Union	0	0	0	0	0	0	0
Canada	0	0	2	0	0	0	2
Total	0	55	70	0	4	0	129
Off the Atlantic coast:							
Bulgaria	44	0	0	6	0	0	50
Canada	6	0	0	0	0	0	6
Federal Republic of Germany	41	0	0	2	1	3	47
France	4	0	0	0	0	1	5
German Democratic Republic	98	74	0	17	3	2	194
Ireland	11	0	0	0	0	0	11
Italy	37	0	0	0	0	0	37
Japan	104	0	54	4	0	0	162
Norway	0	0	1	0	0	0	1
Poland	199	58	0	34	0	2	293
Romania	7	0	0	0	0	0	7
Soviet Union	778	161	265	112	22	9	1,347
Spain	88	31	60	0	0	0	179
Total	1,417	324	380	175	26	17	2,339
Grand total	2,901	1,569	1,635	517	130	51	6,803

*Excludes duplicate sightings within the same month—includes repetitive sightings from 1 month to the next.
[1] Includes all classes of stern factory and stern freezer trawlers.
[2] Includes all classes of medium side trawlers (nonrefrigerated, refrigerated, and freezer trawlers).
[3] Includes fuel and water carriers, tugs, cargo vessels, etc.
[4] Includes exploratory, research, and enforcement (E) vessels.
Source: National Marine Fisheries Service, Law Enforcement Division, NOAA, Department of Commerce.

Source: U.S. HR(1975)

for extended jurisdiction came from these areas.

If U.S. distant-water fleets are taxed by means of a licensing fee that is already a reality, the consumer will bear some of this tax. Generally, the more inelastic the demand curve, the greater the burden will be on the fish consumer. The most recent estimates of demand elasticities for those U.S. fisheries likely to be affected by the license fees are the following:

	Price Elasticity (U.S.A.)
Canned Tuna	- 0.99
Shrimp	- 0.36
Lobsters	- 0.65

According to these estimates, the tuna industry will be hit the hardest because of relatively high price elasticity of demand; shrimp should bear the smallest burden. Of course, if fishing rights are withdrawn completely, severe economic dislocation may occur in these U.S. fishing sectors. This has already happened in the case of U.S. shrimpers fishing off the Mexican coast in the Gulf of Mexico. Some provision should be made for handling vessels and fishermen who are unemployed or underutilized because of extended jurisdiction.

Let us imagine the completely idyllic situation. If our coastal stocks were rebuilt and if our tuna, shrimp, and lobster industries survived with a minimum loss of job and idle capacity, what is the most probable course the Regional Fishery Management Councils (RFMC) would take? Section 304(d) of PL 94-265 limits the charging of fees in excess of administrative costs, and for all practical purposes, this eliminates Limited Entry Scheme 3 (i.e., use of license fees or competitive bidding or both). We have argued that this scheme would, over the long haul, both conserve the resource and make the fishing industry economically progressive. The RFMC may be restricted to Limited Entry Schemes 1 (issuing of licenses) or 2 (fishermen's quotas). The latter has certain advantages if coupled with an excess profits tax when the value of the fishermen's quota (or stock certificate) becomes excessive. However, if the RFMC use the *overall* quota or regulated inefficiency, the 200-mile limit will merely eliminate foreign competition, but will not solve the fundamental problem: *the common property nature of the resource.* Thus, the situation may become less idyllic. With the Bell et al. (1975) and Covin (1975) projections of the likely demands on coastal fishery resources, economic rents will accrue that can only be handled in one of two ways: (1) allow them to dissipate by excessive capitalization, or (2) place a license fee, competitive bidding, or fishermen's quota on the fishing industry to control fishing effort. The former case has led to dwindling productivity and diminished efficiency. And with license fees and competitive bidding effectively outlawed by PL 94-265, the mess of the past may well be perpetuated.

However, the prime movers in determining jurisdictional claims are still the international forces that dictate either the harvesting of fish for domestic consumption or the acquisition of foreign exchange through the export of high-valued fishery products. All countries, including the United States, have a habit of creating "international law" to suit their own objectives. Worldwide population expansion and increased distant-water fishing by the developed countries (see table 4.7) have overcome the principle of the freedom of the seas. A fishery stock can be depleted even more rapidly when conservation controls or extended jurisdiction are anticipated. Pulse fishing has been very effective in beating resource controls to the punch. Francis Christy and Anthony Scott (1972) have called these events the fisheries' "domino effects." They define them as an artificial stimulus to overcapitalization derived from quota systems such as that of the Inter-American Tropical Tuna Commission. As Christy and Scott assert, "In short, we are now on the brink of precipitous action and reaction that could, within a year or two, totally change the character of fishery regimes throughout the world" (1972, p. xv). Hence, the fault is not in the stars, but in ourselves, in failing to properly handle fishery resources as common property that should be transformed as quickly as possible into private property. Our failure to act either multilaterally or unilaterally toward a private property regime (i.e., sometimes called limited entry, as discussed above) has brought us to the point where marine biologists now feel that the global catch of table-grade fish is at or near the maximum sustainable level (see chapter 3). Approximately thirty leading species of commercial-grade fish may be overfished. But where does that leave us with respect to U.S. extended jurisdiction?

We have heard over and over again that the primary purpose of extended jurisdiction is the *conservation of fish.* U.S. legislation states this as an objective. But with international commissions, the concern for conservation has led us away from an emphasis on economic objectives. Consider these facts. Although the biologists, conservationists, and political managers have realized that fishery resources are common property in nature, they have not, except in a few cases (see Limited Entry 1, 2, and 3), come to grips with this problem directly.

In a penetrating paper, Francis T. Christy has severely criticized PL 94-265: "The most important consequence of these provisions is that they provide for the creation of property rights in public resources and then grant these rights, for free, to the domestic fishermen. This is a gift on the order of $5 billion, even though the gift may never by fully utilized" (1976, p. 8). Furthermore, "this is detrimental to the interests of society because no return is received from the use of publicly owned resources" (1976, p. 10). Christy feels that this is not generally in the interest of taxpayers, because research, enhancement, and most importantly enforcement, will be borne by them rather than by the fishermen. Christy also believes that although foreign fishing vessels can obtain permits to fish within the U.S. 200-mile zone, the incentive to exclude foreigners will be great, thereby substituting high-priced domestic production for low-cost foreign production. Thus, consumer interests are likely to be damaged. The pattern Christy predicted is already emerging in the appointments to the RFMCs—95 percent of the appointees have a direct economic interest in the public resource they are regulating. This goes far beyond the typical pattern of conflict of interest already present in so many regulatory bodies outside the fisheries, such as the FAA, FDA, and FEA. Christy concludes his concern with PL 94-265 by saying that "It is readily apparent that I find major flaws in the Fishery Conservation and Management Act of 1976, and that my major area of concern lies with the inability to collect revenues from fishermen. This is . . . my personal predilection for society receiving appropriate returns from public resources. I would, for example, strenuously oppose the free gift of OCS oil lands to the oil industry" (1976, p. 24).

The National Fisheries Plan (NFP) developed by the National Marine Fisheries Service (1976) is closely linked to extended jurisdiction. If you strip the "plan" down, it is merely a simplistic variation on the NACOA recommendation to the NOAA administrator.[13] NACOA calls for an increase in U.S. landings from a 1969-1973 average of 4.7 million pounds to 7.1 million pounds by 1985, or an increase of 3.4 million pounds.[14] No matter what scenario the plan uses, its essential goal is partially to replace foreign imports with domestic landings. With U.S. jurisdiction over additional fishery resources, the fundamental question is what nation or nations can harvest these resources at the lowest cost. Presumably, the NMFS has concluded that the United States can, despite the absence of one single recent economic study to justify this assumption. The mere fact that the resource will be available only to U.S. fishermen is not sufficient reason to say that they can compete with foreign imports. This point was emphasized by Christy (1976). First, most of the alleged substitution is disguised in the rhetoric that the United States will somehow chase all those foreigners off our shores and that this will open up a vast cornucopia for U.S. fishermen. The NFP ignores the fact that only 17 percent of the groundfish is caught in subareas 5 and 6 in the northwest Atlantic, which roughly coincides with a 200-mile fishing zone off the northeastern United States. Hence, the United States does not control all the groundfish resources from which a good percentage of our imports are derived. In Noetzel and Vondruska (1975), we get a more definite statement derived from the NFP on how all this import substitution will take place:

Case (1969–1973 base period)	Result and Causal Factors
1. Status Quo (1985)	0.2 billion pound increase in domestic landings and a 7.4 billion pound increase in imports
2. Extended Jurisdiction (Min. Case by 1985)	0.8 billion pound increase in domestic landings and a 6.8 billion pound increase in imports due solely to extended jurisdiction
3. Extended Jurisdiction (NACOA Case by 1985)	2.4 billion pound increase in domestic landings and a 5.2 billion pound increase in imports due to extended jurisdiction plus *government support programs*
4. Extended Jurisdiction (Max. Case by 1985)	6.1 billion pound increase in domestic landings and a 5.2 billion pound increase in imports due to same factors as in NACOA case plus *increased government assistance to industry*

It is incredible that even with the 200-mile extension, the NMFS should request additional funding under cases 3 and 4 to achieve the objective of import substitution. Under case 2 the largest increase will come in groundfish, but the competitive advantage will still be with foreign fishing fleets in supplying frozen fish, although the fresh fish market might absorb some increases in

U.S. landings over the next ten years. Case 3 (NACOA) is requesting government support in terms of *financial assistance, market development, and tax credits.* Case 4 is based upon bold new "development programs." These are not described in any great detail, except that they will involve underutilized species.

A New World Economic Order Under Extended Jurisdiction. Our discussion in previous sections leads us to project the changing nature of world fisheries. The preoccupation of fairly affluent nations with the sea as common property has had a destructive influence on world fisheries. Conservation has greatly confused the issues. It has been the objective of so many international fishery commissions that little study has been given to the economic consequences of the tools used to conserve the fishery resource.

In the future, the pressure of world population will lead most countries to extend their jurisdiction over these valuable sources of direct food supply or foreign exchange or both. The LDCs will probably, as they have in the past, unilaterally extend their jurisdictions to even more than 200 miles. From an economic point of view, this is a general trend toward the *destruction of the common property nature of fishery resources,* which can only bring benefits to the world economic community, although there will be profound distribution effects in some cases. More specifically, the following new economic system in the fisheries may emerge.

1. The LDCs, with 64 percent of the world's fishery potential off their coasts, may act as sole owners and charge fees roughly equivalent to the economic rents generated by the increasing demand for fish and protein in general. This will apply to those fish over which the LDCs have jurisdiction, but which the LDCs are incapable of harvesting themselves.
2. For those fleets that are forced to pay harvesting fees, the burden will largely depend on the price elasticity for their product. Much of the price increase will be passed on to the consumer, especially in the shellfish area.
3. If the developed countries act as rationally as the LDCs (i.e., charge fees to limit entry), overcapitalization in the world fisheries should gradually be reduced. Fishing costs should also be reduced as redundant resources are used elsewhere in the various economies throughout the world.
4. Anadromous and pelagic fish (primarily salmon and tuna) will remain a problem. For example, we need some coordination among fee-charging LDCs for the use of the tuna

resources that are fished along several South American countries.

In summary, the world fisheries may be at a turning point. Whether conflict or cooperation will result is still quite debatable.

Sociological Aspects of Fishery Management

Dow et al. (1975) concluded in a recent labor mobility study of the Maine lobster fishery that the various management schemes discussed above would have certain socioeconomic effects:

1. License fee levels that would actually displace labor (i.e., reduction in fishing effort) would have a minimum unemployment impact, since the group that would leave is relatively mobile.
2. License fee schemes would considerably reduce the income of those remaining in the fishery, which would be a disadvantage to this proposal.
3. A reduction in the degree of capitalization through any of the management plans would probably raise total revenue produced in the fishery with an increased catch, which would benefit the entire fishing community.
4. From a social point of view, the stock certificate or fisherman quota plan has the fewest disadvantages from the standpoint of the fishing industry and surrounding communities.

The results of our analysis would seem to be in conflict with Limited Entry Scheme 3 and Christy's call for license fees under extended jurisdiction. The charging of a license fee is a dynamic process. As indicated earlier in this chapter, in equilibrium, fishermen are not earning excessive profits (i.e., average cost = price). This is probably the primary motivation behind limiting license fees under PL 94-265—very few U.S. fisheries are earning large "excess" profits (see chapter 9).

Consequently, the management authority must first generate excess profits while limiting effort in order to demonstrate this principle to the fishermen. This is a dynamic process that might be accomplished through a fisherman's quota (i.e., a stock certificate plan) followed by license fees or excess profit taxes or both. Many who work in this area have failed to realize this point.

In addition to a straight assessment of labor mobility, other sociological factors must be considered. Rockwood et al. (1973) indicate that in the Apalachicola Bay (Florida) oyster community, kinship ties are emphasized in order to separate members from outsiders. Thus, kinship ties define who "belongs" in the community. The insularity and clannishness of the in-community is typical of many rural communities. The inhabitants of these communities tend to be very guarded with outsiders, particularly those who come in and ask questions. In oyster communities such as Apalachicola, it would be possible to use large oyster dredges to reduce harvesting employment; however, the social cost of the attendant unemployment might exceed the efficiency benefits. Thus, any fishery management program must allay any fears about the economic impact. Thus, RFMCs would be well advised to consider the sociology of the fishing community as they consider the various forms of limited entry or other management programs. The adjustment or reaction of people to management programs is consistently overlooked in the abstract world of theory.

Demand and Supply Projections and the Increasing Need for Fishery Management

FAO Supply and Demand Forecasts

The tools developed in chapter 2 on demand analysis can be used to make economic projections of the future demand for fishery products. Working for FAO, M. A. Robinson (1973) has made several demand projections to the year 2000. As discussed in chapter 2, two of the most important determinants of the demand for fish, as with many other commodities, are population and per capita income. Robinson assumes that world population will grow to 4.6 billion in 1980 and 6.6 billion by the end of the century and *that the price of fish will remain constant* (so he can isolate the impact of the former two variables). He concludes that merely to maintain per capita food fish consumption at present levels (1970) will necessitate an increased fish supply of some 8 million metric tons by 1980 and 27 million tons by the end of the century. Robinson uses an FAO projection of a rise in world per capita private consumption (i.e., income) of 5.1 percent to 1980 and 4.5 percent from 1980 to 2000. Each country is treated separately because of differences in income elasticities for fish, which according to FAO vary from 1.0 in Indonesia to as low as 0.28 in the United States. Income elasticities are assumed to *remain constant* over the projection period, which conflicts with

Engel's law as discussed in chapter 2. When Robinson adds the income impact to his projections, the demand for food fish will increase 18.5 million metric tons by 1980 and 63 million metric tons by the end of the century. His projections include both fresh and marine species. Food fish demand is projected to be approximately 61 million metric tons by 1980 and nearly 108 million metric tons by the turn of the century. Robinson excludes any allowance for increased demand for fish meal, for which, it is believed (owing mainly to supply limitations), there will be no significant increase in consumption above present levels. In the FAO projections, fish are treated as a homogeneous commodity—apples and oranges again. On the supply side, the world's resources of *marine* fish, crustaceans, and cephalopods can, according to Robinson, sustain an annual catch of about 118 million metric tons—on the *optimistic premise* that fishing effort on each stock is adjusted to take the maximum sustainable yield. Robinson indicates that "it is clear from the comparison [of projected supply and demand] that world fisheries are approaching a critical period" (1973, p. 2057). Clearly, Robinson's projections are too optimistic: much of the increase in supply potential consists of species for which there is presently no demand. Moreover, in light of the current resource crisis, it is clearly unrealistic to assume that all species will be harvested at *MSY*. Robinson's projections indicate a pressing need for effective fishery management (this was the theme of chapter 3). They indicate that the depletion of world fishery resources is likely to increase unless extended jurisdiction mitigates this process. International Fishery Commissions have had only limited effectiveness.

Bell et al.: Supply and Demand Forecast

The FAO projections give only general guidance and suffer from a lack of disaggregation. Bell et al. (1975) identified twelve food fish categories such as shrimp, salmon, and fish meal. For each of these categories, they obtained the relationship between per capita consumption and per capita income and price for all major consuming countries. These relations were added to obtain a world demand curve (i.e., price-quantity relation) for a base year period 1965–1967. Thus, we would have a world demand curve for lobsters, tuna, or groundfish. Using Gulland (1971), the potential supply was obtained for each species category. A supply curve was developed by superimposing a modified version of the Schaefer yield function with a 1965–1967 base point (i.e., estimated point on the yield function). This allowed the development of a *world* supply curve (i.e., the average cost per pound of fish) where overfishing could take place in response to demand pressures. This is completely different from the assumption Robinson used. Figure 4.5 shows an example for lobsters where overfishing is allowed to take place. When such fishery categories as salmon or halibut were considered, it was assumed that management would prevent overfishing. The projections of population and per capita income growth were quite similar to those used by Robinson. However, Bell et al. allowed the income elasticity for the fishery categories to decline as per capita income increased in accordance with Engel's law. The general conclusions were the following:

1. The world demand for various species categories will outstrip the maximum world supply potential before 1985. Table 4.8 indicates the year in which certain species will attain maximum sustainable supply.[15] Critical problems of resource supply are presently or about to occur for groundfish, salmon, halibut, lobsters, crabs, and fish meal. Unless

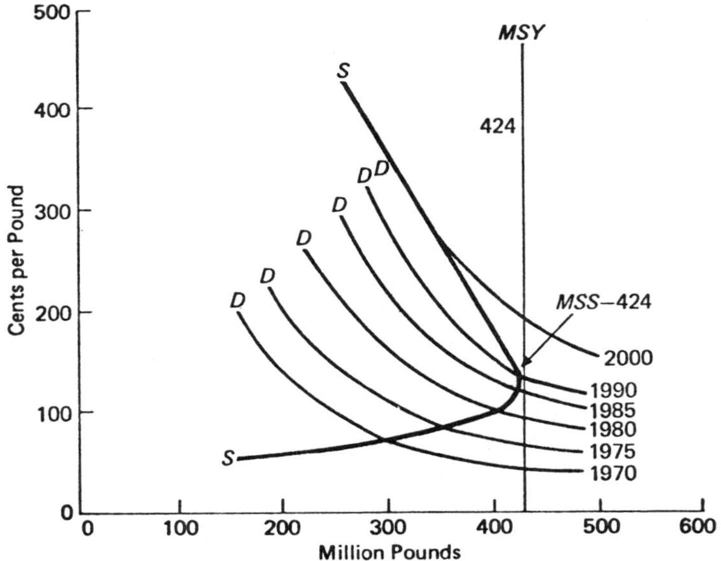

Figure 4.5: World Demand and Supply Functions for Lobster, 1970-2000
Source: Bell et al. (1975)

proper management policies are adopted, overfishing for crabs, lobsters, groundfish, and fish meal is possible *on a widespread basis* by 1985 or sooner. Table 4.9 shows a breakdown of the species categories for the projections of consumption to the year 2000.[16]

2. By 1985, it is projected that demand (and supply) will be in equilibrium or will increase to approximately 78.6 million metric tons and approach 83.5 million metric tons by the year 2000. Let us compare the FAO projections with Bell et al. (in millions of metric tons):

	Actual		Projected	
	1965-1967	1970	1980	2000
1. FAO				
Food		43.9	61.0	107.5
Meal		26.5	33.3	33.3 ?
Total		70.4	94.3	140.8
2. Bell et al.		(Projected)		
Food	36.7	41.2	52.0	68.3
Meal	20.4	22.7	28.4	15.2
Total	57.1	63.9	80.4	83.5

There are three fundamental reasons why the projections of Bell et al. are lower than those of the FAO for both 1980 and 2000. First, the Bell et al. projections allow for the possibility of overfishing and thereby for declines in production. Second, prices are allowed to rise over time as demand pressures are placed upon the resource. This reduces consumption. Third, a declining income elasticity will (as opposed to the FAO assumption) slow up the rise in demand.

One final note is in order. If not controlled, population and per capita income growth over the next twenty-five years will result in widespread depletion of the world's ocean fisheries. The FAO expectations are realistic only if underutilized species are used. This is the subject of chapter 8. Even then, Robinson's assumption of no overexploitation seems unrealistic. Finally, figure 4.6 compares the Bell et al. actual and projected world fish consumption-supply to the year 2000 with FAO's demand projections. There has been no appreciable increase in world catch over the 1970-1976 period, although catch was predicted to increase at a decreasing rate.

Table 4.8: Projected Date of Maximum Sustainable Supply

Species*	Year World Will Reach Maximum Sustainable Supply
Salmon[a]	1970
Halibut	1970
Groundfish[b]	1970
Crabs	1980-85
Fishmeal (i.e., species for reduction)	1980
Lobsters	1985
Tuna[c]	2000
Shrimp	2000
Sardines	2000+
Scallops[d]	2000+
Clams	2000+

*Aquaculture not assumed in these projections

[a]Does not include the possibility of exapanded supply through hatchery operations and stream improvements.

[b]Excludes hake and hake-like fish

[c]Excludes Central Pacific skipjack

[d]Includes recent discovery of calico scallops

Source: Bell et al. (1975)

Summary

All wild stock fisheries are common property natural resources. Since no single enterprise has to pay for the use of fishery resources, they are often overexploited. Unlike the productivity natural resource industries (such as timber and coal), where private property prevails, each fishing firm's *productivity* is influenced by the total number of firms exploiting the common property fishery resource. This is known as a technological externality. With a common property fishery resource, each enterprise acting on its own leads not to a social optimum, as Adam Smith en-

Table 4.9 World Aggregate Consumption of Fishery Products, Projected to Year 2000[1]

	1965-67[2]	1970	1975	1980	1985	1990	2000	Changes 1965-67 to 2000
	Thousand metric tons, round weight							Percent
Food fish								
Groundfish	6,368	6,935	6,940	6,759	5,761	5,262	4,763	-25.2
Tuna	1,291	1,315	1,456	1,556	1,615	1,647	1,657	28.4
Salmon	476	476	481	485	485	485	485	1.9
Halibut	58	58	58	58	58	58	58	0
Sardines	871	1,166	1,464	1,657	1,848	2,013	2,370	172.1
Shrimp	634	894	1.066	1,243	1,347	1,438	1,479	133.3
Lobsters	137	150	174	187	192	186	145	5.8
Crabs[3]	328	395	481	549	517	449	386	17.7
Clams[4]	478	481	535	590	626	658	694	45.2
Scallops[5]	166	209	236	259	281	295	322	94.0
Oysters	777	965	1,218	1,487	1,755	2,015	2,453	215.7
Other fish	25,086	28,123	32,659	37,195	41,504	46,040	53,524	113.4
Total food fish	36,670	41,217	46,768	52,025	55,989	60,546	63,336	86.4
Fish meal	20,440	22,680	27,170	28,350	22,634	19,505	15,196	-25.7
Total (food & meal)	57,110	63,897	73,938	80,375	78,623	80,051	83,532	46.3

1 Under LDR-DIE Assumptions
2 Average of actual
3 Estimated for 1985, 1990 and 2000 based on a gradual decline in the resource base.
4 Without additional aquaculture
5 Includes calico scallops

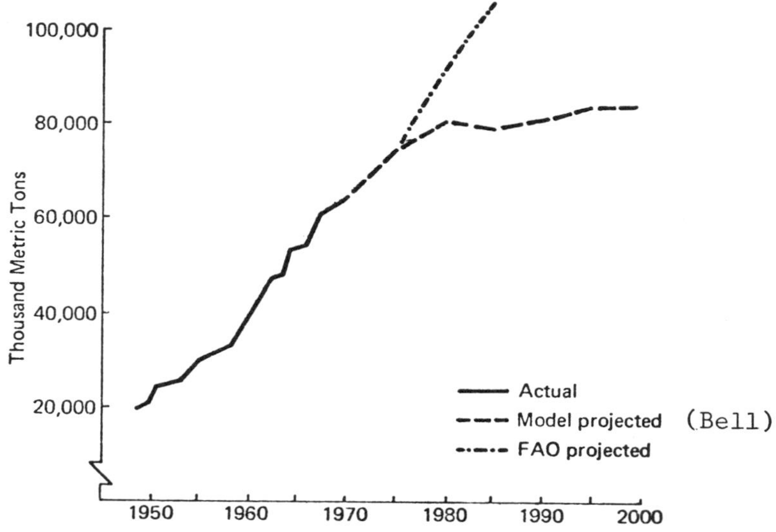

Figure 4.6: Historical and Projected World Aggregate Consumption of Fishery Products

Source: Bell et al. (1975)

visioned, but to a private market failure and thus to overfishing. The fundamental reason why the fisheries need management is economic inefficiency and waste.

Two kinds of management schemes have been explored here. Man can enhance or increase the productivity of the resource through proper mesh selections, closing nursery grounds to fishing and improved hatchery operations, to mention but a few. However, by far the overwhelming concern has been and is management to prevent overfishing or to restore an overfished resource. Historically, regulated inefficiency or technological restriction has been used to reduce the fishing power of a fleet. Of equal importance, total catch quotas for entire fishing fleets have been imposed, resulting in shortened seasons and uneconomic use of vessels and fishermen. These two management schemes have failed to come to grips with the fundamental problem of common property; therefore, many sectors of the fishing industry have remained technologically backward and uncompetitive with other domestic food products and foreign imports. The primary goal of conservation has greatly detracted from good economic management of fishery resources.

Three forms of limited access or limited entry have been considered as alternatives: (1) direct control of fishing effort by licensing, (2) a fisherman's quota, and (3) charging fees for the use of the resource or competitive bidding for fishing rights. The schemes used in Alaska, British Columbia, South Africa, and Japan to control fishing effort have been reviewed. The results indicate that these schemes suffer from a constant battle between fishermen and the regulatory authority to outwit each other as regulations are piled on regulations in order to keep fishing effort in check. The fisherman's quota, in effect, gives a public resource to the fisherman. Rights or quotas may be transferable. Some have attacked this system as a "giveaway" fraught with potential corruption. In theory, charging for the right to fish or competitive bidding is very defensible on economic grounds, but it does suffer from stiff opposition from an industry that may not be earning excessive economic profits. A system combining fishermen's quotas and taxation after the appearance of excessive economic profits seems feasible, but it may be potentially dangerous owing to political opposition to taxation after such profits occur.

The law of the sea has varied historically, but the basic principle of freedom of the seas is being eroded by an international wave of extended fishery jurisdiction. The combination of world population growth and the inability of agricultural production to keep pace has led to greater interest in the sea as a source of food. The less-developed countries have approximately 64 percent of the world's sustainable fishery production off their coasts. Already, countries such as Ecuador and Peru are charging for the right to fish within 200 miles of their respective coasts. Japan and the USSR will be at a serious disadvantage since they have heavily invested in distant-water fishing fleets and greatly depend on the fisheries for food. The Fishery Conservation and Management Act of 1976 extends U.S. fishing jurisdiction to 200 miles; however, it all but prohibits the charging of fees (or, by implication, competitive bidding for fishing rights) to fish. The fear is that the act, or PL 94–265, will lead not to an efficient and equitable limited entry system where the common property problem is met head on, but that *total* catch quotas, arbitrary licensing, or other systems will be used that will not result in a more productive U.S. fishing industry, one that is able to deliver products competitive with foreign imports. The U.S. government's own projections of increased catch and the tremendous cost of management and especially enforcement give rise to some question over whether the 200-mile limit in itself will achieve benefits greater than the costs—at least over the next ten years.

It is also axiomatic that new fishery management plans coupled with extended jurisdiction cannot be successful unless some study is given to the sociological aspects of many rural fishing communities and the economic impact or adjustments that many oppose.

Finally, this chapter has considered the projected future demand and fishery resources' potential supply on a worldwide basis. The FAO projects that the world will fully utilize the wild fishery stocks of approximately 120 million metric tons on a sustainable basis by the year 2000. Other studies, such as that made by Bell et al., are much more pessimistic, especially in light of the increasing depletion of the world's major fishery stocks. The economic projections cry out for fishery management. There is indeed a crisis at sea, and it threatens the world's potential food harvest.

Notes

1. A fishery resource can be used for both food and recreation, which economists consider "commercial use."

2. There could be repercussions if more farms increase aggregate farm output and depress prices. This might influence the profitability of the wheat farm, but not its productivity.

3. The actual yield per trap over the 1964-1966 period was approximately twenty-eight pounds annually; the difference is due to a decline in seawater temperature, which depresses lobster yields even more.

4. The seminal articles in economics on overexploitation of fishery resources were by Gordon (1954) and Scott (1955).

5. Colin Clark (1973) has taken issue with this traditional point of view. He argues that *private* owners may not only overfish, but also render the stock extinct! His argument is based upon the fact that the maximization of rent is not the same as maximization of present value. Quite simply, if the world is literally starving to death, the entire fishery stock may be "mined" by private owners to make an immediate sale. Economists would say the private entrepreneur has a high discount rate—he prefers profits now rather than a continual flow into perpetuity.

6. Five countries—Panama, Ecuador, Mexico, Canada, and Japan—joined the convention, with Ecuador withdrawing in 1968.

7. In *Isakson et al.* v. *Rickey et al.*, the constitutionality of Alaska's limited entry system is now being tested in that state's courts. "In the Superior Court, Judge Thomas Stewart ruled that the limited entry law was constitutional in all respects, placing considerable emphasis on the unrestricted transferability of entry permits in his decision" (1975).

8. For most industries, output will expand in response to demand up to the point where the marginal cost of production (i.e., additional cost of producing one more unit of output) is equal to the price received in the mar-

ketplace. This is considered an efficient level of production. As discussed earlier in this chapter, this condition does not hold for the fishing industry because of the common property nature of the resource coupled with resource limitations. Marginal cost pricing is unstable in fishing, and it is argued by some economists that regulations should be structured so as to achieve this objective.

9. Gates and Norton (1974) conclude that for the yellowtail flounder fishery restricting fishing effort to *MSY* would reduce the fleet by 16 percent and increase economic profit by $14,700 per vessel. If the management agency wished to restrict effort to a level that would be socially optimum (i.e., *MC=P*), the fleet would be reduced to 45 vessels from 132 under free entry and to 111 if *MSY* is used as an objective. Profits (above a normal return on investment) would average $150,000 per vessel. However, it is important to recognize the political and social implications of eliminating a large number of vessels and crewmen from the fishery. The Gates-Norton compensation possibilities for vessel owners who leave the fishery will be discussed later in this chapter.

10. Historically, the length of a cannon shot was thought to divide a nation's territory from the open sea. Galiani calculated this length at three nautical miles, and this principle was generally accepted by the major powers of that era. See Kent (1954).

11. These estimates were provided by Dr. Fred Olson of the NMFS.

12. Much of the catch in the northeast Pacific is beyond 200 miles of the U.S. coast or off the coast of Canada. In 1974 and 1975, the USSR caught 0.7 and 0.61 million metric tons off the Pacific Coast of the United States. Hence, Soviet dependence might be closer to 8-9 percent of its total catch from waters off the U.S. coast in the Pacific. Obviously some of this catch is beyond 200 miles. Conflict in the Pacific between the USSR and the United States over fisheries might be a distinct reality if current practices (1974, 1975) are attempted to be maintained.

13. NOAA = National Oceanic and Atmospheric Administration; NACOA = National Advisory Committee on the Oceans and Atmosphere.

14. The status quo projects U.S. landings to be 4.9 billion pounds by 1985.

15. Maximum sustainable supply differs from *MSY*, since overfishing may take place in some parts of the world—on tuna, for example. Thus the *MSY* of all tuna stock could never be obtained.

16. In chapter 3, sardines were estimated to be fully fished, if not overfished, on a world basis. In making the Bell et al. projections, it was assumed that herringlike fish used for food and industrial purposes would be substituted for sardines such as anchovies.

References

Agnello, Richard J., and Donnelley, Lawrence. P. 1975. Property rights and efficiency in the oyster industry. *Journal of Law and Economics* 13:521-533.

Anderson, Lee G. 1977. *The Economies of Fisheries Management.* Baltimore: Johns Hopkins Press.

Asada, Y. 1973. License limitation regulations: the Japanese system. *Journal of the Fisheries Research Board of Canada* 30:2085–2095.

Bell, Frederick W., and Fullenbaum, Richard F. 1973. The American lobster fishery: economic analysis of alternative management strategies. *Marine Fisheries Review,* paper 994.

Bell, Frederick W., et al. 1975. A world model of living marine resources. In *Quantitive models of commodity markets,* ed. Walter C. Laby. Cambridge, Mass.: Ballinger Publishing Co.

Campbell, B. A. 1973. License limitation regulations: Canada's experience. *Journal of the Fisheries Research Board of Canada* 30:2070–2076.

Centaur Management Consultants. 1975. *Economic impact of the U.S. commercial fishing industry.* Prepared for the National Marine Fisheries Service.

Chapman, Wilbert M. 1968. Social and political factors in the development of fish production by United States flag vessels. In *The future of the fishing industry of the United States,* ed. DeWitt Gilbert. Seattle: University of Washington Press.

Christy, Francis T., Jr. 1976. Limited access systems under the Fishery Conservation and Management Act of 1976. University of Delaware, Sea Grant symposium. This paper has been published in *Economic impacts of extended fisheries jurisdiction,* ed. Lee G. Anderson. Ann Arbor, Mich.: Science Publishers, 1977.

Christy, Francis T., Jr., and Scott, Anthony. 1972. *The common wealth in ocean fisheries.* Baltimore: Johns Hopkins Press.

Clark, Colin W. 1973. The economics of overexploitation. *Science* 181: 630–634.

Commercial Fisheries Entry Commission. 1976. *1975 Annual Report.* State of Alaska.

Covin, Terrence R. 1975. A baseline forecast of the U.S. fishing industry. 22d International Meeting of the Institute of Management Sciences, Japan 1975. Synergy, Inc., Washington, D.C.

Crutchfield, James A., and Pontecorvo, Giulio. 1969. *The Pacific salmon fisheries: a study of irrational conservation.* Baltimore: Johns Hopkins Press.

Crutchfield, James A., and Zellner, Arnold. 1962. Regulation of the Pacific Coast halibut fishery. In *Economic effects of fishery regulation.* FAO fisheries report no. 5.

——. 1963. *Economic aspects of the Pacific halibut fishery.* Fish and Wildlife Service, Fishery Industrial Research Series, no. 1.

Cushing, David. 1975. *Fisheries resources and the sea and their management.* Oxford: Oxford University Press.

Devanney, J. W., III. 1976. *Fishermen and fish consumer income under the 200-mile limit.* MIT Sea Grant Program, report no. MITSG 75-20.

Dow, Robert L.; Bell, Frederick W.; and Harriman, Donald H. 1975. *Bioeconomic relationships for the Maine lobster fishery with consideration of alternative management schemes.* National Oceanic and Atmospheric Administration technical report, NMFS SSRF-683.

Food and Agriculture Organization. 1973. *Yearbook of fishery statistics.* Vol. 36.

Gates, John M., and Norton, Virgil J. 1974. *The benefits of fishery regulation: a case study of the New England yellowtail flounder fishery.* University of Rhode Island Marine Technology Report no. 21.

Gertenbach, L.P.D. 1973. License limitation regulations: the South African system. *Journal of the Fisheries Research Board of Canada* 30:2077-2084.

Gordon, H. Scott. 1954. The economic theory of a common property resource: the fishery. *Journal of Political Economy* 62:124-142.

——. 1958. Economics and the conservation question. *Journal of Law and Economics* 1:110-121.

Gulland, J. A. 1971. *The fish resources of the ocean.* London: Fishing News (Books), Ltd.

——. 1974. *The management of marine fisheries.* Bristol: Scientechnica.

Hardin, Garrett. 1968. The tragedy of the commons. *Science* 162:1243-1248.

Huq, A. M., and Hasey, Harlin I. 1973. Socioeconomic impact of changes in the harvesting labor force in the Maine lobster fishery. National Marine Fisheries Service file manuscript no. 142.

Joseph, James. 1973. Scientific management of the world stocks of tunas, billfishes, and related species. *Journal of the Fisheries Research Board of Canada* 30:2471-2482.

Kent, H.S.K. 1954. The historical origin of the three-mile limit. *American Journal of International Law* 48:537-553.

Meadows, Donella H.; Meadows, Dennis L.; Randers, Jorgen; and Behrens, William W., III. 1972. *The limits to growth.* New York: Universe Books.

Moiseev, P. A. 1969. *The living resources of the ocean.* Translated for the National Marine Fisheries Service.

National Marine Fisheries Service. 1976. *National plan for marine fisheries.* Final draft, October 1975.

Noetzel, Bruno, and Vondruska, John. 1975. *Future investment in U.S. fish harvesting and processing—a discussion of possible alternative requirements through 1985.* National Marine Fisheries Service.

Penn, Erwin S. 1975. Cost analyses of fish price margins, 1972-74, at different functional levels—for management decisions on production, distribution, and pricing policies. National Marine Fisheries Service unpublished manuscript.

Reeves, J. 1969. The lobster industry: its operation, financing and economics. Master's thesis, Stonier Graduate School of Banking, Rutgers University.

Robinson, M. A. 1973. Determinants of demand for fish and their effects upon resources. *Journal of the Fisheries Research Board of Canada* 30:2051-2058.

Robinson, M. A., and Crispoldi, A. 1971. *The demand for fish to 1980.* Rome: Food and Agriculture Organization.

Rockwood, Charles E., et al. 1973. A management program for the oyster resource in Apalachicola Bay, Florida. Florida State University unpublished report.

Rutherford, J. B.; Wilder, D. G.; and Frick, H. C. 1967. *An economic appraisal of the Canadian lobster fishery.* Fisheries Research Board of Canada bulletin 157.

Saila, Saul B., and Norton, Virgil J. 1974. *Tuna—a summary of current status, expected trends, and alternative management arrangements.* RFF Program of International Studies of Fisheries Arrangements. Washington, D.C.: Resources for the Future.

Schaefer, Milner B. 1957. A study of the dynamics of the fishery for yellowfin tuna in the eastern tropical Pacific Ocean. *Inter-American Tropical Tuna Commission Bulletin* 2:245-285.

Scott, Anthony D. 1955. The fishery: the objectives of sole ownership. *Journal of Political Economy* 63:116-124.

Sinclair, S. 1960. *License limitations—British Columbia: a method of economic fishery management.* Ottawa: Canadian Department of Fisheries.

Smith, Adam. 1776. *Inquiry in the nature and cause of the wealth of nations.*

Sokoloski, Adam. 1970. A digest of state commercial fisheries laws in the United States. NOAA/NMFS file manuscript no. 112.

Sysoev, N. P. 1970. *Economics of the Soviet fishing industry.* Translated for the National Marine Fisheries Service.

U.S., Congress, House. 1975. *Marine fisheries conservation act.* Report no. 94-445.

Wahle, Roy J.; Vreeland, Robert R.; and Lander, Robert H. 1974. Bioeconomic contribution of Columbia River hatchery Coho salmon, 1965 and 1966 broods, to the Pacific salmon fisheries. *Fishery Bulletin* NMFS/NOAA 72:139-169.

Wilson, James A. 1973. Economic aspects of fishery management—the northern inshore lobster fishery. Contract to National Marine Fisheries Service no. N-043-30-72.

5. *Environmental Deterioration and Fishery Resources*

The problem of environmental degradation—air and water pollution, solid wastes, threats to wildlife, the destruction of natural resources—is often summed up in the phrase *the environmental crisis.* This chapter is concerned with one aspect of the environmental crisis, the quality of water.

The visual effects of water pollution are obvious enough to all of us. Shellfish areas are closed to fishing because of a danger to public health. Fish kills are numerous and are directly attributable to various pollutants. The *immediate* cause of water pollution also seems generally apparent—it is no farther away than the discharge pipes of the nearest industrial or sewage treatment plant. Industrial and population growth and urban concentrations have made great demands upon water resources. In turn, deterioration in the quality of water has reduced fishing productivity and harvestability. These reductions in both commercial and recreational resources have indirect regional economic impacts as well. Why has all this happened?

Why Pollution Occurs

The answer as to why pollution occurs may seem quite simple. We have just mentioned the *immediate* cause of water pollution—industrial and population growth have placed greater demands on our water resources. However, the real cause of water pollution is much more involved. Pollution means simply that certain materials accumulate where they are harmful or simply undesirable. It occurs because of (1) the existence of common property resources, and (2) the use of these resources as free goods. This should sound familiar. Our discussion in chapter 4 on the fisheries as a common property resource was instrumental in explaining overfishing and technological externalities. In fact, the fisheries literature has been the forerunner of much of the modern theory of pollution. Economists have long recognized that the private market system often produces harmful spillover effects on man's environment. That is, the technical nature of the production process requires the use of such common property resources as air and water. For example, a chemical plant may use water in a production process that results in polluted water being discharged back into the ecosystem; or it may use the water as a way to dispose of pollutants. This is called a "spillover effect," and the result has been deteriorating environmental quality. "Stop the polluters," shout the environmentalists! We can't swim in our lakes, go boating in our rivers, or catch fish in our oceans because of these spillovers. Like fish, the air and water are common property resources. They are *owned* by neither the chemical plant nor the environmentalist. As noted in chapter 4, we have a market failure. Thus, the fundamental cause of pollution is really our system of property rights.

The essence of the problem can be grasped by a very realistic example. Estuaries serve as nursing grounds for shrimp. These estuaries are sensitive ecosystems that are supplied with water from rivers. Suppose a shrimp industry develops rapidly and is based upon the continuing viability of the estuaries. As indicated in chapter 2, the demand for shrimp is placing increasing demands on the shrimp resource. However, the rivers leading into the estuaries are common property resources. There is also a strong demand for industrial chemicals, some of which may be used either indirectly (e.g., for vessel construction) or directly (e.g., paint remover) by the shrimp industry. Suppose Zoopont Chemicals, Inc., decides to locate its plant on a river twenty miles upstream from the estuaries. To produce the chemicals, it discharges large amounts of waste phenols, arsenic, fluorides, and other toxic substances into the river. The river water is free to the chemical plant. Its owners pay nothing for the use of the water even though it is essential to the production process. These toxic chemical discharges into the water find their way into the estuaries and have an adverse impact on the shrimp nursing grounds. As a consequence, the shrimp industry's productivity, or catch per unit of fishing effort, falls. This is called an *external diseconomy*, since the actions of the chemical plant have a negative (it lessens productivity) impact on the shrimpers. Figure 5.1 shows the marginal private cost

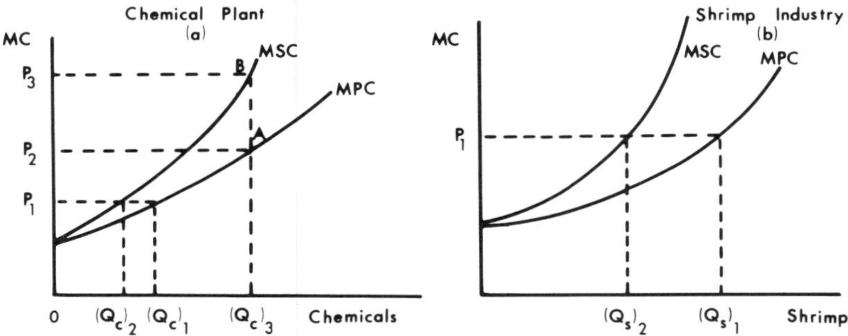

Figure 5.1: Impact of an External Diseconomy on the Shrimp Industry

(*MPC*) of production for the chemical plant and shrimp industry (see chapter 3). The *MPC* shown in panel (b) for the shrimp industry is the cost incurred *before* the operation of the chemical plant. We have labeled this *private* cost, since it is paid by shrimp vessel owners for each resource they use (labor, vessel, gear, gas) in the process of catching shrimp.[1] $(Q_S)_1$ is produced, *assuming the fishery is managed according to marginal cost pricing* (see chapter 3). Chemical pollutants are now introduced as an externality and have the effect of making it costlier to produce shrimp at each level of production (i.e., the productivity of the shrimp resource has declined). The *MPC* curve will shift up and to the left. But have we labeled it incorrectly? What is *MSC? MSC* is marginal *social* cost. What is the difference? If you think back, you will remember that the chemical plant did not pay for the use of the water. It certainly paid for the labor, insurance, machinery, and raw materials necessary to the production process. Well, water is necessary to the production process!

Indirectly, the shrimp industry is forced by the circumstances to bear the cost of producing chemicals. The external impact raised their cost of production, so we call the new curve marginal social cost. Why social? From society's standpoint, the shrimp industry is forced to bear part of the cost of producing chemicals. The net result is that the chemical industry produces too much—because obviously it does not pay for the free resource, and its costs are therefore lower—and the shrimp industry produces too little. An economist would say that the chemical industry should be made to "internalize" the full cost (private and social) of producing chemicals. That is, the *MPC* of producing chemicals understates the full cost of their production. For example, if the chemical plant were forced to install pollution abatement equipment to reduce the effluent discharges, the real cost (social cost) of chemicals would increase. Thus, we can conceive of another marginal cost curve that includes all the costs of chemical production, both private and external. This alternative would be called the marginal social cost of producing chemicals.

In figure 5.1 the vertical difference between the *MPC* and *MSC* for the chemical plant at any given quantity measures the external cost per extra unit of output. For example, the *MPC* of producing the $(Q_C)_3$ unit of chemicals is P_2 dollars, but the *MSC* of this unit is P_3 dollars. The difference, $P_3 - P_2$ dollars, is the marginal external cost. We have drawn the marginal cost curves for the chemical plant so that the external cost increases with production, since it is more realistic to assume that proportional increases in production (and hence chemical toxins discharged into the water) increase water pollution with a more than proportional damage to the productivity of shrimp nursing grounds. The limit of such a case would be the level of chemical production at which the chemical toxins destroyed the entire shrimp fishery.[2]

Unless some action is taken, the chemical plant will continue to produce $(Q_C)_1$ units of output at a price of P_1 for chemicals; after all, the water is free. The shrimp industry will produce $(Q_S)_2$ units of output, something less than that based upon its *MPC*.

When no charge is made for use of the common property of an ecosystem, firms, municipalities, and individuals can reduce their own expenditures by using the air and water as a receptacle for waste. From a social point of view, there is too large an output of commodities that are produced without meeting all costs. In part, this explains why affluent nations are well supplied with chemicals, plastics, cans, and plumbing while coastal fishery production is on the decline. Now that we know why pollution occurs, what can be done about it? What are our options? We shall look at the solutions, theoretical and actual, later in this chapter. However, we should first investigate the interaction between various classes of pollutants and water resources.

The Impact of Waterborne Pollutants

Degradable Pollutants

At the outset, we should make the critical distinction between discharges of *degradable* and *nondegradable* materials. A widespread degradable discharge is domestic sewage; however, industry (e.g., pulp and paper mills and food processing plants) produces greater amounts of degradable organic effluents. Once a degradable substance is discharged into a stream of "clean water," *aerobic degradation* begins. In so-called clean water, certain bacteria are always present, bacteria that feed on degradable waste, breaking it down into *inorganic* forms of nitrogen, phosphorous, and others. Two things happen in this process: (1) some of the oxygen dissolved in the "clean" water is used by the bacteria, and (2) the inorganic material thus created becomes fertilizer (i.e., phosphorous) for plant life. The ecosystem will normally be able to handle reasonable amounts of organic (biodegradable) discharges through reoxygenation by the air and through photosynthesis by plant life in the water. But here is the rub! If the degradable organic waste discharged into the water becomes excessive (if the ecosystem is overloaded), the process of degradation may exhaust the dissolved oxygen. High temperatures accelerate degradation. They also decrease the saturation level of dissolved oxygen in the water. This is one of the effects of thermal pollution—from a nuclear power facility, for example. Thus, degradable effluents place certain oxygen demands on the water. A conventional sewage treatment plant mimics this process. Under usual circumstances, treatment plants are capable of reducing the *BOD* (biochemical oxygen demand) in waste effluents by 90 percent.

Stretches of water that persistently carry less than 4 or 5 ppm (parts per million) of oxygen will not support the higher forms of fish life. Remember that in the process of degradation, fertilizer for plant life is created. Up to a certain point, algae growth because of the free fertilizer is not harmful and may even increase fishery productivity. This would be a good description of the "carrying capacity" of the ecosystem. But what if the biodegradable effluents exceed the carrying capacity? First, plant growth becomes excessive. We have "algae bloom"! Second, the dissolved oxygen decreases. Fish and animal life are endangered; after all, fish require oxygen in the water. This whole process is really the death of a body of water and is called *eutrophication.* A good example is Lake Erie. Discharges are measured in terms of *BOD* to assess the impact on the ecosystem.

Nondegradable Pollutants

Many pollutants cannot be broken down chemically by the bacteria in the water. The water body, or ecosystem, cannot "purify itself" of nondegradable pollutants. Such *inorganic* substances as DDT, detergents, and heavy metals (i.e., mercury, lead, zinc) are good examples, as is the pesticide, endrin, which is toxic to fish even in minute concentrations. The public health problem centers on the *chronic* effects of prolonged exposure to very low concentrations. DDT and mercury, for example, get into fish flesh. Above certain concentrations (see below), fish containing these substances are banned from sale for reasons of public health. Radionuclides are persistent nonbiodegradable effluents.

As Barry Commoner states (of nondegradable pollutants), "the new technology has an appreciably greater environmental impact than the technology which it has displaced, and the postwar technological transformation of productive activities is the chief reason for the environmental crisis" (1972, p. 271). Commoner says, "the growth pattern has been counter-ecological." Figure 5.2 shows the increase in the world's use of heavy metals, the U.S. release of nuclear waste, the global production of crude oil, and the increase in required pesticides for Africa, Latin America, and

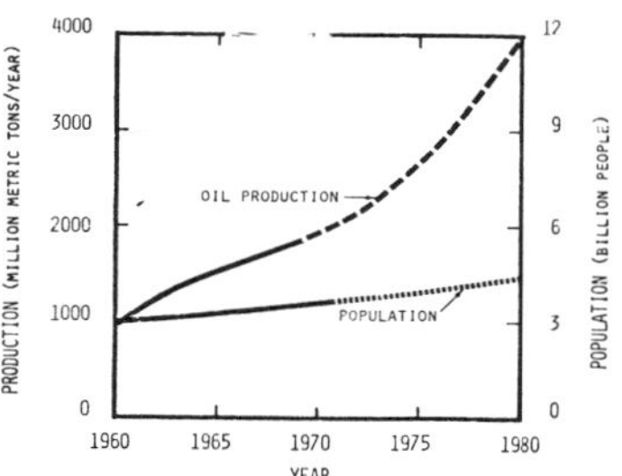

Actual and projected global crude oil production and human population, 1960–1980

Sources: Oil production data from SCEP 1970, p. 266; population data from U.N. 1969, p. xxvii.

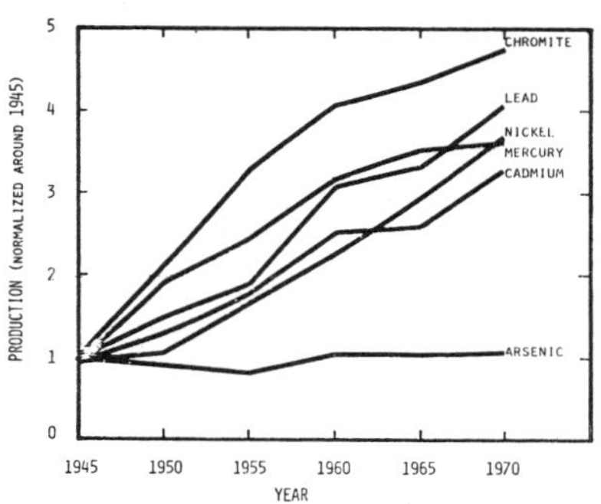

Growth in the global production of six toxic heavy metals, 1945–1970

Source: Production data from *Commodity Yearbook* 1950–1970.

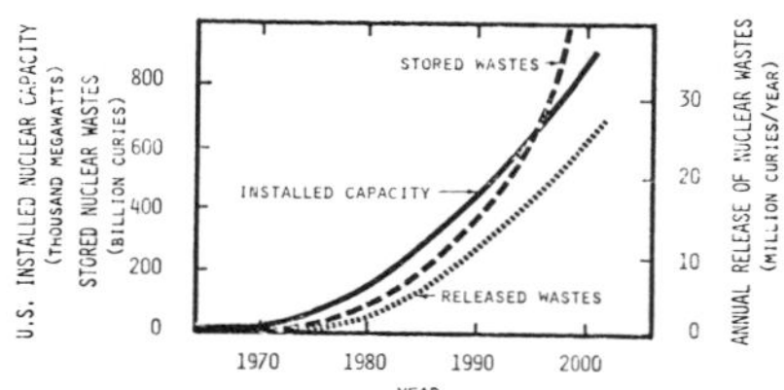

Projected generation of radioactive wastes from the operation of U.S. nuclear power plants, 1970–2000

Sources: Installed capacity to 1975 from AEC 1971; installed capacity to 2000 from Starr 1971; stored nuclear wastes calculated from specifications for 1.6-thousand megawatt plant in Calvert Cliffs, Maryland.

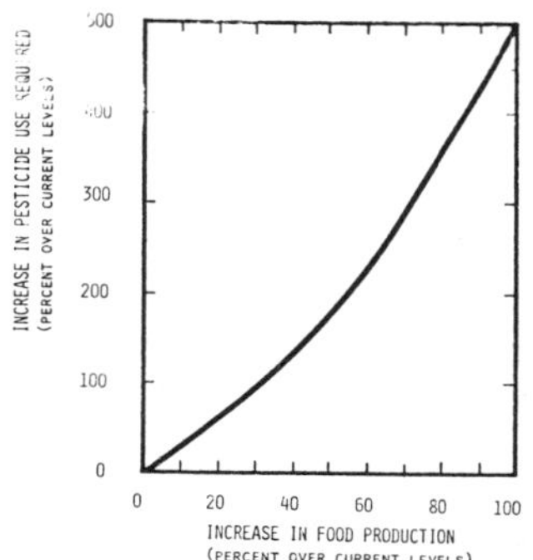

Pesticides required to increase food production on land now under cultivation in Africa, Latin America, and Asia (except mainland China and Japan)

Source: Data from SCEP 1970, p. 282.

Figure 5.2: Trends in Potential Pollutants Impacting Marine Life
Source: Taken from Meadows et al. (1974)

Asia. Typically, there is a delay between the moment that persistent pollutants such as DDT, PCB (polychlorobiphenyl), mercury, and radioactive waste are released into the environment and the moment they have negative effects on the ecosystem. Now let us look at some of the economic losses resulting from the impact of water pollution on U.S. and foreign fisheries.

The Effect of Pollution on Fishery Resources: The United States Experience

The purpose of this section is to evaluate, with the fragmentary data available, the historical record on the impact of pollution, primarily on U.S. coastal fisheries.[3] Coastal fisheries are used for both commercial and recreational purposes; therefore, pollution affects both sectors.

Published estimates of *commercial* fishery losses from pollution center on localized problems, although there are at least three national estimates. Most damage estimates deal with landing revenue lost to fishermen rather than retail value. Figure 5.3 indicates the dollar value and geographic area of many published estimates for 1970 (Tihansky, 1973). Other estimates illustrate the wide range of effects that waste discharges have on a number of marine species ranging from clams to salmon (Howard 1973; Council on Environmental Quality 1970; Federal Water Pollution Control Administration 1970). Over the past sixty-five years, environmental changes in Connecticut have resulted in a loss in shellfish revenue of $1 billion (Federal Water Quality Administration 1970). Recently, it was estimated that the fishery economic losses from water pollution in the area near Pensacola, Florida, were over $3 million (not shown in figure 5.3) for 1972 (Terrebonne 1973). Furthermore, the initial effects—lost fishermen's wages—are multiplied throughout the economy.

National shellfish loss estimates are based on the proportion of shellfishing areas closed because of pollution. The National Marine Fisheries Service estimated losses of $12 million in 1970 (Bale 1971). This estimate assumes that only clams and oysters are affected, since they are immobile and harvested primarily within bays and estuaries. The Council on Environmental Quality (1970), on the other hand, assumes that all shellfish, including lobsters, shrimp, and crabs, are affected by contamination. Its 1970 estimate was $63 million, a figure based on the closure of one-fifth of the nation's shellfish beds and a corresponding loss of potential revenue. It should be stressed that this brief review of fisheries

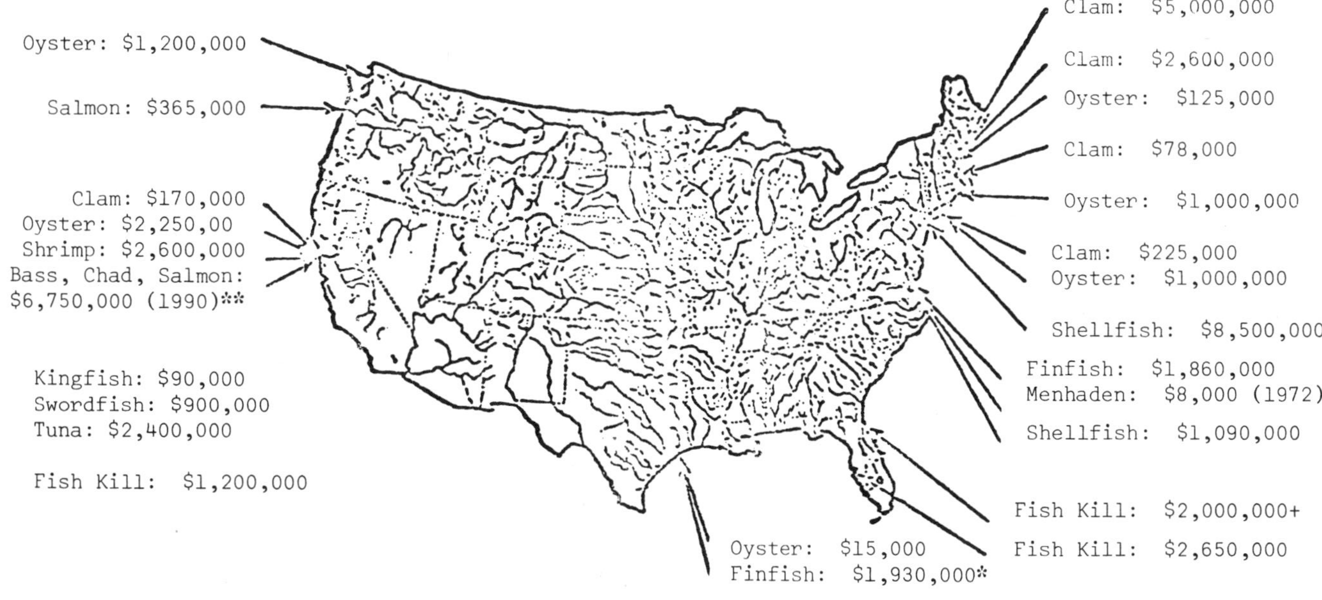

Figure 5.3: Commercial Fishery Revenue Loses From Marine Water Pollution in the United States, 1970

Source: Tihansky (1973)

literature reveals that studies deal with a variety of species in a variety of regions. In many cases, the losses were only temporary. In addition, techniques to measure losses differ and, of course, not all areas were systematically studied. Finally, Tihansky (1973) estimated that for 1970, commercial fishery losses could be as much as $37.5 million.

We shall update these analyses and focus both upon specific coastal areas and particular species. In the following sections, we shall (1) identify the main sources of recreational and commercial fishery economic losses from pollution, (2) evaluate the impact of pollution on the marketability of shellfish, (3) evaluate the impact of pollution on the marketability of finfish, (4) assess the relation between pollution and lethal, sublethal, and extinction impacts on fish stocks (both finfish and shellfish), and (5) discuss the incidence of water pollution and its effect on areas that potentially could have been used for aquaculture. That is, economic losses from actual lost harvests and potential harvests of shellfish and finfish will be roughly assessed. Let us first look at the *sources* of economic losses that the U.S. coastal fishery resources may suffer from water pollution.

Potential Sources of Economic Losses from Pollution

In this section, we shall identify the circumstances in which pollution reduces the harvest of fishery resources (i.e., commercial harvest and the harvest available to the recreational fishermen).

Reduced Marketability. If shellfish harvest areas have excessive coliform counts (a measure of the presence of pathogenic bacteria), they are legally closed. That is, even though the fish or shellfish are probably not biologically affected at some levels of water or species contamination, shellfish from these areas are rendered unsuitable for human consumption because of the bacteria they contain. The biomass for shellfish with excessive bacterial levels may actually be growing, but it is not legally available for harvest. Finfish may also be contaminated by mercury salts, DDT, and PCDs. In this case, the finfish in question is usually banned from sale or seized by the appropriate authorities, such as the Food and Drug Administration. In terms of data sources, we must distinguish between shellfish and finfish, because they are controlled or regulated as to their safety by different government agencies.

Shellfish are relatively immobile and are under the control of state departments of health. The U.S. Public Health Service periodically publishes a *National Shellfish Register,* which shows areas closed owing to pollution (the Environmental Protection Agency now publishes it). No species breakdown is available from this publication, although the closed areas probably contain mainly oysters and clams.

Finfish are more mobile, and control by individual states is more difficult. Moreover, finfish in general spend more of their lives in the open sea, where pollution may be less severe, as contrasted to shellfish, which are relatively immobile and harvested near shore, where bacterial organisms are more localized. The finfish themselves may also have been relatively less affected by the contaminant(s), such as mercury salts, they are carrying. However, once government health guidelines for human consumption are exceeded, the finfish are excluded from the national fish harvest. The Federal Drug Administration has usually carried out this banning function.

Reduced Biological Productivity. Pollution may *directly* influence both finfish or shellfish by *outright mortality* or *reduced capability of biomass growth* (sublethal effects) or both. Both factors obviously influence catch rates or the harvest size and thereby have a profound economic impact in terms of lost protein. Statistics on the subject are scarce. The Federal Water Quality Administration collects data on pollution-caused fish kills. However, there is no published breakdown by species. In addition, biologists are presently working on the question of reduced capability of biomass growth, but these studies must be tabulated and translated into catch impacts. This has not been done on any systematic basis, and we cannot therefore draw definitive historical conclusions. In fact, the National Commission on Water Quality and other agencies have commissioned such studies. However, we shall discuss the question of reduced biological productivity for both finfish and shellfish *given available historical data.*

Reduced Opportunities for Protein Production: Aquaculture. In contrast to world catches of fish from the world's wild stock, aquaculture has been growing throughout the world at a phenomenal rate. In this country, most commercially aquacultured species are *freshwater* species, such as minnows, catfish, trout, and crayfish. In chapter 8, we will discuss water quality as a deterrent to other forms of potential aquaculture, such as shrimp, pompano, salmon, redfish, abalone, lobster, scallop, and mullet. These species are presently in the experimental stage for U.S. coastal areas, and it is not generally believed that water quality has significantly influenced economic feasibility. Over the next ten years, however, water quality may enter as a significant variable either affecting marketability or biological productivity.

Shellfish Marketability: Trends in Closures

Part of the growing concern over pollution has been a concern about the injuries that rising levels of contaminants have inflicted on the public health and the commercial fishing industry. Contaminants in rivers, estuaries, and coastal waters could threaten the human use—indeed, the very existence—of fish resources. The contaminants that appear in water systems may be divided into three broad classifications: heavy metals and chemical compounds, pesticides, and bacterial organisms. Other major types of pollution, which apparently have a relatively localized effect, are thermal pollution and the dumping of radioactive substances.

Commercial shellfish areas are most often closed because of bacterial pollution. Improperly treated sewage dumped into rivers deposits enteric viruses along with the sewage on coastal and estuarine bottoms. Certain diseases, especially infectious hepatitis, have broken out when raw shellfish contaminated by such sewage deposits have been eaten. A positive correlation has been established between a high coliform bacterial count in water and such infection of shellfish. These health problems dictate shellfish closure by public health authorities if there is a high coliform count. All states have adopted the National Shellfish Sanitation Program Manual of Operations (1965), which states that an area is classified as prohibited if a sanitation survey indicates any of the following:

1. The area is so contaminated with fecal material that consumption of the shellfish might be hazardous;
2. The area is so contaminated with radionuclides or industrial waste that consumption of the shellfish might be hazardous; or
3. The median coliform MPN (Mean Probable Number, or number of coliform organisms per sample) of the water exceeds 70 per 100 ml, or more than 10 percent of the samples ordinarily exceed an MPN of 230 per 100 ml for a five-tube decimal dilution test, in those portions of the area most probably exposed to fecal contamination during the most unfavorable hydrographic and pollution conditions.

Those commercial shellfish areas closed by public health authorities are reported by state and EPA regions for 1966 and 1971 in table 5.1. These figures were taken from the *National Shellfish Register* (*NSR*). The "conditional" closings are those areas closed during the summer, when bacterial levels are at their peaks. These areas are fished later in the year and hence are not considered "closed" for the purposes of our analysis.

According to the *NSR,* there was an upward trend in closings in all regions between 1966 and 1971 and in all states except New York and Mississippi, which experienced declines in acreage closed. A total shellfish area of 2,001,500 acres was "closed" in 1966, and 3,297,991 acres were closed in 1971, an increase of 61 percent.

Although many species of estuarine shellfish are affected by water pollutants, clams and oysters are probably the most affected by shellfish closures. Because of their feeding mechanism, clams and oysters are particularly susceptible to pollutants. All bivalves are filter feeders—i.e., capable of filtering fine materials from water for food. In a contaminated environment, a filter feeder is exposed to large amounts of pollutants because of the huge quantity of water it filters.

In order to derive an estimate of the economic value associated with the closings, the quantities, values, and prices, of clams, oysters, and "other" shellfish were collected by state. These data are reported in table 5.2. The first two columns show for the twenty-three coastal states the estimated pounds lost because of shellfish closure (based upon data published in the *National Shellfish Register*). Column three shows the pounds lost (based upon the independent survey by Bell and Canterbery [1976]). The term *designated by the state* indicates that shellfish and fishery administrators in each state were asked what species of fish are found in closed areas. They tended to mention oysters and clams most frequently. The estimated potential catch from closed areas increased from over 350 million pounds in 1966 to nearly 629 million pounds in 1971. As might be expected, oysters and clams were the species most often closed to harvesting. The Bell and Canterbery survey indicates no real change in potential catch from 1971 to 1975, due to the fact that closed areas did not increase over the 1971–1975 period. The potential value of these unharvestable resources did increase, from approximately $28 million to $38 million. These estimates are conservative: Bell and Canterbery found that other species such as lobsters and crabs were being caught in open areas adjacent to the closed areas. Therefore, they thought it reasonable to assume that closed areas will not only

Table 5.1: Changes in Acreage of Shellfish Waters, 1966-71

STATE	OPEN		CONDITIONAL		CLOSED		RESTRICTED	TOTAL		% SURVEYED FOR SHELLFISH	
	1966	1971	1966	1971	1966	1971	1971	1966	1971	1966	1971
ME	291,600	942,501		6,723	61,300	88,913	6,728	352,900	1,044,865	33.8	100.0
MA	25,100	307,533	6,100	220	7,900	32,658	4,255	39,100	344,666	4.9	43.0
RI	64,800	97,551	11,800	10,846	19,700	18,464		96,300	126,861	75.9	100.0
CT	67,100	270,146	1,300	1,796	14,500	46,557		82,900	318,499	21.2	81.3
TOTAL	448,600	1,617,731	19,200	19,585	103,400	186,592	10,983	571,200	1,834,891	24.1	77.3
NY	820,300	477,279		266	163,800	151,058		984,100	628,603	96.4	61.6
NJ	293,500*	255,611	4,100	5,354	63,500	114,364	19,399	361,100*	394,728	91.5	100.0
TOTAL	1,113,800	732,890	4,100	5,620	227,300	265,422	19,399	1,345,200	1,023,331	95.0	72.3
DE	209,700	205,153		153	6,200	28,251		215,900	233,557	78.8	85.2
MD	1,122,400	1,236,735		101,499	75,200	118,764		1,197,600	1,456,998	78.2	95.2
VA	1,352,600	1,352,505	1,100	724	58,200	82,975		1,411,900	1,436,204	93.9	95.5
TOTAL	2,684,700	2,794,393	1,100	102,376	139,600	229,990		2,825,400	3,126,759	85.4	94.5
NC	927,700	1,464,448			45,200	519,153		972,900	1,983,601	45.9	93.6
SC	133,500	198,237	100	347	49,500	76,664		183,100	275,248	66.5	100.0
GE	119,100	123,528			44,700	80,439		163,800	203,967	80.3	100.0
FL	606,600	663,834		83,334	797,100	1,025,023		1,403,700	1,772,191	61.9	78.2
AL	147,500	81,937	700	187,484	256,700	85,618		404,900	355,039	108.0	94.7
MS	27,500	76,232	8,500		87,300	32,842		123,300	109,074	31.6	27.9
TOTAL	1,961,900	2,608,216	9,300	271,165	1,280,500	1,819.739		3,251,700	4,699,120	57.8	83.5
LA	997,800	1,584,384	28,600		5,900	198,812		1,032,300	1,783,196	29.3	50.6
TX	683,200	820,043	19,600	11,251	231,200	275,653		934,000	1,106,947	57.3	67.9
TOTAL	1,681,000	2,404,427	48,200	11,251	237,100	474,465		1,966,300	2,890,143	38.1	56.1
CA	2,100	14,747	3,000		3,400	264,154		8,500	278,901	1.6	53.1
OR	1,600	12,323	3,200		8,500	15,766		13,300	28,089	15.9	33.6
WA	43,200	177,263	200	224	1,700	41,863		45,100	219,350	2.2	10.8
TOTAL	46,900	204,333	6,400	224	13,600	321,783		66,900	526,340	2.5	19.9
GRAND TOTAL	7,936,900	10,361,990	88,300	410,221	2,001,500	3,297,991	30,382	10,026,700*	14,100,584	48.8	68.7

*New Jersey data adjusted due to publication error in 1966 register

Source: National Shellfish Register, (1971)

Table 5.2: Estimated Potential Catch and Value Harvestable From Closed Shellfish Areas for Twenty-three Coastal States, 1966, 1971, 1975*

Species	1966 NSR Poundage Loss at 1966 Yields	1971 NSR Poundage Loss at 1971 Yields	1975 FSU Poundage Loss at 1971 Yields	1966 Value Lost (Ex Vessel $)	1971 Value Lost (Ex Vessel $)	1975 Value Lost (Ex Vessel $)
	---Designated by the State---					
Clams	80,839,652	146,645,600	132,849,220	3,453,699	10,380,749	10,536,317
Oysters	270,595,310	479,235,890	471,695,891	7,002,932	17,545,482	24,047,246
Crabs	46,610	1,550,734	4,522,619	3,729	170,581	2,984,929
Lobsters	0	0	0	0	0	0
Mussels	176,663	925,234	852,600	3,688	31,722	30,206
Scallops	0	287,094	0	0	0	0
Total	351,658,235	628,644,552	609,920,419	10,464,048	28,128,534	37,598,698

*Average yields were obtained by dividing landings of each species for each respective year by the area open (acres) to shellfish harvesting in that year. The average yeilds for each species multiplied by the area closed to shellfish harvesting gives the poundage loss for that species. Poundage loss was multiplied by the unit value for each species to arrive at value lost. Unit values were obtained by dividing the value of landings by the quantity landed for each species for each year. The reader should note that the total water area available for shellfish production according to the NSR may be considerably less than the actual area occupied by shellfish, especially lobsters and crabs which are somewhat mobile. Therefore, when catch of these species are divided by NSR shellfish areas, yields (i.e., catch per acre) may be biased upward and hence value.

Source: Bell and Canterbery (1975)

limit sedentary fish production, but also the production of fish that are more mobile but may be contained in closed areas. In fact, there is some evidence that areas with high coliform counts may actually attract lobsters and crabs; therefore, Bell and Canterbery felt that these estimates should be given in order to establish an upper boundary of loss. Table 5.3 shows the estimated losses for those species designated by the state and other potential species for 1975. Estimated losses climb from $38 mil-

Table 5.3: Estimated Potential Catch and Value From Closed Shellfish Areas by Species for Twenty-three Coastal States, 1975

Species	1975 Catch Loss Using 1975 Yields* (pounds) (round weight)	1975 Value of Catch (dollars)
(1) Designated by State		
clams	132,849,220	$10,536,317
oysters	471,695,980	24,047,246
crabs	4,522,619	2,984,929
lobsters	0	0
mussels	852,600	30,206
scallops	0	0
Subtotal	609,920,419	$37,598,698
(2) Other Potential Species		
clams	1,976,130	156,728
oysters	9,608	490
crabs	57,338,615	35,691,837
lobsters	12,459,977	19,063,764
mussels	1,958,171	69,375
scallops	5,595,900	1,228,498
Subtotal	79,338,401	56,210,692
(3) Total	689,258,820	93,809,390

SOURCE: Bell and Canterbery (1973)

lion (lower boundary) to nearly $94 million (upper boundary) when all possible shellfish are considered. Although the increase in poundage was not great with addition of the "other potential species" category, such species as lobsters and crabs have a high unit value.

For 1975, the following states suffered the greatest loss because of closed areas for those species *designated by state officials:*

	Loss (in millions)
1. California	$ 7.3
2. Louisiana	6.0
3. New Jersey	6.0
4. Florida	2.5
Total	$21.8

These four states alone account for over 57 percent of estimated total national losses.

Finfish Marketability: FDA Actions

In the past several years there has been increased concern that the content of mercury, other heavy metals, and pesticides in fish will have deleterious effects on humans. For example, the Food and Drug Administration has set the standard for seizing mercury-contaminated fish at 0.5 ppm. This poses serious problems, since mercury has important industrial and agricultural uses, namely, in the production of chlorine and alkali compounds and as an element in fungicides.

Mercury enters fish primarily through the food chain and is deposited in the fatty tissues in amounts that vary directly with the size of fish. Consequently, fish having large amounts of fat within the muscles, such as mackerel, tuna, and swordfish, tend to accumulate large quantities of mercury. While data on the economic losses incurred owing to contamination are scant and incomplete, an indication of the potential contamination problem is given in Bale (1971) and Bell et al. (1970). The species for which estimates of losses are made are listed in table 5.4. With the exception of *swordfish, tuna,* and *mackerel,* the values represent potential income loss if the FDA were to ban the harvesting of such species.

The FDA totally banned landings of swordfish in 1970, which caused the market practically to disappear. According to Bale (1971), this ban on landings resulted in an economic loss of $500,000. Tihansky (1973), on the other hand, estimates the

Table 5.4: Estimated and Potential Economic Losses Due to Sub-Lethal Contamination

Species	Cause	1970 Landed Value (mil. $)
Swordfish*	Mercury	0.5
Anchovies	DDT	1.4
Bonito	Mercury	0.8
Bluefin, Yellowfin, Albacore Tuna*	Mercury	3.43
Jack Mackerel*	DDT	.32
Pacific Halibut	Mercury	8.0
Red Snapper	Mercury	4.0
Atlantic Mackerel	Mercury	0.4

*estimated loss due to FDA ban
Source: Harvey E. Bale, Jr. (1971) and Bell (1970)

1970 monetary loss for swordfish in the Pacific Ocean alone at $900,000. Due to DDT contamination, landings of mackerel were curtailed for a two-month period in 1969. Bell et al. (1970) have evaluated this loss in revenue at $320,000. Between December 1970 and February 1971, the FDA threatened seizure of tuna (mainly yellowfin). The National Canning Association has informed the writer that the manufacturers undertook a survey and found that 180,400 cases of tuna exceeded the FDA standard. If these cases are valued at the Cannery Association average price of $19.00 each, the cost of this particular recall—due to excessive mercury content—was $3,427,600. An FDA release in February 1971 indicated that 3.6 percent of all imported tuna exceeded mercury guidelines. Since there is no way to estimate the number of cans marked with a particular code, the monetary loss in this instance cannot be calculated. In summary, the only actual losses for finfish because of unsuitability for market (i.e., FDA actions) are an estimated $4.25 million for swordfish, tuna, and jack mackerel.

Reduced Biological Productivity

Direct Mortality: Fish Kills. One of the major sources of data on fish kills in the United States is the *Fish Kills Caused by*

Pollution series published by EPA's Office of Water Planning and Standards. Although this series does indeed contain useful data, it is inherently difficult to analyze.

There have been thirteen bulletins in the series, with state cooperation apparently increasing during the 1960–1972 period. The low point of reporting was in 1962, when thirty-seven states submitted data; in 1972 all fifty states responded. Table 5.5 presents a history of pollution-caused fish kills for the period from June 1960 to December 1972.

It is obvious from the total reported number of fish killed that no clear trend is evident for this period. Consider some recent evidence. In 1970, over 22 million fish were reported killed, but in 1971 this number tripled. By 1972, however, the total number of fish killed had dropped substantially—to about 18 million.

In order to verify that fish kills have been random over the 1960–1972 period, we carried out some statistical analyses. The hypothesis for this analysis is that fish kills occur randomly. The alternative hypothesis is that fish kills are on the increase and are the result of chronic water pollution. To test these hypotheses, a linear trend for each coastal state was fitted to published (EPA) fish kill data for the years 1960–1972.[4] The results indicated that for twenty-one of the twenty-three coastal states, fish kills are primarily random and are probably due to specific accidents in the majority of cases.

We should note, however, that all of the data in the fish kill bulletins must be interpreted with the utmost caution. Several criticisms are necessary here. First, only aggregate fish kills are reported; freshwater and saltwater fish are lumped together. Second, no consideration is given to regional factors, except for the most recent year (1972). Third, no distinction is made among the types of fish killed. Low-value fish are weighted equally with high-value fish when considered on this aggregate level. Perhaps the most devastating criticism is the fact that the reports do not always distinguish between unpredictable environmental accidents and chronic and identifiable "point source" and "non point source" pollutants.[5] The latter, in most cases, are harder to identify and are usually omitted altogether.

In order to supplement the gaps in fish kill data, Bell and Canterbery (1976) used extensive interviews and survey questionnaires. The water quality administrators for the twenty-three coastal states were surveyed, and their responses provide a more detailed analysis of the entire fish kill problem.

Table 5.5: Historical Summary of Pollution-Caused Fish Kills
June 1960-December 1972

	1960[2]	1961	1962	1963	1964	1965	1966	1967	1968	1969	1970	1971	1972
Number of states responding	38	45	37	38	40	44	46	40	42	45	45	46	50
Number of reports	289	413	421	442	590	625	532	454	542	594	635	860	760
Reports which state number of fish killed	151	265	246	304	470	520	453	364	469	492	563	759	697
Total reported number of fish killed	6,035,000	14,910,000	44,001,000	6,937,000	22,914,000	12,140,000	9,614,000	11,291,000	15,815,000	41,166,000	22,290,000	73,670,000	17,717,000
Average size of kill[1]	2,925	6,535	5,710	7,775	5,490	4,310	5,620	6,460	6,015	5,860	6,412	6,154	4,639
Largest kill reported	5,000,000	5,387,000	3,180,000	2,000,000	7,887,000	3,000,000	1,000,000	6,549,000	4,029,000	26,527,000	3,240,000	5,500,000	2,922,000
Number of reported incidents for each pollution source operation													
Agricultural	79	74	51	84	131	114	88	87	77	117	108	132	113
Industrial	103	169	209	199	193	244	195	139	177	199	213	231	189
Municipal[3]	24	52	33	60	120	125	87	91	122	84	120	162	167
Transportation	0	0	1	17	26	27	27	23	39	32	28	52	56
Other	33	58	47	27	17	23	38	35	23	33	28	64	72
Unknown	50	60	80	55	103	92	97	79	104	129	138	219	163
Total reports	289	413	421	442	590	625	532	454	542	594	635	860	760

Number of reports and fish killed by size grouping	1960 No. reports	1960 No. fish (millions)	1961 No. reports	1961 No. fish (millions)	1962 No. reports	1962 No. fish (millions)	1963 No. reports	1963 No. fish (millions)	1964 No. reports	1964 No. fish (millions)	1965 No. reports	1965 No. fish (millions)	1966 No. reports	1966 No. fish (millions)
1,000,000 or more	1	5.0	4	12.6	2	41.0	1	2.0	5	16.9	3	5.4	2	2.0
100,000 to 1,000,000	3	0.53	5	0.85	9	1.69	12	2.68	15	3.82	17	4.62	23	5.48
10,000 to 100,000	15	0.31	45	1.05	38	1.01	54	1.82	59	1.65	63	1.42	58	1.53
1,000 to 10,000	64	0.18	107	0.34	89	0.30	134	0.41	167	0.49	202	0.59	185	0.55
0 to 1,000	68	0.02	104	0.03	108	0.03	103	0.03	224	0.07	235	0.07	185	0.05
No size reported for incident	138		148		175		138		120		105		79	
Average duration of kill in days		**2.95**		**2.64**		**2.59**		**3.18**		**2.44**		**2.57**		**2.71**

Number of reports and fish killed by size grouping	1967 No. reports	1967 No. fish (millions)	1968 No. reports	1968 No. fish (millions)	1969 No. reports	1969 No. fish (millions)	1970 No. reports	1970 No. fish (millions)	1971 No. reports	1971 No. fish (millions)	1972 No. reports	1972 No. fish (millions)
1,000,000 or more	1	6.5	3	6.1	4	35.1	5	11.4	28	63.0	6	8.97
100,000 to 1,000,000	7	2.66	30	7.44	9	3.15	26	7.44	26	6.37	27	5.43
10,000 to 100,000	49	1.58	64	1.79	81	2.06	91	2.73	124	3.33	81	2.60
1,000 to 10,000	143	0.46	153	0.48	165	0.52	198	0.62	266	0.86	216	0.62
0 to 1,000	164	0.05	219	0.06	233	0.06	243	0.07	315	0.10	367	0.09
No size reported for incident	90		73		102		72		101		63	
Average duration of kill in days		**3.34**		**2.99**		**3.11**		**3.25**		**3.35**		**3.40**

[1] Derived after excluding reports of 100,000 kills or more as being unrepresentative.
[2] Reporting system in effect for last six months of 1960.
[3] Municipal operations include electric power-generating stations.

Source: EPA (1972)

Of the twenty-three coastal states surveyed, detailed, usable answers were available for only thirteen. The rest had nothing material to add about fish kills. Three features were apparent: (1) municipal waste is a major source of water pollution in five of the thirteen states surveyed; (2) menhaden is the most frequently affected fishery; and (3) the nature of the other water pollutants was a function of the economic structure of the respective areas.

EPA data indicate three major sources of water pollution in induced fish kills for *all* states: industrial, municipal, and agricultural wastes. These data are generally consistent with the Bell and Canterbery survey, which indicates numerous fish kills associated with municipal sewage.

Table 5.6 indicates that for the first time since the annual report was started in 1960, more fish were reported killed in estuarine waters in 1971 than in fresh or salt water. The 1971 estuary fish kills also increased considerably over the two previous years, but declined in 1972. 1972 fish kills show a considerable decline over 1971; however, 42 percent were in salt or estuarine water.

The increase of fish killed in estuarine water may be a matter of great national concern, since estuaries serve as nursery grounds for many species of marine fish. Before interpreting this increase as a national trend, however, it should be noted that the large increase of fish kills in estuarine water in 1971 resulted from a number of localized massive kills, principally in the Escambia Bay, Florida, and Galveston Bay, Texas, areas. The fish kills in Escambia Bay, Florida, were largely menhaden, a very valuable industrial fish.

Although the annual reports do not identify the species of fish killed, later attempts have been made to give a reliable estimate of

Table 5.6: Pollution Caused Fish Kill Summary by Type of Water: 1969-1972

			Reported Fish Killed[1]			
Type of Water	1963	1964	1969	1970	1971	1972
Fresh[2]	5.5	15.3	35.0	12.0	15.2	10.7
Salt[3]	1.2	2.5	.6	.5	2.0	.4
Estuary[4]	.1	.0	5.0	9.8	56.4	7.0
Total	6.8	17.8	40.6	22.3	73.6	17.7

[1]In millions. Reports specifying number of fish killed.
[2]Fresh water includes any inland water upstream of tidal action.
[3]Salt water means water beyond the coastline.
[4]Estuary means the water of inlets, bays, or river mouths that are affected by tidal action.

Source: EPA (1971)

the number of fish of commercial value that have died. Owing to insufficient data, 1968 is the latest figure available. Of the 15 million fish killed in 1968, the Federal Water Quality Administration estimated that two-thirds had commercial value (Bale 1971). If fish kills are evaluated in the usual manner (i.e., each fish counted is worth 10 cents), a conservative estimate of the commercial economic loss in 1968 is $1 million. If two-thirds of the fish kills in 1971 had commercial value, this would be a loss of approximately $5 million because of pollution-caused fish kills.

It must again be emphasized that the annual report figures of pollution-caused fish kills grossly underestimate the number of actual fish kills. Tihansky (1973) reports that in 1970 the commercial fishery revenue losses from fish kills caused by marine water pollution exceeded $6.95 million. Since Tihansky valued fish kills at 10 cents each, he must have estimated fish kills in excess of 69 million.

Our investigation of fish kills caused by water pollution has revealed several salient features. The major pollution sources include municipal wastes, agricultural runoffs, industrial by-products, and some other minor sources. It is interesting to note that these tend to reflect the prevalent industrial base in the respective regions. Municipal sewage alone is closely correlated with population density and inadequate treatment facilities. The obvious randomness of the pollution source is also overwhelming. Only in a few states were persistent pollution-induced fish kills significant or identifiable.

Sublethal Effects. Most of the emphasis in data collection has been on outright fish kills or mortality. Obviously, these are more sensational and relatively easy to identify. However, pollutants may have more subtle effects, because they are often difficult to disentangle from other factors. A poor year or a decline in catch (or catch per unit of effort) is often attributed to natural factors or intensive fishing effort. Relating various pollutants ceteris paribus to a decline in *CPUE* is difficult, and research on this problem has just begun. Pollutants can influence growth and the incidence of *natural* mortality, both of which play a part in the population dynamics of the fishery.

However, historical data on these relations have not been collected and may never be collected. There simply are not enough baseline data, such as levels and kinds of pollution. Figure 5.4 shows the usual relation between catch and fishing effort and a hypothetical interaction with other variables, such as water pollution.

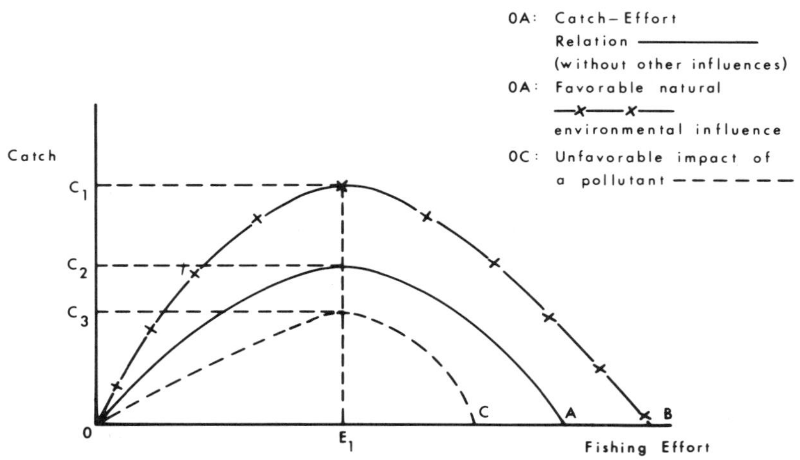

Figure 5.4: Determinants of Fish Catch

As one can see, we must deal with three major factors at one and the same time: (1) changes in fishing effort, (2) changes in the natural environment, and (3) changes in man-made pollution. Catch may be expected at level C_2 under "normal circumstances." Unfortunately, data over long time periods are not available on pollutants that might have sublethal effects.

Extinct Species from Various Areas. The Bell and Canterbery survey indicated eleven species regarded as *endangered, extinct,* or *no longer present* in U.S. coastal waters as a direct result of pollution. The species it mentioned were clams, short-nosed sturgeon, shad, Atlantic salmon, herring, tarpon, snook, and chinook salmon. The major pollutants linked to the absence of these species were (1) landfill, (2) low dissolved oxygen, (3) wood fibers, (4) sulfides, and (5) phenols. In general, these species occur in limited numbers and were mentioned for a limited number of areas. We must conclude that complete extinction is not widespread; however, pollution does reduce catch owing to public health considerations, sublethal effects, sporadic fish kills, or unsuitability of the water for aquaculture.

The Effect of Pollution on Fishery Resources: The International Experience

The main reference on international experience with pollution is *Marine Pollution and Sea Life* (1972). With the increasing use of the sea-lanes for commerce, the increased size of cargo ships, and the use of the ocean for mineral extraction, the threat of pollution to the marine environment on an international level is growing.

North Sea

Pollution has severely damaged several estuarine and coastal areas; indeed, the fishless estuary is a feature of all industrialized areas bordering the North Sea. A great deal of sewage reaches the North Sea, much of it untreated. Often the use of long pipelines to the open sea to avoid contamination of estuaries and tourist beaches results in the discharge of large quantities of waste with a high BOD and volumes of persistent metallic wastes on offshore fishery nursery areas or spawning grounds.

Since 1968 the countries that border on the North Sea have used less of persistent and highly toxic organochlorine pesticides such as dieldrin, aldrin, endrin, and DDT. However, industrial waste must still be disposed at sea since there are no extensive unpopulated land areas available. The mixing characteristics of the North Sea are extremely favorable for the dispersion and dilution of pollutants. Probably the greatest threat to the North Sea is oil pollution because of the concentration of shipping at the approaches to the major ports. As H. A. Cole states: "Although it is easy to point to estuaries in the North Sea where fisheries for shellfish and migratory species of fish have been lost, . . . it is not possible to demonstrate that the North Sea as a whole is less productive than it used to be" (1972, p. 7).

The Baltic Sea

In the Baltic Sea, the oxygen content of the deeper water layers has decreased considerably during recent decades. Figure 5.5 shows the decline in oxygen over the last seventy years in the Baltic along with the growth in phosphate content. The former is usually attributable to an increased nutrient supply to the water, presumably from pollution, which has increased the biochemical oxygen demand (BOD) in deeper water. The latter may be due to increased human population, the use of washing powders containing phosphorous, and artificial fertilizers. This increase of phosphate values has seriously polluted many coastal areas. Signs of eutrophication can be observed in the Baltic areas around large cities, especially around Stockholm and Helsinki. The Baltic is a stagnant basin similar to several of the Great Lakes in the United States.[6]

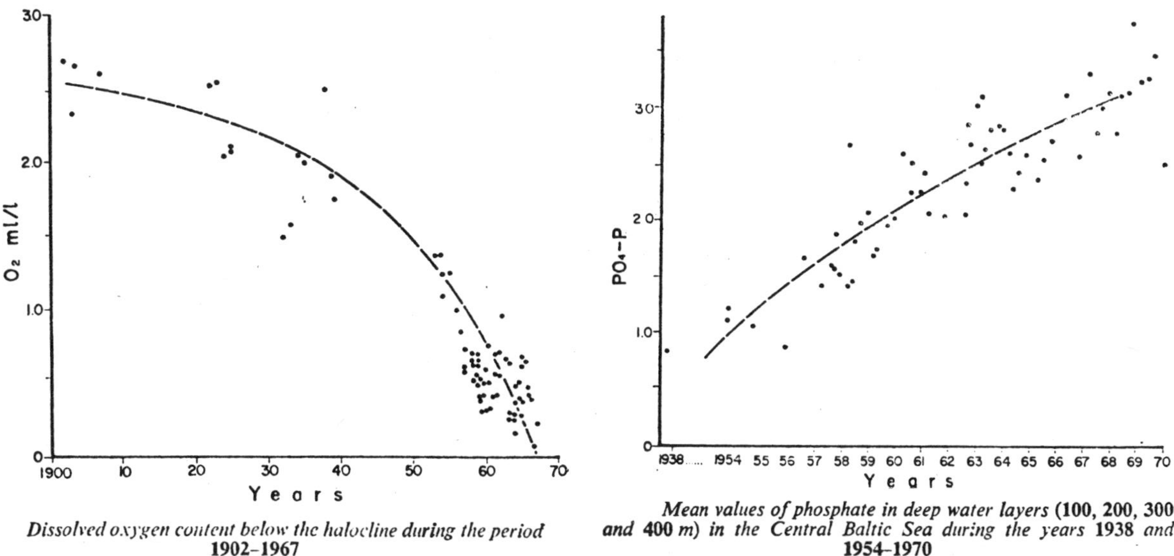

Dissolved oxygen content below the halocline during the period **1902–1967**

Mean values of phosphate in deep water layers **(100, 200, 300** ***and*** **400** *m) in the Central Baltic Sea during the years* **1938** *and* **1954–1970**

Figure 5.5: Decline in Dissolved Oxygen (DO) and Increase in Phosphates (PO_4) in Baltic Sea

Source: Fonsollus (1970)

As oxygen deficiency increases, the ecosystem becomes increasingly unfavorable for the fisheries.

Mediterranean Sea

This enclosed sea has only minor exchange with the open ocean; its increasing receipt of waste that cannot be handled naturally by the ecosystem will sooner or later become a real danger. Generally, sewage is not treated before discharge into the sea. Most Mediterranean beaches are heavily polluted by oil—up to 80 percent of the coastline is so polluted. Oil damage to fishing gear is becoming more frequent. It is also becoming increasingly difficult to sell fish caught in some areas of the Mediterranean. The conflict between the fisheries and the two major pollutants, sewage and oil, is becoming increasingly apparent.

Coastal Areas of Japan

All rivers in the developed regions of Japan are polluted. Unlike inland pollution, marine pollution occurs only in limited areas along the coast, but it is increasing. For example, a dark green coloration in oysters, caused by a high content of copper and zinc, is known as (what else!) the "green oyster." Fish become tainted with foreign odors in harbors and near engineering and petroleum plants. Osaka harbor has this problem. Because the Japanese coastal cities are highly industrialized, eutrophication and other forms of pollution are becoming increasingly harmful to fish life.

Ocean pollution had attracted little public attention until the *Torrey Canyon* incident. The *Torrey Canyon,* an oil tanker, was one of the ten largest ships in the world. Despite warning signals, it went aground off the coast of England and tore a 500-foot gash through six of its eighteen oil storage tanks. The oil spread up the English Channel and into large areas of the Atlantic. This turned the world's attention to ocean pollution. Though tanker accidents have given us a most vivid image of ocean pollution, 90 percent of the polluting hydrocarbons come from normal tanker operations, refineries, industrial plants, and offshore wells, not to mention *air* pollution, which indirectly affects the sea. Figure 5.6 shows marine pollution around the world. The reader can identify those areas that are continually polluted, those that are intermittently polluted, and those that may soon be polluted. Finally, in contrast to our study of the United States, the international pollution losses have not been assigned dollar values. Let us now look at some possible solutions to these pollution problems.

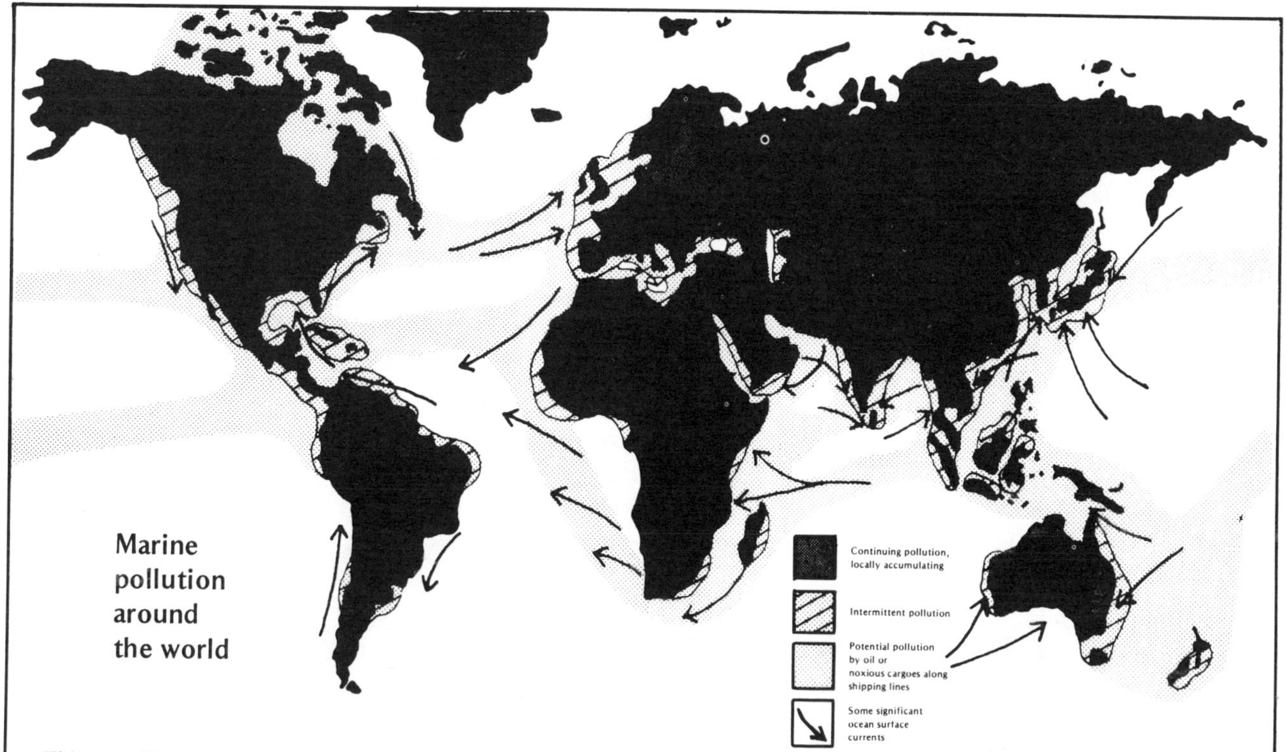

This map illustrates present and potential pollution round the world in relation to main surface currents indicated by arrows. Heavy shading and dots show continuing pollution locally accumulating; light shading with sloping strokes indicates intermittent pollution; and the light clear shading the potential pollution by oil or noxious cargoes along shipping lines.

Source: M. Waldichuk, Fisheries Research Board of Canada, and L. Andren, FAO.

Figure 5.6: Areas of Marine Pollution

Solutions to the Water Pollution Problem

Our section on why pollution occurs indicates that the private market fails to allocate resources properly. This section will look at alternative mechanisms that allow society to correct, partially or fully, for the allocation distortion caused by externalities or spillovers. Before we begin our analysis, let us consider environmental quality as an economic good. For example, we might use a water quality index to measure the state of environmental quality. When we consider environmental quality as an economic good, we are really asking whether people attach any value to changes in environmental quality, changes they in effect consume. It is projected (Water Resources Council 1968) that by the year 2000 U.S. citizens will demand 4.7 billion recreational swimming days, 1.0 billion recreational fishing days, and over 1.3 billion boating days. These demands imply a certain degree of environmental water quality. However, there is a price associated with increased environmental quality: the cost of pollution abatement.

Pollution cost can be broken down into two classifications: (1) the expenditures (i.e., pollution abatement equipment) undertaken to avoid pollution damage (such as that occurring to the shrimp industry, as discussed earlier) and (2) the welfare damage of pollution. The latter is associated with the more subtle consequences of pollution, those that result from deterioration of physical assets and human health and that are not offset by increased expenditures on maintenance and repair. Thus, we can look at the pollution problem in terms of the benefits derived or the sum of the reduction in pollution costs—both damage avoidance cost and welfare damage. Let us consider the various theoretical solutions.

Market Solution

Remember our example at the beginning of this chapter. The chemical plant was polluting the water and reducing the productivity of the shrimp resource. The chemical plant may treat its water discharge in degrees, thereby creating various levels of water quality. The cleaner the water, the greater will be the cost of treatment. Shrimp fishermen may, as a group, offer to bribe (or pay) the chemical plant to reduce discharges or improve the water. The horizontal axis of figure 5.7 may be viewed as a water quality index representing levels of water treated ranging from zero at point 0 (i.e., no treatment) to *K*, which is the maximum capacity of the plant (i.e., all water used is treated). Note that there is a

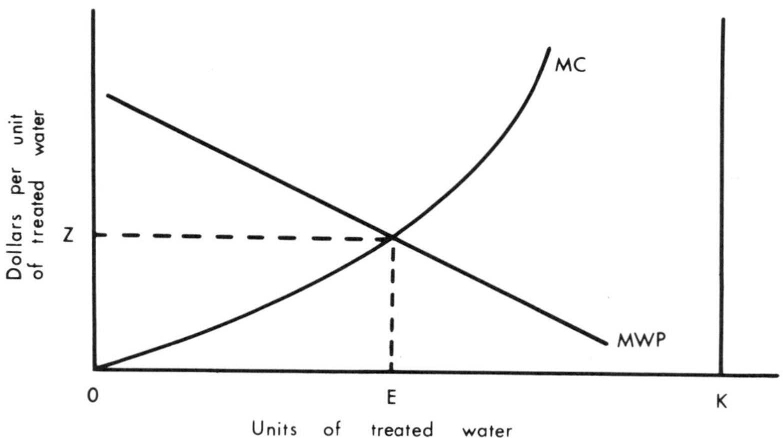

Figure 5.7: A Market in Water Quality

marginal cost curve associated with more and more units of water treated. The marginal cost of abatement rises owing to the rapidly increasing total cost of cleaning up all the water. The marginal willingness to pay curve (*MWP*) is the demand for water quality on the part of the fishermen. At zero units of water treatment, the shrimp resource may be in danger of extinction. Thus, fishermen would be willing to pay a considerable amount for some reduction in pollution (i.e., first units treated). As the chemical plant gradually cleans up the water, the additional cost (i.e., marginal) increases, and the damage to the resource diminishes; hence, fishermen will be willing to pay less and less for incremental improvements in water quality. The fishermen will pay *Z* dollars per unit of treated water, and *E* units of treated water will, in effect, be sold to the fishermen. This arrangement is difficult to set up, especially when polluters and pollutees are rampant throughout the economy. More important, what gives the chemical plant the right to make a market in pollution? In effect, the chemical plant is seizing the common property resource and charging others to clean it up. If you were a swimmer or recreational fisherman, you would object if you had to bribe a polluter to stop polluting the water. After all, the polluter does not own the water!

Regulation by a Tax on Pollutants

Under this system, the government regulates the polluter by charging an effluent tax. Consider figure 5.8. The marginal damage

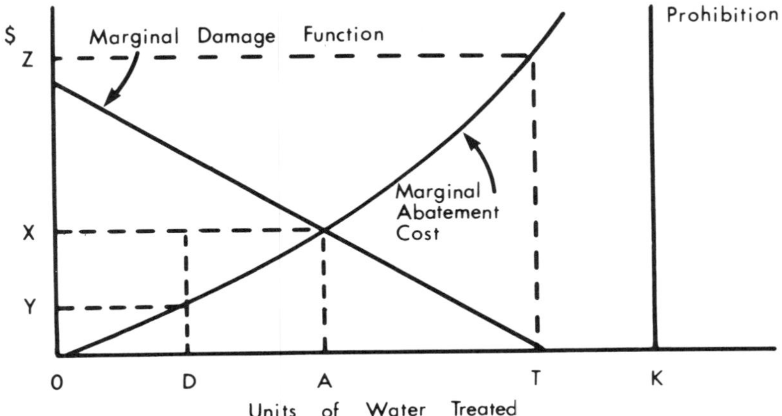

Figure 5.8: Economic Impact of an Effluent Charge to Reduce Pollution

function is similar to what we earlier called the "marginal willingness to pay curve." If little or no water is treated, the chemical plant's use of this common property resource will *damage* the shrimp fishery. As more and more units of water are treated, we reach *OT*. At this point, enough effluent has been eliminated so that it has no appreciable impact on the shrimp nursery grounds (i.e., marginal damage is zero beyond *T*). The ecosystem is either handling the pollutant and cleansing itself, or the pollutant (e.g., nonbiodegradables) is present in insufficient concentrations to influence shrimp *productivity* or *marketability*. Remember our earlier discussion of the ways pollution can affect fishery resources. The marginal cost of abatement increases as more and more units of water are treated. If the plant treated all the water it used, it would treat *OK* units. This is a very important point, which we shall consider below. From society's point of view, as long as the marginal damage is greater than the marginal cost of abatement, it will be in the general interest to treat more units of water. In this situation, the benefits—increased shrimp production—exceed the cost, namely, the additional cost of chemicals due to the abatement cost incurred by the chemical plant. It is sometimes argued that if the government could ascertain the damage and abatement functions, it could levy a *tax on effluents* equal to *OX* per unit of water treated. The chemical plant would then have an option: *either pay the tax and continue to pollute or install abatement equipment.* You are the president of the chemical plant. Should you pay a tax and discharge *OD* units of *un*treated water? Economically, this would be a bad decision; it is cheaper to treat *OD* units of water with a marginal cost of *OY* rather than pay *OX* for the right to pollute. Beyond *OA*, it would be cheaper to pollute and pay the tax. To eliminate *all* pollution damage, the government would have to raise its effluent tax to *OZ* with *OT* units of water treated. However, at *OT* level of treatment, society is allocating more resources (i.e., abatement equipment) to clean up the water than what is lost through damage. *OX* would thus be the optimal effluent tax from a cost-benefit point of view. Many will say, "the environment is not for sale!" Effluent charges are not in wide use, except in Germany, and have been called a license to pollute. From a cost-benefit approach, however, effluent taxes are at least in theory the best way of using scarce economic resources.

Regulation by Prohibition

In our last section, we implied that there is a good economic argument that there is an optimal amount of pollution. That is, beyond a certain point (which is operationally very difficult to determine), the cost of reducing pollution exceeds the benefits of reduction. How much environmental quality can you afford? How much are you willing to pay to eliminate all pollution?

In the United States, the Federal Water Pollution Control Act of 1972 (PL 92-500) is administered by the Environmental Protection Agency. First, point-source polluters must obtain permits specifying (1) allowable amounts, (2) constituents of effluent, and (3) a schedule for achieving compliance. The individual states issue the permits under the EPA's supervision and review. The act established the following timetable:

1. 1977: municipal plants must provide "secondary treatment" by this date, and industrial facilities must have complied with certain control guidelines.
2. 1983: limitation of all pollutant discharge for classes of nonmunicipal facilities for which it is "technologically and economically achievable."
3. 1985: elimination of all pollutants (i.e., prohibition).

In general, the act is designed to move public and private polluters toward what is called EOD, or end of discharge. Realistically, this cannot happen overnight: there are limits to known abatement technology, and there may be an adverse economic impact on many firms, firms that employ literally millions of workers. In terms of figure 5.8, PL 92-500 is ultimately directed at *OK* units of water treated. A study by the National Commission on Water Quality (1975) indicates that EOD will be extremely costly relative to the benefits (i.e., the marginal cost of abatement is greater than the marginal benefits from reducing pollution). The United States has chosen a method of regulation that moves us toward complete prohibition. However, we may now be treating less than *OA* units of water, and the legislation may be modified as the environment is cleaned up (i.e., beyond *OA*). Because of the numerous polluters and damage functions in the economy, PL 92-500 may be the most pragmatic way of dealing with environmental pollution. This will, of course, increase fishery production by opening up closed shellfish areas, by increasing aquatic productivity, by allowing more fish to be marketed (e.g., absence of DDT or mercury), and by enhancing the potential for coastal aquaculture.

Summary

Although the visual effects of water pollution are quite obvious to us all, the fundamental cause is that water is common property. For many industries, the technical nature of the production process requires water inputs, which are free. For example, a pulp and paper mill uses water and discharges substances into the water. If these substances are harmful to the ecosystem or other water users, they are defined as pollutants. Polluters, in essence, do not pay the entire social cost of production, since any harmful effect will probably raise the cost of production in, for example, fishing. Thus, polluters produce too much (since they are not required to pay for polluting the water) and pollutees produce too little (since they must bear the cost or damage from pollution). Economists call this a private market failure, or spillover effects.

Waterborne pollutants can be classified as degradable and nondegradable. The ecosystem can break down reasonable amounts of degradable substances such as raw sewage through reoxygenation by the air and through photosynthesis by plant life in the water. On the other hand, nondegradable substances such as DDT, detergents, and heavy metals cannot be broken down by the natural environment and become threats to animal life and man. Degradable substances often overload the ecosystem and cause many detrimental effects.

Studies of the impact of various pollutants on fishery resources are limited. National shellfish loss estimates are based on the proportion of shellfishing areas closed by pollution. The most recent comprehensive study for the United States indicated an annual loss of $38 million dollars to commercial fisherman, which does not include losses to recreational fishermen. The only actual losses for finfish because of unsuitability for the market was $4.25 million over the 1969–1971 period. Fish kills seem to occur on a random basis in the United States.

The North Sea, Baltic Sea, Mediterranean Sea, and the coastal areas of Japan, for example, have experienced extensive pollution, which has seriously damaged fish life. For example, the level of dissolved oxygen, which is crucial to fish survival, has reached almost zero throughout the Baltic Sea.

Finally, the possible solutions to the pollution or environmental quality problems were explored. Economists generally favor taxing polluters up to the point where the marginal cost of abatement is equal to the marginal reduction in damages. In this way, polluters are forced to pay for the free resource, water, and the adverse effects of pollution would be significantly reduced. By and large, this is a theoretical solution, one that would be difficult to implement. The United States and other governments of the world have chosen regulations that apply increasingly stringent standards to the discharging of pollutants. The object of the Federal Water Pollution Control Act is to end discharge by 1985. Although this objective may not be attained, it does force polluters in the right direction, which will relieve the commercial and recreational fisherman of bearing the social cost of pollution. If the goals of the act are met many closed shellfish areas can be reopened, and improved water quality will increase the productivity of existing fisheries.

Notes

1. It may be argued that shrimpers do not pay for the common property fishery resource. This is true. However, it in no way influences our illustration or basic argument.

2. The chemical plant's social and private costs are identical at zero units of output since no pollution or external diseconomy is imposed on the shrimp industry. In addition, the shrimp industry's MPC and MSC coincide at zero units of shrimp landings, but diverge as the chemical plant increases its level of production.

3. We shall confine our emphasis to marine fisheries; however, some freshwater pollution impacts will be discussed.

4. A linear trend merely tests for randomness versus secular increases.

5. Point source pollution means we can identify the specific entity that is discharging effluents. Non point source pollution is not specific and may originate from agricultural runoff, municipal runoff, or the general environment.

6. The fish catch in Lake Ontario has declined for herring, blue pike, whitefish, and lake trout. Fish catches have declined by factors of 100 to 1000 for most species. See Beeton (1970).

References

Bale, H. E., Jr. 1971. Report on the economic costs of fishery contaminants. National Marine Fisheries Service file manuscript no. 80.

Beeton, A. M. 1970. *Statement on pollution and eutrophication of the Great Lakes.* University of Wisconsin special report no. 11.

Bell, Frederick W., and Canterbery, E. Ray. 1976. *An assessment of the economic benefits which will accrue to commercial and recreational fisheries from incremental improvements in the quality of coastal water.* National Commission on Water Quality, NTIS, U.S., Department of Commerce PB-252172-01-12.

Bell, Frederick W., et al. 1970. The future of the world's fishery resources. National Marine Fisheries Service file manuscript no. 65-1.

Cole, H. A. 1972. North Sea pollution. *Marine Pollution and Sea Life,* ed. Mario Ruivo. London: Fishing News (Books), Ltd.

Commoner, Barry. 1972. The environmental costs of economic growth. In *Economics of the Environment,* ed. Robert and Nancy Dorfman. New York: W. W. Norton.

Council on Environmental Quality. 1970. *Ocean dumping: a national policy.* Washington, D.C.: Government Printing Office.

Environmental Protection Agency. 1969, 1971. *National shellfish register.* Annual reports.

——. 1972. *Fish kills caused by pollution in 1971.* Washington, D.C.

Federal Water Pollution Control Administration. 1970. *The national estuarine pollution study, II.* U.S., Department of Interior, Washington, D.C.

Federal Water Quality Administration. 1970. *New Haven harbor: shellfish resource and water quality.* U.S., Department of Interior, Needham Heights, Mass.

Fonsollus, Slig H. 1970. Stagnant sea. *Environment* 12:2-11, 40-48.

Food and Agriculture Organization. 1972. *Marine pollution and sea life.* United Nations, Surrey, England.

Howard, G. V. 1973. National Marine Fisheries Service, Terminal Island, California, personal communication.

Meadows, Dennis L., et al. 1974. *Dynamics of growth in a finite world.* Cambridge, Mass.: Wright-Allen.

National Commission on Water Quality. 1975. *Staff Draft report.* Washington, D.C.

Public Health Service. 1971. *Fish kills caused by pollution: 1960, 1963, 1964, 1969.* Washington, D.C.

Spencer, S. L. 1970. *Monetary values of fish.* American Fisheries Society, the Pollution Committee, Montgomery, Alabama.

Terrebonne, Peter R. 1973. The economic losses from water pollution in the Pensacola area. *Florida Naturalist,* October 1973, pp. 21–26.

Tihansky, Dennis P. 1973. An economic assessment of marine water pollution damages. National Marine Fisheries Service file manuscript no. 174. Rockville, Maryland.

Water Resources Council. 1968. *The Nation's Water Resources.* Washington: Government Printing Office.

6. *Fisheries for Recreation Use: The Sleeping Giant*

Outdoor Recreation and the Fisheries

Outdoor recreation includes such diverse activities as swimming, fishing, sailing, hunting, bird watching, and camping, to mention but a few. Man has always felt the need for more than food and shelter—the need to escape from toil and worry, which finds relief in various forms of recreation. Available statistical information shows that millions of people in the United States are engaged in outdoor recreation, and the numbers are rising each year. For example, attendance at major types of outdoor recreation sites, such as state parks and national forests, has been rising 8–10 percent a year over the 1910–1963 period. Population growth over the same years was generally below 2 percent; hence, the per capita visitation rate rose considerably. In 1972, U.S. per capita consumption expenditures on all goods and services was $3,479, of which $122, or 3.5 percent, was spent on *all* forms of outdoor recreation. This amounted to $25.4 billion dollars.

This chapter will focus on fishery resources as one form of outdoor recreation. We have earlier referred to recreational fishing as a "sleeping giant" in terms of economic importance. Up to this point, the fisheries have been considered as a food resource; however, this chapter will place recreational fisheries within the proper perspective of overall resource use. The ocean's magnetic attraction for human enjoyment and especially recreational fishing has created a large industry that by many measures dwarfs commercial fishing. Robert R. Nathan Associates (1974) estimated that ocean-related outdoor recreational expenditures per capita were $24.90 in 1972, or $5.18 billion dollars altogether. According to the *National Survey of Fishing and Hunting* (*NSFH*) (1970), saltwater recreational fishermen spent $1.2 billion in 1970.[1] Although our figures for ocean-related recreational expenditures are gross approximations, it is apparent that saltwater recreational fishing is a major component. Table 6.1 shows the *relative* importance of saltwater, as compared to freshwater, recreational fishing for 1970. *Saltwater* recreational fishermen and their corresponding expenditures accounted for approximately one-quarter of all fishermen and all money spent on recreational fishing in the United States. According to table 6.1, freshwater fishermen tend to spend more recreational days per fisherman yearly, but less expenditure per day, than their saltwater counterparts.

National surveys of marine (i.e., saltwater) recreational fishermen have been few, and they are not considered too reliable. A national survey of marine recreational fishing constructed for five-year intervals (1955–1970) by the Bureau of the Census for the Fish and Wildlife Service showed that the number of marine recreational fishermen increased from about 4.6 to 9.5 million, or over 100 percent (7 percent annually). Table 6.2 shows the trend in expenditures, miles traveled, and per capita values over the 1955–1970 period. Real expenditures (i.e., 1967 dollars) *per angler* and *per recreational day* have shown no trend over the period for which we have data. Thus, the rise in *total* real expendi-

Table 6.1: A Comparison of Fresh Versus Salt Water Recreational Fishing* 1970

Measure	Salt	Fresh	Total	Percent Salt of Total
	---thousands---			
Number of Fishermen	9,460	29,363	33,158	28.5
Expenditures	$1,224,705	$3,734,178	4,958,883	24.7
Recreational Days	113,694	592,494	706,187	16.1
Expenditures Per Fisherman	$129	127		---
Expenditures Per Recreational Day	10.77	6.30		---
Recreational Days Per Fisherman	12.02	20.18		---

* finfish only

Source: National Survey of Fishing and Hunting (1970)

tures and days fished is due primarily to increasing numbers of anglers.

Anglers as a percent of the U.S. population have been increasing steadily since 1955, but the increase seems to have slowed somewhat over the 1965–1970 period. As more and more Americans participate in saltwater recreational fishing, they put more and more pressure on ocean fishery resources. This often leads to policy questions over alternative uses, and conflicts between user groups (such as commercial and recreational fishermen) are by no means uncommon. The potential user groups are not only commercial and sports fishermen, but also developers, industrialists, and other business interests, whose activities are sometimes completely incompatible with the ecological system in which the fishery resource exists. This topic was discussed in chapter 5 dealing with environmental deterioration and fishery resources.

As with commercial fishing (see chapter 3), the increasing fishing pressure of marine recreational fishermen results in declining catch per unit of effort. The usual measure of *both* quantity demanded (i.e., the recreational experience) and fishing effort is the recreational day. The statistics gathered by the *1970 Salt-*

Table 6.2: Number, Expenditures, Recreational Days and Passenger-miles traveled by Automobiles for Salt Water Recreational Fishermen* (thousands)

	1955	1960	1965	1970	Percent Change (1955–70)
Number	4,557	6,292	8,305	9,460	108
Expenditures	$488,939	629,191	799,656	1,224,705	150
1967 Dollars	$609,649	709,347	846,197	1,053,567	73
Days Fished	58,621	80,602	95,837	113,694	94
Miles Traveled	2,904,001	3,404,945	4,138,307	5,459,276	88
Expenditures (1967 dol) Per Angler	$134	113	102	111	-17
Expenditures (1967 dol) Per Day Fished	$10.40	8.80	8.83	9.27	-11
Days Fished Per Angler	12.86	12.81	11.54	12.01	-7
Angler ÷ U.S. Population	2.7%	3.5%	4.3%	4.6%	-

* finfish only

Source: National Survey of Fishing and Hunting (1970)

Table 6.3 Marine Sport Fishing Statistics, 1960–1970

	1960	1965	1970	Percent Change (1960–1970)
Weight of Fish[1] (pounds)	1,380,301	1,474,353	1,576,823	14
Anglers[2]	6,292	8,305	9,460	108
Days Fished[2]	80,602	95,837	113,694	94
Weight/Angler	219	178	167	-24
Weight/Day	17.1	15.4	13.9	-19

1. Finfish only (SWAS).
2. Table 6.2.

Water Angling Survey (*SWAS*) (1973) and the 1970 *NSFH* are shown in table 6.3. We are here dealing with total pounds caught by saltwater recreational fishermen, not economic value—the apples and oranges problem again! However, these data do show some broad trends that are inconsistent neither with studies of commercial species (see chapter 3) nor with some recent detailed studies of sport fisheries (see below). Over the 1960–1970 period, fishing pressure on U.S. sports fishery resources increased around 50 percent (using anglers or days fished) but the physical catch increased a mere 14 percent, resulting in a 24 percent decline in catch per angler and a 19 percent decline in catch per recreational day.[2] Detailed population dynamics studies of sports fisheries are startling few relative to such studies of commercial fisheries. This is particularly interesting since all indications are that recreational fisheries have a much larger economic role in the U.S. economy than commercial fisheries. Let us consider a recent study of some California sports fisheries.

According to MacCall, Stauffer, and Troadec (1974), the California barracuda partyboat pounds per angler dropped from fourteen in 1935 to approximately two by 1970. The authors categorize the California barracuda as a depleted sports resource. In the 1930s fewer than 50 percent of the barracuda were caught for recreational purposes; however, the commercial fishery declined to such an extent that in 1973 96 percent of the barracuda catch was taken by sport anglers. Similarly, MacCall, Stauffer, and Troadec indicate declining abundance (*CPUE*) for partyboat bonito (i.e., tuna) and white sea bass. From our initial discussion, it is apparent that recreational ocean fisheries suffer from many of the problems—such as rising demand and limited resource potential—from which commercial fisheries suffer. Unlike commercial fisheries, however, sport or recreational fisheries suffer from an additional problem: the difficult task of valuation. Although we have data on gross expenditures by anglers, the true value of the recreational experience is more complicated.

How Much is a Recreational Experience Worth? The Valuation Problem

Recreation as an Economic Good

The output of a sport fishery is fishing, not fish. Fishing is a recreational experience: that is, it embraces both subjective evaluations and intrinsic characteristics of recreation sites. For example, Crutchfield and MacFarlane (1968) estimate that in 1962 recreational fishermen spent $50 million dollars in the state of Washington fishing for chinook, coho, and pink salmon. These fishermen caught an estimated 7,358,246 pounds of salmon, which works out to $6.80 per pound. The 1962 market value of salmon was approximately $0.31 per pound at dockside. The retail value of salmon was less than a dollar per pound. This simple example illustrates that recreational fishing involves far more than the capture of food. Individual reactions to recreational experiences are personal and highly variable. So why do people fish? A survey of saltwater sport fishing for salmon along the east coast of Vancouver Island, British Columbia, was conducted to determine why anglers were making their current fishing trips. The purpose was to test the assumption that there is more satisfaction from sport fishing—recreating—than the mere catching and eating of fish. Table 6.4 shows the results of this survey. The two most important motivations for fishing trips were "to take it easy" and "to be outdoors." At least four motives (2, 4, 7, and 8 in table 6.4), representing 38 percent of all reasons selected, are a form of "escapism." Only four motives (3, 5, 10, and 11), or 27 percent, are "fish-oriented." Thus, in the recreational experience, the consumer or recreationalist is willing to pay for many components other than fish for food. The personal values in recreational fishing are reflected in what people are willing to give up to obtain them. In choosing to use parks, forests, or fishery resources and to spend time, money, and trouble in doing so, people behave in a way that is not fundamentally different from the way they purchase other

Table 6.4 Motives of British Columbia Salmon Anglers for Their Current Fishing Trips (1973)

Motives*	Responses	Percent
1. To be outdoors	290	22.3
2. To take it easy and get rid of tension	230	18.0
3. To eat fresh fish	166	13.1
4. Change from working pressures	153	12.1
5. The experience of the catch	126	9.9
6. To take family and/or friends out	69	5.4
7. Change from home pressures	53	4.1
8. To do something different	47	3.8
9. Solitude	37	2.9
10. Good fishing available	23	1.8
11. Fair fishing available	21	1.7
12. To enjoy the scenery	20	1.6
13. To travel to and from fishing site	7	.5
14. Other	36	2.8
	1,278	100.0

* The above list of specific motives represents the responses of 524 anglers who were interviewed, and selected two or three motives each on the average.

Source: SFI Bulletin (1974)

items. However, unlike most goods such as tires, sugar, or spark plugs, a unit of recreational fishing is difficult to define. The whole recreational experience involves planning, anticipation, travel, experience on the site, and by no means least, recollections. The last may, in retrospect, turn out to be a valuable nostalgic experience. As we shall see below, the most pragmatic way of approximating a unit of recreation is by defining the experience in terms of time, or more specifically, a unit-day measure.

Not only is it difficult to define a unit of recreation, but outdoor recreation is what economists call "extra or nonmarket activities." That is, it is very difficult directly to estimate the *value* of the sport fishery, because the product is rarely marketed commercially in the United States. In chapter 4 we said that a fundamental problem in fisheries is the common property nature of the resource. This leads to inefficiency and overfishing for *commercially* caught fish. But we did not call commercially marketed fish extra or nonmarket activities! Why not? The basic reason is that sports, as opposed to food fishing, has no discernible market price. In chapter 2, we were able to derive the demand curve for shrimp by looking at the observed relationship between market prices and quantity demanded. But where are the market prices for sports fishing? In isolated cases they do exist, for example, where the owner of a private lake will charge a fee for a day of sport fishing.[3] According to the 1970 *NSFH,* only 1.27 percent of expenditures by saltwater fishermen (see table 6.5) were for

Table 6.5 Expenditures of Salt-Water Fishermen in 1970

The total number of salt-water fishermen (12 and over) in the United States was 9,460,000

Expenditure item	Number of spenders	Percent of all salt-water fishermen	Total spent	Average spent per fisherman
	Thousands		*Thousands*	
United States, total	8,445	89.3	$1,224,705	$129.46
Food and lodging:				
Food	4,366	46.1	135,109	14.28
Lodging	970	10.3	35,654	3.77
Transportation:				
Automobile	6,083	64.3	109,288	11.55
Bus, rail, air, and water	168	1.8	13,213	1.40
Auxiliary equipment:				
Special fishing clothing	192	2.0	3,751	.40
Tents	56	.6	2,157	.33
Boats	172	1.8	211,664	22.37
Motors	123	1.3	51,791	5.47
Other equipment	1,076	11.4	92,426	9.77
Fishing equipment:				
Fresh-water rods	181	1.9	3,063	.32
Fresh-water reels	162	1.7	3,072	.32
Salt-water rods	956	10.1	20,600	2.18
Salt-water reels	840	8.9	20,374	2.15
Lures	1,259	13.3	9,661	1.02
Lines	1,394	14.7	7,065	.75
Other fishing equipment	2,564	27.1	29,474	3.12
Licenses, tags, and permits:				
Licenses	1,403	14.8	5,619	.59
Privilege fees and other:				
Annual lease and privilege fees	76	.8	5,131	.54
Daily entrance and privilege fees for fishing	315	3.3	6,898	.73
Special government fees	4	0	25	0
Bait, guide fees, and other trip expenses:				
Bait	5,501	58.2	146,078	15.44
Guide fees	21	.2	972	.10
Head and charter fees	1,717	18.1	82,158	8.68
Alcoholic beverages	1,960	20.7	64,754	6.84
Rental equipment	928	9.8	24,892	2.63
Other trip expense	2,625	27.7	108,125	11.43
Magazines	858	9.1	5,982	.63
General club dues	101	1.1	3,033	.32
Special club dues	91	1.0	3,136	.33
Boat launching fees	436	4.6	9,544	1.01
Other	455	4.8	9,994	1.06

Source: NSFH (1970)

the privilege of fishing (such as annual lease, entrance, or other privilege fees), which indicates that there are no real market prices for the privilege of recreational fishing. Most marine recreational fishing is done by (1) private or rented boat, (2) party or charter boat, (3) bridge, pier, or jetty, and (4) beach or bank. The recreational fisherman is not faced with a "price" for the use of the resource. In general, the principal difference between most forms of outdoor recreation and marketed goods and services is the pricing mechanism. Sponsors of athletic contests, concerts, movies, and the like base their fees on supply and demand. The lack of a conventional market pricing system for recreational marine fishing greatly complicates the valuation process. Let us turn to the alternative techniques used to place a value on recreational fishing.

Techniques of Valuation

Generation of meaningful values for sport and commercial fisheries has been often requested, but seldom accomplished. In alternative-use debates on water, economic evaluations in the form of traditional benefit-cost analysis may be important in decisions on resource allocation. Thus, in many instances, it is essential for fisheries managers to have estimates of the value of fishery resources.

Gross Expenditures. The value of sport fisheries has often been estimated on the basis of how much anglers spend on fishing. Their expenditures fall into two general classifications: (1) *transfer or travel costs* (sometimes called variable costs), including expenses (such as transportation, food, lodging) that are incurred while traveling to, using, and returning from a fishery resource; and (2) durable expenditures, those for fishing equipment (tackle, boats, unique clothing) that can be used over a period of years. Table 6.5 was taken from the 1970 *NFHS,* which shows the *gross expenditures* of saltwater fishermen in the United States (see also table 6.2). This is probably the most common technique used in recreational fishery evaluation. It is fairly easy to distinguish between *travel* and *capital* (durable goods) expenditures. For example, *auxiliary equipment* and *fishing equipment* constitute 48.2 percent of gross expenditures. Thus, 51.8 percent are transfer or travel costs. In 1970 gross expenditures per recreational day were $10.77. Is this what a recreational day is worth? Let us look at the question of valuation from another point of view.

Net Market Benefits. Although many use it, most economists consider gross expenditures a fallacious method of valuation. The rationalization for using this method is that individuals or groups who make such gross expenditures as shown in table 6.5 must have received a value corresponding to the expenditures or else they would not have made them. Crutchfield has argued that the actual value of a fishery resource is a full charge for the right to fish, which is omitted from gross expenditures:[4] "Fishing costs [gross expenditures] in the usual sense really cover only what it costs to get to the fishing area, properly equipped. If fish were so abundant that anyone could fish without restrictions . . . only this would be required. If . . . the catch of the more desirable game fish must be rigidly limited . . . , the "price" of fishing is below the equilibrium level where supply and demand are equated" (1962, p. 148). Crutchfield goes on to admit that gross expenditures are an expression of the recreationalist's desire to fish, but he asks, What if there were no recreational fisheries? All the money spent on recreational fisheries might then be spent on other goods and services. Now here is the key to the critique of gross expenditures. Loss from the shift to a second choice—to which the sport fishermen would be forced because no sport fishery exists—would not be gross expenditures but the *net value,* which the gross expenditures method failed to measure. From this point of view, the *margin above gross expenditures* on the fishery measures the real monetary value that would be lost if the *fishery were to disappear.* Crutchfield calls this *margin* the "net value" of a sports fishery, or the value of a fishing right. An example might clarify this concept. Suppose the fishery resource was *not* common property but was instead owned by a corporation. The corporation would charge a price to anglers for use of the resource, a price consistent with maximizing the corporation's profit. If the corporation were not in close competition with other recreational fishery resources, it would act as a monopolist and vary the price or fee it charges in order to see how angler days "purchased" respond to fees charged. Behold! We would have a demand curve (see chapter 2) for fishing rights.

The concept of net market value is sometimes difficult to grasp. What Crutchfield is really saying is rather simple. Suppose an individual decides to visit an amusement park—a recreational site. He will incur travel cost and parking fees. He may buy popcorn and candy canes at the park. Assume that the individual in question has come with the express purpose of riding the Ferris wheel. The Ferris wheel is a capital good owned by the proprietor, so he will charge a fee for its recreational use. Hence, we have a direct analogy between a Ferris wheel and a fishery resource—

except that the latter has no owner. If we ask, what was the value of the recreational experience of riding the Ferris wheel, most people will quite correctly say that it is the fee per ride multiplied by the number of rides purchased. They will not include the travel cost to the Ferris wheel or popcorn eaten—i.e., gross expenditures. Thus, the concept of net market value allows us to compare alternative uses of fishery resources. For example, who would pay higher fees for the right to fish—recreational or commercial fishermen? Clearly, net market value would be the measure by which alternative uses would be economically compared.[5]

But how can one measure net market value for a recreational resource such as sport fishing? The most promising method of measurement has evolved from an ingenious suggestion by Harold Hotelling in his letter (1949) to the director of the National Park Service. Hotelling indicated that it would be possible to establish approximate measures for evaluating the services of national parks to the public. His technique was based upon establishing concentric zones around each park so that the cost of travel to the park from all points in one of these zones is approximately

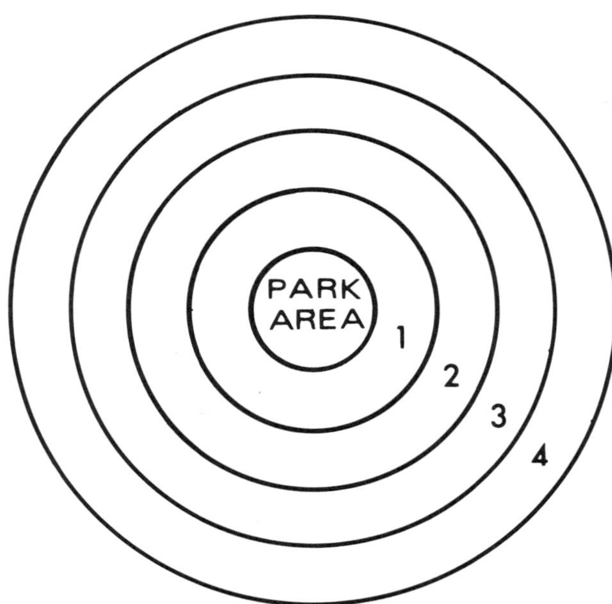

Figure 6.1: Concentric Zones around a Park

constant. This is shown in figure 6.1. If we compare the cost of coming from a given zone with the number of people who do come from it (per thousand of population), we will be able to plot one point for each zone on a Hotelling demand curve for the services of the park. Let us look at the example given in table 6.6.

Table 6.6 Derivation of the Hotelling Demand Curve for a National Park

Concentric Zone	Visits to Park	Zone Population	Visits per 1000 Population	Cost Per Visit
1	500	1,000	500	$1
2	800	2,000	400	$2
3	1,200	4,000	300	$3
4	1,000	10,000	100	$5
	3,500			

Note that we cannot relate the visitors from each zone to the cost per visit, since the size of the population may vary from zone to zone. Thus, we express the visitors per unit of population in order to eliminate the population size effect. Figure 6.2 shows the Hotelling demand curve for a national park's recreational services.

An interesting application of Hotelling's concentric travel zones was made by Trice and Wood (1958). They used data obtained from visitors to three similar areas in the Sierras. They obtained the

1. number of persons in each recreational party
2. city or county of origin of each party
3. number of days spent by each party in the recreational area
4. number of days the party spent on the entire trip

Using these four statistics, Trice and Wood computed the average cost of travel per visitor day. They estimated travel costs at 6.5 cents per mile. Thus, in deriving a demand curve for recreation they expressed the quantity variable as visitor days per unit of population.

Marion Clawson (1959) has made the most extensive application of the Hotelling principle. He first computed what he called an approximation to the demand for the recreation experience as a whole. The demand schedule or curve was measured by plotting the estimated costs per visit as a function of the number of visits per 100,000 population in a zone in a given distance range. Clawson assumed that the visit to the recreational site was the central purpose of the trip. Two examples are taken from his book

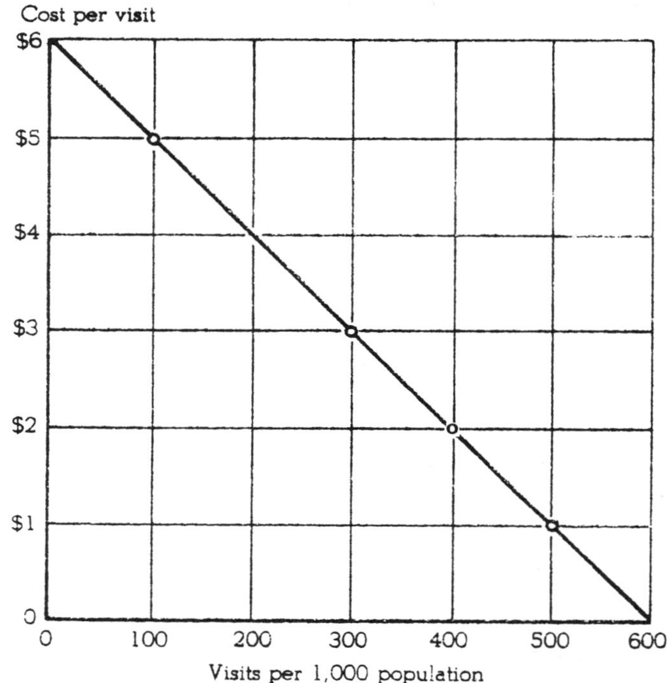

Figure 6.2.: Demand Curve for Whole Recreational Experience for Hypothetical Recreational Area
Source: Clawson and Knetsch (1966)

(with Jack L. Knetsch), *Economics of Outdoor Recreation* (1966), and are shown in figure 6.3. Of course, as we saw in chapter 3, a demand curve relates price to quantity consumed. How do we go from visits per thousand in each zone to a total "market" demand curve for recreation? This can be *simulated* by raising the cost per visit in each zone (not necessarily concentric, but geographically drawn so that cost will be the same in each zone). In effect, we are attempting to see what the response will be to an entrance or users fee imposed on the use of the site. Table 6.7 shows these computations. We note that without any added cost (table 6.6), total attendance is 3,500. We can now see the impact of a price rise of $1. Those in zone 1, who had been paying $1 per visit, must now pay $2. They have been going to the area at a rate of 500 per thousand, but the Hotelling demand curve (see figure 6.2) for this area indicates that people will attend at the rate of 400 per

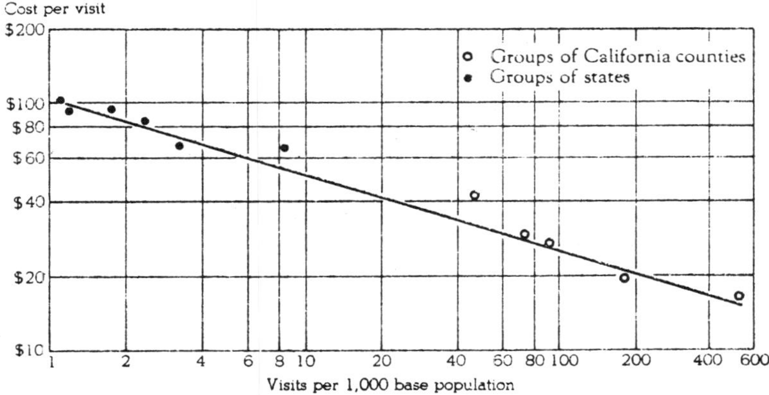

Estimated costs ("shared") per visit to Yosemite National Park in relation to number of visits per 1,000 base population, 1953.

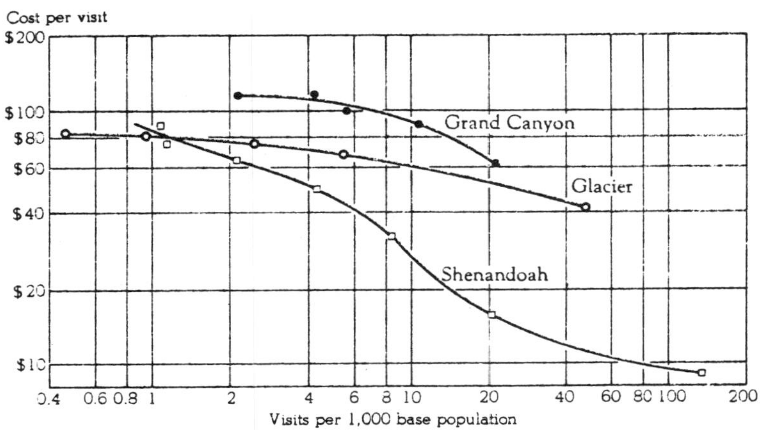

Estimated cost per visit ("shared") related to number of visits per 1,000 base population, Grand Canyon, Glacier, and Shenandoah national parks.

Figure 6.3: Clawson Demand Curves for Selected Sites

Source: Clawson and Knetsch (1966).

Table 6.7 Impact of Increases in Cost on Number of Visits to the Recreation Park

Zone	Number of visits at added cost per visit of*					
	$0	$1	$2	$3	$4	$5
1	500	400	300	200	100	0
2	800	600	400	200	0	0
3	1200	800	400	0	0	0
4	1000	0	0	0	0	0
Total Attendance	3,500	1800	1100	400	100	0

*In addition to regular travel cost

Derived from Table 6.6 and Figure 6.2

thousand when faced with a $2 per visit. Working with the Hotelling demand curve and the data in table 6.6, we can establish how an increase in fees from zero (i.e., the original state) to higher values affects the demand for the recreational park. The relation between increased cost and attendance is shown in figure 6.4 (i.e., from table 6.7). Thus, Clawson has really *two* demand curves. Figure 6.2 uses the basic Hotelling techniques to arrive at the reaction of recreational visits to increased travel cost (i.e., first demand relation), where the second demand curve (figure 6.4) could be used by the resource owner—if there were one—to see the responsiveness of demand for his resource to increased fees. Using this technique, Clawson has calculated the relation between fees and visits for four actual outdoor recreational areas. These are shown in figure 6.5.

The Clawson technique has also been applied—using travel costs—to recreational fishing. If fisherman A visits a recreational fishery five times and his travel cost is $5 per trip and if fisherman B visits two times and his travel cost is $12 per trip, the travel cost approach concludes that if a fee of $7 per trip were imposed on fisherman A, he would decrease his trips from five to two. When is such a conclusion fully warranted? When "price" is the same, A should behave like B only if they have the same tastes, income, face the same price for substitutes, and only if all other demand determinants are the same for each.[6] Thus, the Clawson technique is subject to considerable qualification. Of course, these other factors do not have to be equal, provided the analysis suitably accounts for individual variations. For a good discussion of the criticisms of travel cost market simulation, see the NMFS's *Workshop in Fishery Economics* (1973, pp. 22-23).

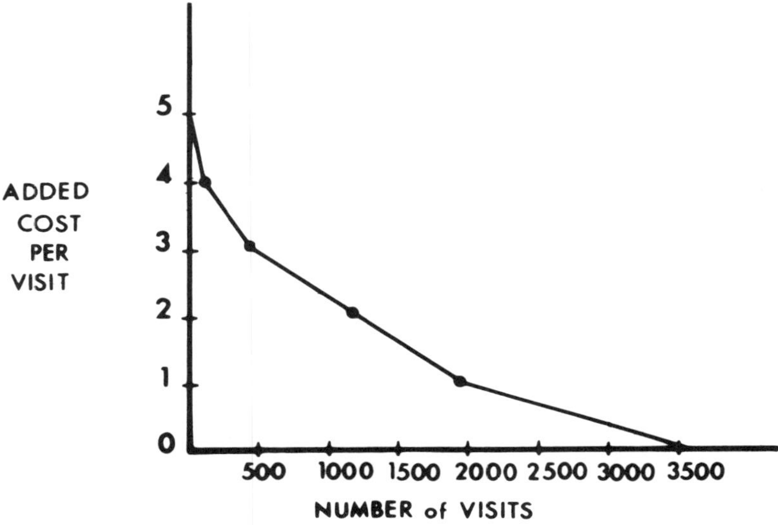

Figure 6.4: Relation Between Added Cost Per Visit and Number of Visits for a Park

Source: Table 6.7

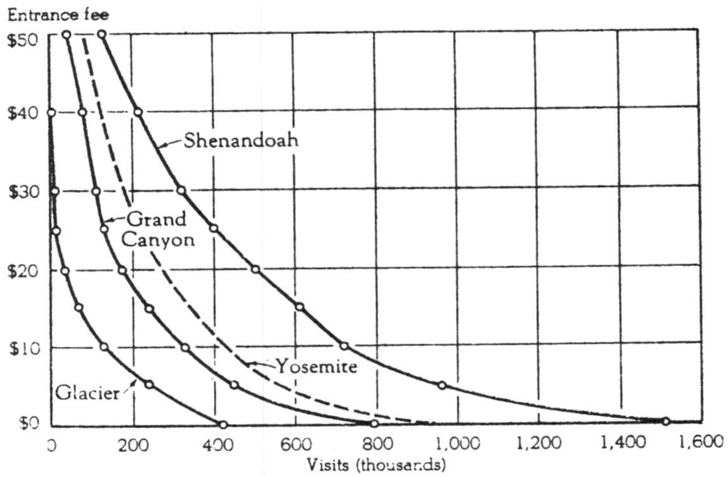

Estimated visits under various entrance fee schedules, Yosemite, Glacier, Grand Canyon, and Shenandoah national parks.

Figure 6.5: The Total Demand Curve for Selected Sites

Source: Clawson and Knetsch (1966)

Using the Clawson method for recreational fishermen, we can project days fished (i.e., derived from days fished per unit of population in each zone and travel cost per day) at increased levels of daily costs as shown in figure 6.6. Point *A* represents the total number of days fished at present costs (table 6.7 shows 3,500 visits at present or zero added cost). Point *XY* is the maximum *net market value* that can be obtained by the state (a monopolist over the fishery resource) and the number of fishing days bought. Some economists, such as Knetsch (1963), propose a concept called "consumer surplus" to reflect the true value of the fishery resource.

Consumer surplus is the area under the demand curve but above the present variable cost (i.e., travel cost at zero user fee).

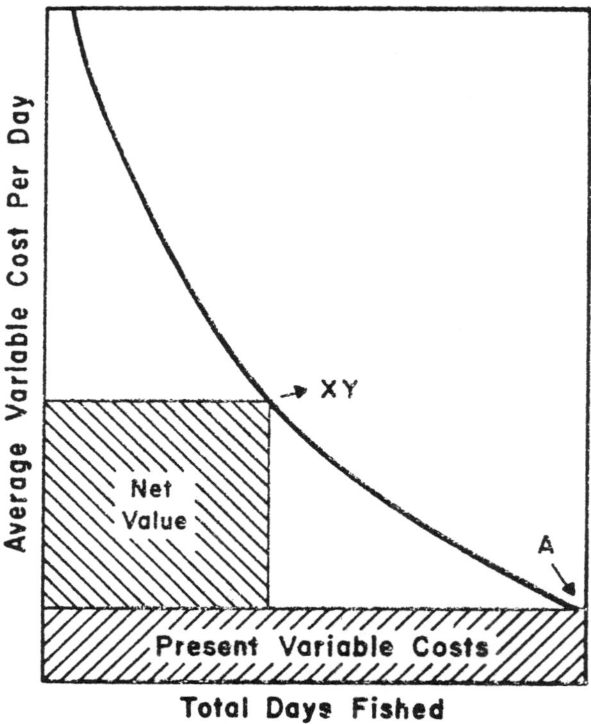

Figure 6.6: Relation of Average Variable Cost and Total Days Fished

Source: Gordon (1973)

If no fees are charged, more angler days will be devoted to recreational fishing. This is the nonmarket condition now prevailing with most fishery resources. The simulated demand curve tells us that people would be willing to pay higher prices (i.e., above present variable cost) for the use of the resource; however, the present charge or fee is zero. Therefore, consumer surplus is the free value received by all anglers who currently use the resource.

Let us look at some applications of the Clawson technique for recreational fisheries. Singh (1965) used this technique to measure the net market value of the salmon–steelhead trout (S-S) sport fishery in Oregon. Singh divided Oregon into five main zones, each based primarily on the average distance traveled for S-S angling. The zone closest to S-S fishing is the coastal zone. Families in this zone averaged only thirty-seven miles per S-S fishing trip, since they lived close to the ocean and coastal rivers where most of the S-S fishing is done. The zones are shown in figure 6.7. Singh's actual statistics are shown in table 6.8. Variable or travel cost per S-S day has been plotted against S-S days per unit of population and is shown in figure 6.8. This function corresponds to what Clawson calls the demand curve (i.e., Hotelling) for the

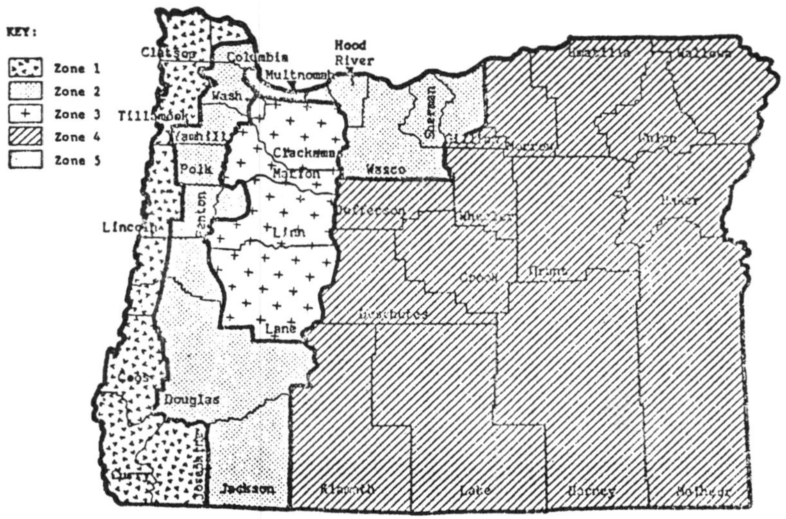

Figure 6.7: Geographic Location of the Five Distance Zones in Oregon

Source: Singh (1965)

Table 6.8 Average Number of Miles Per S-S Fishing Trip, Variable Cost Per S-S Fishing Day, Population, Total S-S Fishing Days in Sample, and Per Capita S-S Days X 10^4

1	2	3	4	5	6
Distance zone	Average No. of miles per s-s fishing trip	Average variable cost per s-s fishing day	Zone population	Sample s-s fishing days	s-s fishing days X 10^4 / zone population
1	37	$ 4.02	184,147	455	24.71
2	105	6.14	455,923	721	15.81
3	140	6.00	473,861	704	14.86
4	220	12.00	229,786	808	6.27
5	120	6.71	481,421	144	16.78

Source: Singh (1965)

recreational experience as a whole. As pointed out above, Clawson's procedure can be used to predict the estimated number of S-S fishing days per zone as S-S variable costs (i.e., fees charged) are increased for each zone. The projected number of S-S days taken at several assumed increased levels of daily costs are calculated using the equation in figure 6.8 and the population in each zone. Remember, in our hypothetical illustration, the graph shows the visits (or days) per unit of population corresponding to increased cost. In the usual case, however, recreational days per unit of population are calculated from the equation fitted to the "Clawson demand curve." Table 6.9 shows the projected total of S-S fishing days for Oregon, and figure 6.9 plots increased S-S variable cost (e.g., fees, entrance charges) per day against thousands of S-S fishing days. In table 6.10 we see how a fishery commission acting in the state's interest can obtain some idea of how much more revenue it will get if it increases the daily charge per angler.

Table 6.9 Predicted Number of S-S Days Taken By The Five Distance Zones with Assumed Increases in S-S Fishing Costs Per Day

Distance zone	$0[1]	Total s-s days taken at Δp: $1	$2	$4	$6	$8
	(s-s days)	(s-s days)	(s-s days)	(s-s days)	(s-s days)	(s-s days)
1	171,429	144,800	122,900	88,500	63,400	45,500
2	298,587	253,500	213,600	153,700	110,400	79,100
3	317,617	269,100	227,700	163,500	117,400	84,300
4	56,921	48,200	40,900	29,200	21,100	15,100
5	286,839	242,800	206,500	147,800	105,900	76,100
Total	1,131,392	958,300	811,500	582,800	418,200	300,000

Source: Singh (1965)

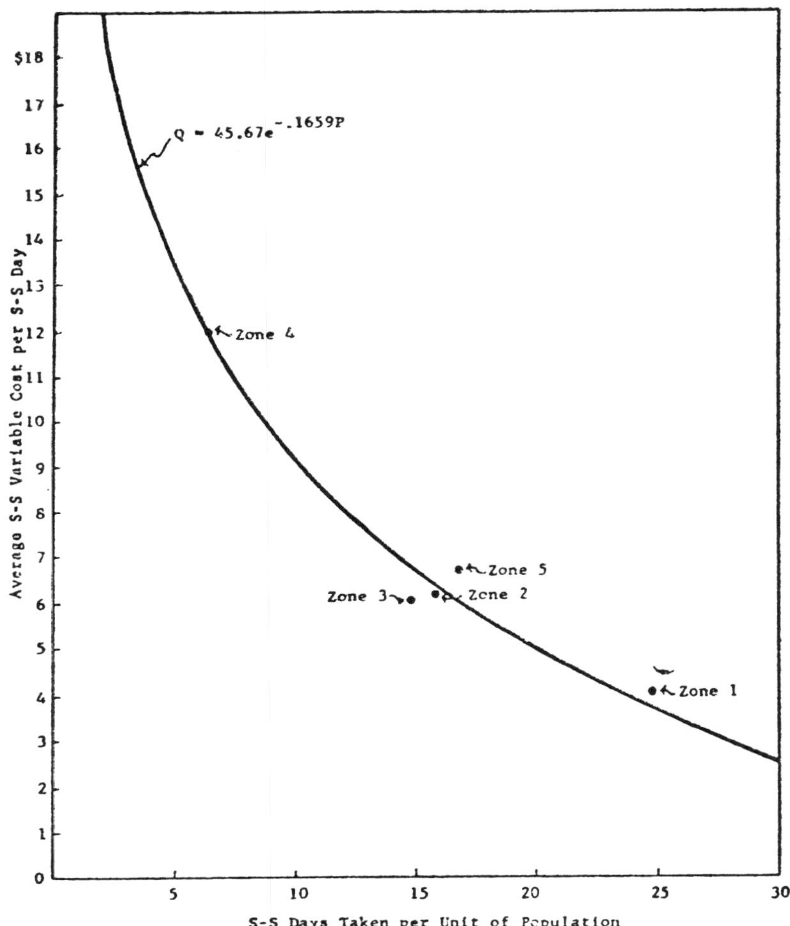

Figure 6.8: Relationship Between Average Cost per Fishing Day and the Number of S-S Days Taken per Unit of Population by the Five Main Distance Zones in Oregon

Source: Singh (1965)

According to the calculation, the maximum *net market value* is about $2.5 million per year if each angler is charged around $6 for each day of S-S fishing. Singh estimated that *gross expenditures* were approximately $17.5 million dollars per year. Most economists would argue that the economic value of the S-S recreational fishery resource is $2.5 million, not $17.5 million; generally, gross

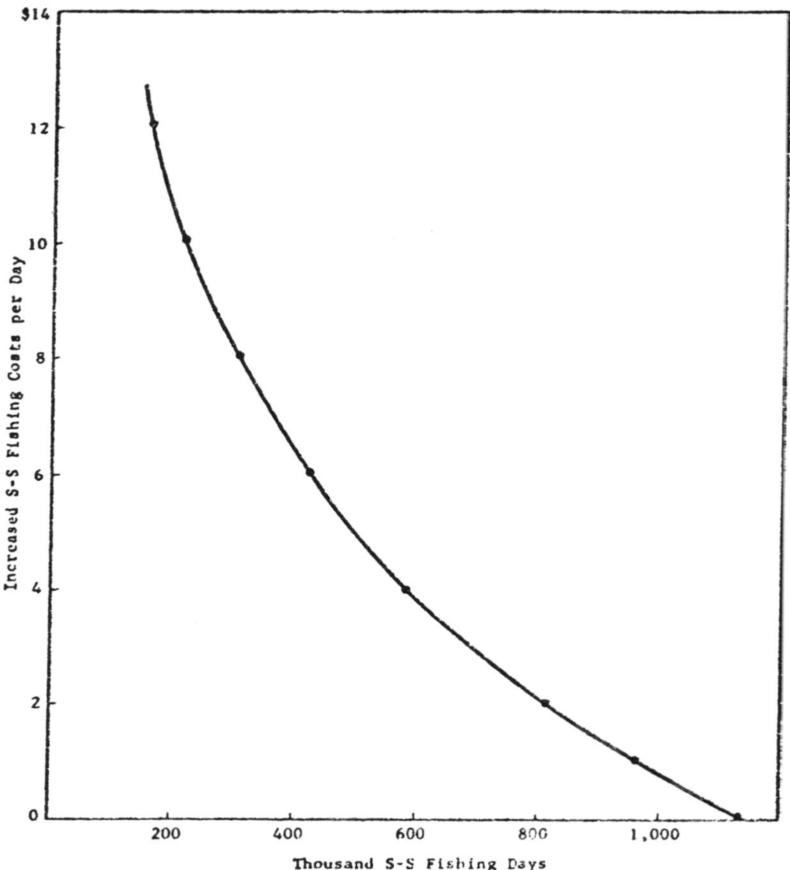

Figure 6.9: Projected Effect of Increased Cost per S-S Day on Number of S-S Fishing Days Taken by Anglers (Projection by Clawson Model)

Soruce: Singh (1965)

expenditures will overstate the economic value of a fishing right. Of course, one can argue that once a fishery resource is "priced" by a license fee, a derived demand for other goods (travel expenses, food, tackle) will be set up. Remember, the net market value is in addition to gross expenditures at any level of daily charge per angler even though part of gross expenditures was used as input to the Clawson procedure. In our example 1,131,392 recreational S-S fishing days would be taken at zero additional cost (see table 6.9).

Table 6.10 Relation of Angler Day Charges to S-S Days Taken

Daily charge per angler per day		Predicted S-S days taken	Predicted possible annual revenue
$1		958,300	$ 958,300
2		811,500	1,623,000
4		582,800	2,331,200
5		493,500	2,467,500
6	Max. Revenue	418,200	2,509,200
7		354,100	2,478,700
8		300,000	2,400,000

Gross expenditures were $17.5 million, or $15.47 per day. If a $6 charge were placed on the fishery, we would have the following:

Net market value	=	$2.5 million
Derived benefits from the existence of the fishery	=	6.5 million[1]
		$9.0 million

1. 418,200 days x $15.47.

Once again, the derived benefits are subject to debate, since they could be spent on something else if the fishery did not exist. However, there are definite regional impacts of our derived benefits or gross expenditures. Gordon, Chapman, and Bjornn (1973) and Piehler (1972) have recently applied the Clawson method to freshwater and saltwater recreational fisheries, respectively.

The User-Opinion Approach to Sport Fishery Valuation. Another approach researchers use to determine the value of sport fish caught is to ask the fisherman valuation questions. One type of question, roughly phrased, is, What is the largest amount you would be willing to pay for the right to fish at a certain location, for a certain period of time, given a certain rate of catch? Or, how much greater would your costs have to be before you would stop fishing? This second questions is akin to asking how much the respondent would have to receive to give up his or her "right" to fish. If these questions are phrased properly, a usable estimate of total value of the fishery can presumably be obtained (assuming of course that the respondents understand the question and answer it honestly). The obvious criticism to the user-opinion questionnaire approach is: ask a hypothetical question, and you get a hypothetical answer.

Horvath (1974) has made one of the most comprehensive user-opinion surveys—for the southeastern United States. Three of the most important questions asked were:

1. What benefits did you receive from a day of fishing expressed in dollars?
2. If you participated, what amounts of money would you have required to give up fishing?
3. If you did not participate, but wanted to, what amount of daily benefit expressed in dollars would you have assigned to fishing?

These questions can be broken down as (1) benefits received, (2) benefits assigned, and (3) value demanded to give up. Horvath's results are shown in table 6.11.

Table 6.11 Value of Recreational Fishing Days

Fishing	(1) Average daily value received by participant	(2) Average daily value to give up by participant	(3) Average daily value assigned by nonparticipants
Saltwater	$59.80	$74.47	$43.69
Warm-Freshwater	40.84	49.28	17.83
Cold-Freshwater	35.58	39.83	23.35

Given the number of recreational days, saltwater fishing recreational participants in the southeastern United States received benefits of $2.4 billion. A saltwater fishing day brought more benefits than warm or cold freshwater fishing. Those that participated in saltwater fishing (column 2) would have to be paid $74.47 to give it up. Finally, nonparticipants valued saltwater fishing much lower than those that participated ($43.69 per day). These values do seem quite high. Horvath says of fishing, hunting, and wildlife enjoyment that "A comparison of expenditures—actual outlays for one year by the population (with capital items amortized for one year)—with the monetary benefits received and required to give up, reveals that for every dollar of cash outlay, a total of $5.90 benefits is received, and for every dollar placed on giving up activity, a total value of $7.68 is received" (1974, p. 192). Do not confuse gross expenditures, the user-opinion values, and net market value. Each is a distinct measure of fishery valuation. User-opinion values would be quite similar to consumer surplus, a concept discussed earlier.[7]

One objection to user-opinion surveys is that the angler's response may depend on why he is being asked. For example, the question about giving up the fishing experience may be a signal to the respondent that someone may want to take away the fishery. As in the Horvath study, this question brought the highest value response. Other researchers—Hammock (1969), Mathews and Brown (1970), and Pearse (1968)—have used the user-opinion method on outdoor recreational resources with varying degrees of success. These results merely keep the user-opinion questionnaire approach in the running as a possible useful technique.

Fallacious Techniques of Valuation. Several techniques have been used but are unacceptable. The *cost method* asserts that the value of a recreational facility is its costs. The National Park Service used this method from 1950–1957. It has the dubious distinction of making every fishery or recreational project self-justifying. It may tell us the magnitude of the sports fishery, but not the loss sustained if it should disappear (nor the net gain from the new project). The cost method is contrary to the valuation based on net market value (see above), since it includes money or expenditures that would be shifted to other goods and services if there were no fishery.

Another method uses the gross national product (GNP). This method attempts to measure the contribution of recreation to GNP—the market value of all goods and services produced in the economy—assuming that *recreation*—like capital, labor, or land—is a factor of production. For example, the GNP for the United States in 1974 was $1,397 billion dollars, and a population was 212 million. This procedure estimates the "value" of a recreational day by dividing GNP by the product of population and the number of days in a year. For 1974, a recreational day was "worth" $18.05. If we then multiply the number of recreational days during the year by $18.05, we have the "value of recreation." Treating recreation as a factor of production fails to quantify what is intended to be measured. It equates work (i.e., hours of time producing goods) with leisure. The computed figure also does not discriminate among different kinds of recreation.

Among fishery biologists a very common approach to recreational fishing evaluation is the *commercial value of fish.* This method imputes to sport fishing the market value of fish caught, calculated at the dockside price for commercially caught fish. The major problem, as indicated above, is that fish for food and fish for recreation are two entirely different economic goods (see NMFS 1974). The dockside price is only one element (i.e., food) in the "price" of the entire recreational experience. If we view the angler as a competitive user of fish, such as salmon and steelhead trout, sport landings represent a net reduction of the potential commercial production and at a minimum must have that value. This method confuses *fish* with *fishing,* a distinction we made earlier in this section (see also table 6.4 for the motivations behind sport fishing).

Finally, many support the idea that no value can be placed on sport fishing. The beauty of nature and the outdoors as a source of healthful sport have infinite value. This is a position analogous to that discussed in chapter 5, that is, some would argue that the environment is not for sale at any price. To argue that the intangibles associated with sport fishing cannot be measured monetarily is to misunderstand the role of prices in a market economy as a measure of human choice. The common denominator of all satisfaction (beauty, sustenance, etc.) is what one is willing to sacrifice for more or less of any good. This satisfaction, or consumer choice, is measured in terms of willingness to buy at a price. If the fishery were owned, consumers would choose—through prices—a mix between sport and commercial fishing.

Policy Implications of Valuation

Policy decisions on sport fish are basically of two kinds. Action can be taken that ultimately increases the catch, action such as hatchery programs, water quality standards, or restrictions on commercial catch. Such programs as resource enhancement and water quality standards were discussed extensively in chapters 3 and 5, respectively. Or, policies like river projects can be promulgated that decrease the catch. The competing demands take the form of municipal and industrial use, hydroelectric power generation, and irrigation. Ideally, if we knew how the total value to fishermen changes as the number of fish caught changes, these policies could be evaluated. No study has yet established in functional form an empirical connection between recreational value and physical catch.

Two studies illustrate the various types of policies that decisionmakers may adopt. Richards (1968) has evaluated the effectiveness of government expenditures (costs) in providing benefits associated with the Columbia River anadromous fish programs. Dams and economic development in the Columbia River basin have threatened the continued existence of anadromous fish, especially salmon. Dams retard upstream migration, block access to upstream spawning areas, and flood downstream spawning areas. The federal government has cooperated with state and private agencies to initiate programs to help the dams and anadromous fish to coexist. So here is the question: does society place a sufficiently high value on the products of these fish resources to offset the additional cost of coexistence? Richards first estimated the cost of the government program at various levels and related this to commercial landings and sport fishing. The commercial fishery was converted to incremental dollar benefits, and the net market value (i.e., the Clawson approach) of the sport fishery was used. Richards found that for 1965 the net economic benefits of the commercial catch were \$8 million and that the net value of the sport catch was \$5.8 million, or approximately \$13.8 for both fisheries. Annual governmental costs *subject to control and alteration* for facilities constructed or under construction were \$9.5 million dollars. Thus, the 1965 benefit-cost ratio was 1.45. In evaluating government programs, however, time is an important factor, and the flow of benefits over time should be compared to costs.

One of the most heated and controversial areas is the direct sports-commercial conflict for fishery resources. Common charges of one group against the other are as follows: (1) damage to habitat, (2) loss of food organisms, (3) destruction of eggs and young, (4) space competition on the fishing grounds, (5) depletion or reduction of stocks, (6) wasting fish, (7) sale of sport fish, (8) optimum yield as *MSY*, and (9) biased regulations. Cleary (1971) has prepared a table (see table 6.12) indicating these conflicts for various sport fish. The most frequent charge by far is that of depletion or reduction of stocks. The second most frequent charge is biased regulation. In addition to domestic fishermen, foreign fleets (notably the Soviet and Japanese) have been charged with diminishing the quantity and average size of several sport fish species through intensive fishing off our coasts. Table 6.13 compares sport catch and commercial catch (domestic plus foreign) for some of the most important finfish (shellfish data are not available). At first glance, one might infer that sport fisheries are mainly for sportsmen and that commercial fisheries appear to have relatively minor sport catches. However, these aggregate figures are for the coastal United States. Although sport fishermen account for only 7 percent of the total salmon catch, conflicts have arisen mainly over depletion or reduction of stocks. The committee on sport-commercial conflicts (1967) reported forty-five marine species and ten freshwater species of fish and shellfish that are

Table 6.12 Summary of Sport/Commercial Conflicts for Specific Fisheries

Sport Fish Species	Charges										
	Damage to Habitat	Loss of Food Organism Preditor/Prey Relationship	Destruction of eggs or young	Depletion or Reduction of Stocks	Wasting Fish	Incon-siderate Behavior	MSY vs. Large Sport Fish	Catching of Game Fishes Incidental Catch	Specie Com-petition on Fishing Grounds	Sale of Sport Fish	Biased Regula-tion
Pacific Coast											
Salmon (coho Alaska)				X				X			
Salmon (king, sockeye)				X				X			X
Salmon (chinook, coho)				X							X
Salmon, Steelhead trout (Columbia River System)				X			X				X
Steelhead trout				X							X
Striped bass, American shad					X						
Herring		X									
Anchovy		X		X							X
Marlin, Sailfish, Blue fin tuna, and Mackerals				X			X				
Razor Clam				X							X
Broad bill swordfish				X		X					
Gulf Coast											
Speckled trout				X							
Texas-Flounders, Red drum, Squeteague, and Mackeral				X			X	X		X	
Atlantic Coast											
Blue fin tuna				X	X	X	X		X		
Tuna				X	X	X			X		
Bluefish				X				X			
Swordfish and other Billfish				X	X	X		X		X	X
Flounder				X							X
Striped bass		X	X	X		X			X	X	X
Saltwater smelt											X
Menhaden			X	X		X		X	X		X
N. Lobster											X
Shrimp			X	X							
Sharks, probeggle				X							
Trout, croaker, spot, and whiting			X	X							

Note: All information contained in this table was obtained from Report of the Committee on Conflicts Between Sport and Commercial Fishermen, October 16, 1967, U.S. Dept. of Interior; and from articles appearing in periodicals such as The Salt Water Sportsman, Sport Fishing Institute Bulletin, and others. This table is not exhaustive nor is it necessarily representative of the national pattern of sport/commercial conflicts.

Source: Cleary (1971)

Table 6.13 Sports and Commercial Catch by Various Fisheries in 1970 of the Most Important Species and Species Groups in the U.S. Marine Fishery Resources

Species or Group	U.S.A. Sport	Commercial (Foreign & U.S.A.)	Total	Sport as Percent of Total
	--------millions of pounds------------			
A. Primarily Sport				
Catfish (marine)	72.5	.5	73.0	99
Bluefish	120.8	6.9	127.7	95
Sea Trouts	152.6	13.3	165.9	92
Croackers	216.0	21.6	237.6	91
Striped bass	83.8	10.4	94.2	89
Groupers	41.0	7.9	48.9	84
Scup	55.5	11.0	66.5	83
Mackerels	157.3	143.7	301.0	52
B. Primarily Commercial				
Cod	36.9	194.8	231.7	16
Flounders	90.5	729.4	819.8	11
Tunas	40.6	469.0	509.6	8
Salmon	30.7	417.7	448.4	7
Mullet	2.3	31.2	33.5	7
Haddock	2.5	40.3	42.8	6
Halibut	2.8	76.6	79.4	4
Jack Mackerel	1.2	47.1	48.3	3
Ocean Perch	13.8	588.7	602.5	2
Sablefish	1.0	54.5	55.5	2
Whiting (silver hake)	2.1	143.7	145.8	1
Pollocks	5.6	5,643.7	5,649.3	0-1
Hakes	.9	486.1	487.2	0-1
Anchovy	0	192.4	192.4	0
Herring	0	1,785.8	1,785.8	0
Menhaden	0	1,813.8	1,813.8	0
Total Finfish	1,172.1	12,939.3	14,111.4	

Source: NMFS (1974).

either harvested by both sport and commercial fishermen or are used for food or for bait for predator sport fish. One of the most notable cases is the California anchovy, which serves as bait for recreational fishermen and is an important food for carnivorous game fish. Under the pressure of sport fishermen, commercial catches are limited by quota set every year by the California state regulatory authorities. The size of the spawning biomass off California and Baja California has been estimated at 4-6 million metric tons. If this estimate is correct, current commercial exploitation is fairly nominal.

Recreational Fishing in the United States: How Big Is It?

The Recreational Dimension

The subject matter of this chapter is fisheries as a recreational resource. However, it is often overlooked that as a by-product of "recreating" a considerable amount of food is harvested. This relates to our earlier chapters that deal with the population/food imbalance throughout the world. Thus, we will now deal with *food* from recreational fishing.

In 1973, the National Marine Fisheries Service collected information on recreational boats in the United States. A summary article was published by Ridgely (1975). Of over 8 million privately owned recreational boats, approximately 1 million are used for saltwater fishing. In addition to the 1 million recreational saltwater fishing boats, 2,500 are operated for commercial purposes (e.g., partyboats). By and large, therefore, anglers own their own boats for recreational saltwater fishing. Earlier in this chapter, the *NSFH* reported (table 6.5) that 22 percent of expenditures by saltwater fishermen in 1970 were for boats and motors. This represents an annual increment to the stock of existing recreational fishing boats of over *one-quarter of a billion dollars.*

The National Marine Fisheries Service (1975) has also made a survey of participation in marine recreational fishing in the northeastern United States during 1973–1974. Unlike the *Salt-Water Angling Survey,* it included participation in shellfish recreational saltwater fishing. In the fourteen northeastern states (including the District of Columbia), the survey estimates that there are nearly 5 million recreational fishing households and nearly 11 million participants. In 1970, the *Salt-Water Angling Survey* (*SWAS*) reported only 3.4 million participants. The latter survey (1) *excluded* persons under twelve years of age; (2) *excluded* those fishing less than three days or spending less than $7.50; (3) *excluded* recreational shellfishermen; and (4) appeared four years before the former surveys; increased participation therefore accounts for some of the difference. From 1965–1970, the number of participants in saltwater recreational fishing grew at 2.5 percent annually in the United States. If we extrapolate for four years the 3.4 million saltwater anglers who fished from Maine to Cape Hatteras, North Carolina (Northeast), and who were reported by the *SWAS,* we have 3.74 million anglers. The ratio of reported participation of 11 million to 3.74 million is 2.94. If this ratio holds nationally, we estimate that 30.6 million people took part in saltwater fishing during 1973–1974 (i.e., a one-year interval), or 14 percent of the entire U.S. population. Although most participants with residences in the Northeast fished in that region, the one noteworthy exception is Florida, where nearly 1.5 million people with residence in the Northeast fished. Thus, marine recreational fishing is a significant aspect of Florida's tourism-based economy. Centaur Management Consultants (1977) estimate that over 50,000 man-years are associated with marine recreational fishing, with sales of $1.8 billion current dollars to anglers in 1975. Finally, Bell and Canterbery (1975) have recently estimated gross expenditures and nonmarket value to the consumer for saltwater recreational fishing to be $3.1 billion in 1974 for the twenty-three U.S. coastal states. Stroud (1977) reports that in 1975 saltwater anglers spent $2.1 billion dollars on fishing goods and services. The data presented here are quite fragmentary, but they do give some notion of the considerable economic significance of marine recreational fisheries to the U.S. economy. Now let us look at the food by-product of saltwater recreational fishing.

Food from Recreational Fishing: A Neglected Area

In 1973, the National Marine Fisheries Service reported that the direct per capita consumption of seafood averages 12.6 pounds in the United States. This figure represents edible fish that enter the usual commercial channels, that is, about 2.6 billion pounds of fish at edible weight. The commercial figures ignore the very substantial contribution that fish harvested by recreational fishermen make to the national diet. The 1970 *SWAS* put the 1970 catch by anglers of saltwater fish at 1.58 billion pounds (round weight). If we calculate the edible weight to be about 40 percent of round or live weight, the corresponding edible weight was about 0.632 billion pounds. This would be equivalent to about one-fourth (24.3 percent) of the amount of edible fish that enter the national diet through commercial channels. An estimated 0.308 billion pounds, the edible weight of freshwater fish taken by anglers, should also be added. The latter would be equivalent to 11.8 percent of the commercial catch. Moreover, the commercial catch includes shellfish, which are not included in the recreational catch statistics. Assuming that shellfish are 15 percent by weight of the total marine anglers' catch, table 6.14 provides a breakdown of aggregate food from the sea. Thus, the recreational catch may be as much as 28 percent of the total food fish consumed in the United States. Furthermore, we have not included fish used for reduction purposes.

Projections of the Demand for Recreational Fishing

Increasing population, higher real per capita incomes, shorter workweeks (fewer hours and fewer working days), and longer

Table 6.14 Food from Commercial and Recreational Fishing

		Aggregate (billions of pounds)	Per Capita (pounds)
Commercial		2.600	12.6
Recreational		1.052	5.0
A. Marine			
Finfish	.632		
Shellfish	.112		
Total	.744		
B. Fresh Water	.308		
Total (commercial plus estimated recreational)		3.652	17.6

vacations mean more time and money for leisure and more recreation, including marine-related recreation. According to a study by Cicchetti, Seneca, and Davidson (1969), recreational days of water-related activities are expected to be 42 percent above the 1965 base by 1980 and 152 percent above by 2000, as shown in figure 6.10. Though the demands for sport fishing are projected to increase significantly, sport fishing opportunities are in fact expected to decrease because of increasing demands for sport fish waters and habitat to be used for industrial, residential, and waste deposition purposes, because of difficulties in increasing the sport fishing resources (mentioned above), and because of continually increasing demands for sport fish as commercial food products. Future disequilibriums between demands and opportunities for sport fishing are expected to result. Thus, recreational fishing was projected to increase by only 37 percent over the 1965–2000 period, or a little over 1 percent annually measured in terms of recreational days fished (saltwater and freshwater combined). This is in sharp contrast to the more than 5 percent annual growth rate in recreational fishing days, both saltwater and freshwater, for the 1955–1970 period. Adams, Lewis, and Drake (1973) estimated that recreational fishing days will increase by 1.75 percent annually over the 1972–1978 period. Thus, this short-term projection is lower than historical rates of growth. The effect of increasing demand with constant or dwindling supply will most certainly raise the real value of a fishing recreational experience. It will also increase public pressure for government intervention to allocate scarce natural resources properly.

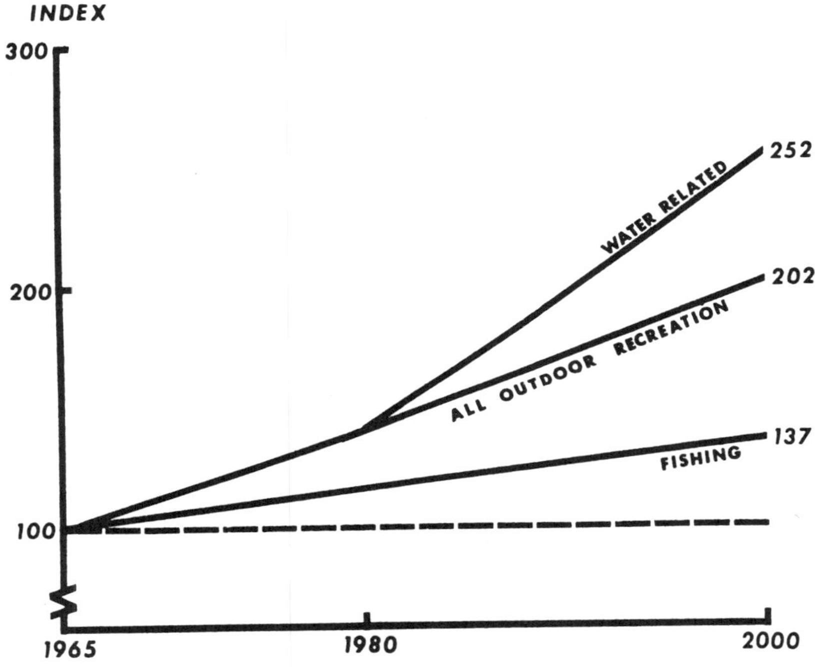

Figure 6.10: Projections of Recreation Days Fishing, Water Related and All Outdoor Recreation from 1965 to 1980 and 2000

Source: Cicchetti et al. (1969)

Summary

Too little attention has been given to fishery resources used for recreational purposes. In 1970, gross expenditures on saltwater recreational fishing were $1.2 billion, and approximately three times as much was spent on freshwater recreational fishing, for a total of $4.96 billion. In contrast, the 1973 *commercial* domestic landings are estimated to have a retail value of $3.0 billion. From an economic standpoint, therefore, recreational fisheries are indeed a "sleeping giant"; they should be given more recognition by both the state and federal governments. The saltwater recreational fishing sector expanded rapidly over the 1955–1970 period, but future projections are less optimistic owing to limitations on resource supply. There are indications that the resource productivity of marine recreational fisheries is declining. However, more documentation is needed in this area.

By and large, individuals do not recreate for fish alone, but principally to "take it easy and be outdoors," although the fishery itself is certainly necessary to the recreational experience. One principal problem with recreational fishing is that it is a nonmarket activity, where no "price" or user's fee is charged for the use of the fishery resource. Most studies emphasize gross expenditures as the "value" of a recreational fishery. If there were no fishery, these expenditures would undoubtedly be reallocated to other forms of recreation. Therefore, it is often argued that the real value of the fishery is the net market value, or what people would pay for the use of the fishery resource if fees were imposed. The Clawson technique has been extensively used to estimate the demand curve for the use of the resource. Several examples were discussed in this chapter. In addition, user-opinion surveys to ascertain the "value" of the fishery resource have yielded estimates of the consumer surplus (i.e., what people would be willing to pay or benefit—in dollars—derived from the use of the resource). Several fallacious techniques of valuation were discussed in order to point out their lack of economic sense.

Sport fisheries are faced with many problems, such as depletion, commercial conflicts, and destruction of habitats. Like commercial fisheries, most sport fisheries are common property in nature; therefore, government intervention is necessary to solve the failure of the private market mechanism. Because of difficulties in valuation, sport fisheries remain a difficult and complex area for government action.

Notes

1. The national income "accounts" prepared by the Department of Commerce include (among numerous consumer expenditures) those made for "recreation." Expenditures as used here (i.e., Robert Nathan Associates) are total sales less purchases, or the "value added" by capital and labor. Thus, $5.18 billion dollars were spent on ocean-related outdoor recreation. This figure *cannot* be directly compared with the $1.2 billion reported by *NSFH,* since the latter includes purchases. The ratio of total sales to value added for recreational fishing expenditures is very roughly 2.4; therefore, the comparable figure is $0.5 billion ($1.2 billion ÷ 2.4). That is saltwater fishing is about 10 percent of total ocean-related outdoor recreation expenditures (i.e., $0.5 billion ÷ $5.18 billion).

2. The U.S. Sport Fishing Institute doubts the statistical validity of the data base used to conclude there is declining catch per unit of effort in sport fisheries. The reader should see Ronald M. North (1976) for a debate on this issue. Thus, the catch per angler and per angler day over the 1960–1970 period is far from clear using aggregate data. The reader may draw his own conclusions.

3. Market-priced recreation is presently limited, such as fishing for Atlantic salmon (Canada and Northern Europe). The extremely high *rentals* paid for land on salmon streams in the Canadian Maritimes demonstrates that sport fishing can be priced directly. This is also true in England.

4. As we indicated earlier, gross expenditures do include some actual charges for the right to fish, but these are negligible.

5. The capital to build the Ferris wheel was allocated on the basis of the same principle. Capital is "scarce" (i.e. not free), and the use of capital to offer Ferris wheel services was determined on the basis of its net market value. If building *more* roller coasters at the expense of less Ferris wheels would *increase* net market value, the price mechanism would act to make this allocation. Unfortunately, there is no price charged for recreational fishing, and the concept of market allocation breaks down.

6. A further qualification is that the travel cost method will equal the value of the fishing right if the only aspect of the trip is the fish. This does not agree with table 6.4. Expenditures of time and money produce a recreational experience composed of a bundle of experiences, such as the pleasure of travel, lodging away, and natural surroundings.

7. The present user of the resource does not have to pay. With the Clawson method, we derived a demand curve, or willingness-to-pay curve. This is still implicitly in the consumer's mind; therefore, the user-opinion survey might measure the total area under the potential demand curve, or consumer surplus.

References

Adams, Robert L.; Lewis, Robert C.; and Drake, Bruce H. 1973. *Outdoor recreation: a legacy for America; appendix A: an economic analysis.* USDI Bureau of Outdoor Recreation.

Bell, Frederick W., and Canterbery, E. Ray. 1975. *An assessment of the economic benefits which will accrue to commercial and recreational fisheries from incremental improvements in the quality of coastal waters.* National Commission on Water Quality, NTIS, U.S. Department of Commerce. PB-252172-01-12.

Centaur Management Consultants. 1977. *Preliminary Report on the Economic Impact of Marine Recreational Fishing in 1975.* Prepared for the National Marine Fisheries Service.

Cicchetti, Charles J.; Seneca, Joseph J.; and Davidson, Paul. 1969. *The demand and supply of outdoor recreation—an econometric analysis.* Bureau of Outdoor Recreation USDI contract no. 7-14-04-4.

Clawson, Marion. 1959. *Methods of measuring the demand for and value of outdoor recreation.* Washington, D.C.: Resources for the Future.

Clawson, Marion, and Knetsch, Jack L. 1966. *Economics of outdoor recreation.* Baltimore: Johns Hopkins Press.

Cleary, Donald P. 1971. Sport/commercial conflicts and the National Marine Fisheries Service's role in fishery management. National Marine Fisheries Service file manuscript no. 76.

Crutchfield, James A., 1962. Valuation of fishery resources. *Land Economics* 38:145–154.

Crutchfield, James A., and MacFarlane, Douglas. 1968. *Economic valuation of the 1965–1966 salt-water fisheries of Washington.* State of Washington research bulletin no. 8.

Fish and Wildlife Service (USDI). 1970. *National survey of fishing and hunting.* Resource publication 95.

Gordon, D.; Chapman, D. W.; and Bjornn, T. C. 1973. Economic evaluations of sport fisheries—what do they mean? *Transactions of the American Fisheries Society* 102:293–311.

Hammock, J. 1969. Toward an economic evaluation of a fugitive recreational resource: waterfowl. Ph.D. dissertation, University of Washington.

Horvath, Joseph C. 1974. Economic survey of southeastern wildlife and wildlife-oriented recreation. Paper read at 39th North American Wildlife and Natural Resources Conference, Wildlife Management Institute, April 1974.

Hotelling, Harold. 1949. Letter cited in *The economics of public recreation: An economic study of the monetary evaluation of recreation in the national parks.* Washington, D.C.: U.S. Government Printing Office (National Park Service).

Knetsch, Jack L. 1963. Outdoor recreation demands and benefits. *Land Economics* 39:387–396.

MacCall, Alec D.; Stauffer, Gary D.; and Troadec, Jean-Paul. 1974. Stock assessment, fishery evaluation, and fishery management of Southern California recreational and commercial fisheries. National Marine Fisheries Service contract no. 03-4-208-160 to California Department of Fish and Game.

Mathews, S. B., and Brown, Gardner. 1970. *Economic evaluation of the 1967 sport salmon fisheries of Washington.* Washington Department of Fisheries technical report no. 2, Olympia, Washington.

Nathan, Robert R., Associates. 1974. *The economic value of ocean resources to the United States.* Report submitted to the Congressional Research Service, Library of Congress, Washington, D.C.

National Marine Fisheries Service. 1973a. *1970 salt-water angling survey.* Current fisheries statistics 6200.

——. 1973b. *Workshop in fishery economics.* Contract no. N208-0350-72.

——. 1974. *The United States marine fishery resource.* MARMAP contribution no. 1.

——. 1975. *Participation in marine recreational fishing: northeastern United States, 1973–74.* Current fisheries statistics 6236.

North, Ronald M. 1976. Economic values for marine recreation fisheries. In *Marine recreational fisheries,* ed. Richard H. Stroud and Henry Clepper. Washington, D.C.: Sport Fishing Institute.

Pearse, Peter H. 1968. A new approach to the evaluation of non-priced recreational resources. *Land Economics* 44:87–89.

Piehler, Glenn R. 1972. An investigation of the Massachusetts marine sport fisheries. Ph.D. dissertation, University of Massachusetts.

Richards, Jack A. 1968. An economic evaluation of Columbia River anadromous fish programs. Ph.D. dissertation, Oregon State University.

Ridgely, John. 1975. Selected information on recreational boats in the United States. *Marine Fisheries Review*, paper 1122.

Singh, Aimer. 1965. An economic evaluation of the salmon-steelhead sport fishery in Oregon. Ph.D. dissertation, Oregon State University.

Sport Fishing Institute. 1974. Why do people fish? *Sport Fishing Institute Bulletin*, no. 258.

Stroud, Richard H. 1977. Recreation fishing. In *Wildlife in America*. Council on Environmental Quality.

Trice, Andrew H., and Wood, Samuel E. 1958. Measurement of recreation benefits. *Land Economics* 34:195–207.

7. *Aquaculture: A Food Panacea?*

Aquaculture, the farming and husbandry of freshwater and marine organisms, is a very old and highly productive management practice. The culturing of oysters, to take one prominent example, has been known at least since Roman times, and the culturing of Chinese carp may have a history at least as long as that. The history of aquaculture shows that it originated and developed under different social and economic conditions in various parts of the world. The greatest concentration of aquaculture is in Asia and the Far East, particularly China. Aquaculture currently provides an estimated 10 percent of the world's water-derived protein—a harvest of between 5 and 6 million metric tons valued in excess of $2.5 billion (Caton, Moss, and Urano 1974). As noted in chapter 4 the FAO estimates that the demand for food fish (excluding industrial demands) will increase to over 107 million metric tons; however, the overexploitation of traditional wild stock species (chapter 3) has created an interest in aquaculture as a source of increased supply of fish protein in this food-starved world. This is especially true for a number of high-priced species, the stocks of which appear to be substantially depleted by overfishing.

The heavy concentration of aquaculture in certain areas of the world, principally the Indo-Pacific region, suggests that the practice might be more greatly emphasized in other areas of the world than is now the case. The fact that aquaculture output has approximately doubled over the last five years and the fact that some countries already rely upon aquaculture for up to 50 percent or more of their total fisheries production—both portend an important future for aquaculture.

The reasons for the recent and projected rapid growth of aquaculture activities are (1) the growing and incessant world demand for protein; (2) the rising costs of commercially caught fish as maximum sustainable yields of more and more wild stock species are approached or exceeded; and (3) aquaculture's more favorable feed conversion rates and higher productivity per hectare than in traditional agricultural methods and products.

It is clear that world aquaculture is well beyond the purely experimental stage. Profitable technologies exist that can and are being transferred within and among various countries. The world protein demand and food supply problems are bound to force the continuation of aquaculture technology transfer and technology enhancement.

The purpose of this chapter is to give a brief survey of current world aquaculture and to introduce the reader to the fundamental biological and economic factors that interact to produce a successful or profitable fish farm. We shall also ask whether the waters of the earth may still be a panacea for the continuing imbalance between population and food supplies, which has been a recurring theme of this book.

Present World Aquaculture: Extent and Distribution

A World Survey

The exact extent of world aquaculture is not fully known. The problem is partly definitional. At what point does wise and careful management of wild natural resources shade off into the farming of freshwater and marine organisms? But the problem is mostly one of measurement. Many aquaculture operations are carried out in remote areas, and the product is either consumed on the farm or bartered locally and in any event never included in national or world production figures.

On the basis of FAO data collected from thirty-six countries (1970), world aquaculture output of finfish in 1970 was estimated to have been over 3.6 million metric tons. The estimated production of mollusks for that same year was put at 1 million tons, and that of seaweeds at about one-third of a million tons. Thus total world aquaculture output in 1970 seems to have been approximately 5 million tons, exclusive of the production of sport fish and bait fish, ornamental fish, and cultured pearls. The precise extent of sport and bait fish culture apparently is not known, but it is obviously a large and growing activity. Ornamental fish culture must also be large, but it, too, has not been measured. It is estimated that the United States alone imports over $80 million worth of ornamental fish annually. Pearl culture, too, is a large and expanding activity. In 1969 Japan produced over 100 tons of pearl oysters valued at more than $500 million (Furukawa 1971).

In a more recent study, Pillay (1973) calculated finfish aquaculture production in forty-two countries, including mainland China, at about 3.7 million metric tons. Table 7.1 compares aquaculture finfish production as a percent of total catch (finfish plus shellfish) for these forty-two countries. It is of interest that over one-third of China's fish harvest came from finfish aqua-

Table 7.1: A Comparison of Aquaculture Finfish Production with Total Catch by Selected Countries, 1970 (metric tons)

Country	Total Catch	Aquaculture Fin Fish Production	Aquaculture Fin Fish % of Total Production	Aquaculture Fin Fish Area (hectares)
China (Mainland)	6,880,000	2,240,000	32.6	700,000
India	1,845,000	480,000	26.0	607,915
U.S.S.R.	7,340,000	190,000	2.6	126,666 (est.)
Indonesia	1,249,700	141,075	11.3	266,300
Philippines	1,049,700	94,573	9.0	164,414
Thailand	1,571,600	87,764	5.6	58,509 (est.)
Japan	9,994,500	85,000	0.9	508
Taiwan	650,200	56,135	8.6	39,234
U.S.A.	2,766,800	40,200	1.5	28,300
Pakistan & Bangladesh	416,500	37,540	9.0	30,780
Malaysia	390,300	25,648	6.6	90,473
Hungary	26,000	19,697	75.8	22,000
Italy	391,200	18,000	4.6	12,000 (est.)
S. Vietnam	587,200	16,500	2.8	2,500
Yugoslavia	49,840	15,840	32.2	9,747
Ceylon (Sri Lanka)	87,700	15,000	17.1	10,000
Rumania	68,700	12,000	17.5	6,400
Denmark	1,400,900	11,000	0.8	7,333 (est.)
Poland	517,700	10,909	2.1	62,791
Czechoslovakia	13,000	10,641	81.9	42,798
Israel	28,200	10,220	36.2	4,904
Brazil	515,400	9,967	1.9	6,644 (est.)
Mexico	402,500	9,026	2.2	12,650
Khmer	171,100	5,000	2.9	3,333 (est.)
Germany, East	323,100	3,669	1.1	2,446 (est.)
Germany, West	507,600	2,627	0.5	11,824
Burma	442,700	1,494	0.3	2,920
Zaire	145,800	1,406	1.0	4,058
Bolivia	1,200	1,400	---	25,502
Austria	4,000	780	19.5	3,000
Hong Kong	114,100	690	0.6	829
Zambia	39,300	689	1.8	459 (est.)
Uganda	137,000	670	0.5	410
Madagascar	48,000	615	1.3	1,280
Norway	3,074,900	600	.02	400 (est.)
Singapore	15,000	554	3.7	890
Nigeria	155,800	127	0.1	85
Kenya	35,000	122	0.3	610
Spain	1,498,700	50	.003	33 (est.)
S. Korea	1,073,700	40	.003	76
Ghana	220,400	30	.01	204
Puerto Rico	57,700	25	.04	135
	69,400,000	3,657,373	5.3 (avg.)	2,643,551

Source: Based upon Pillay (1973) and FAO Fisheries Yearbook (various) and estimates of total catch by the National Marine Fisheries Service.

culture. Hungary and Czechoslovakia derived over three-quarters of their catch from finfish aquaculture. In terms of aggregate physical production of finfish through aquaculture techniques, China, India, the USSR, and Indonesia are the leading producers. Of the forty-two countries surveyed, nearly 90 percent of the aquacultured finfish production came from Asian and Indo-Pacific countries. This area of the world is the major center for aquaculture industries; other nations that aspire to develop aquaculture as a source of increased protein can learn much from it. The United States is still in its infancy as far as aquaculture is concerned; only 1.5 percent of the catch is from finfish aquaculture sources.

Pillay also estimated shrimp and prawn aquaculture production for seven countries at 14,000 tons, oyster culture in eleven countries at 710,000 tons, mussel production in four countries at 180,000 tons, cultured clam output in four countries at 56,000 tons, and miscellaneous mollusk production in one country at 20,000 tons; seaweed culture in two countries was estimated at 373,000 tons. This is shown in table 7.2. Total measured aquaculture output for these selected countries and these products was thus slightly above 5 million metric tons annually.

The total area used in world aquaculture production also is not known, but statistics on finfish production indicate that the average production per hectare is about 1.5 metric tons per year.

Table 7.2: Estimated World Production Through Aquaculture Exclusive of Fin Fish Production (metric tons),1970

Country	Shrimps and Prawns	Oysters	Mussels	Clams	Other Mollusks	Seaweeds
Australia		9,800				
Canada		4,100				
France		34,200	39,800			
India	3,800					
Indonesia	3,328					
Italy			13,700			
Japan	1,800	194,600		10,800		357,000
Korea		45,700	16,800	16,800	20,000	16,000
Malaysia	250		28,600			
Mexico		43,500				
New Zealand		10,700				
Philippines	2,500					
Portugal		2,900				
Singapore	120					
Spain		1,800	109,700			
Thailand	2,500					
Taiwan		12,700		60		
United States		350,500				
	14,298	710,500	180,000	56,260	20,000	373,200

Source: Pillay (1973)

If this estimate is accurate, there are more th[illegible]
tares of water impoundments used in world a[illegible]
tion. This sum is exclusive of the area used [illegible]
seaweed production, which is most often an o[illegible]
estaurine area.

Current trends in aquaculture production [illegible]
ascertain than current production statistics. Ac[illegible]
one source (Caton, Moss, and Urano 1974), aquaculture production now is double that of five years ago. Whether or not this estimate of growth is reasonably precise, it is evident that a very high rate of growth in aquaculture production could be sustained for a considerable period merely by extending present technology to additional areas and by raising the output of existing areas to currently feasible levels. The prediction of Caton, Moss, and Urano that aquaculture output can reach at least 50 million metric tons by the year 2000 seems quite feasible: after all, much current aquaculture uses a relatively simple technology, which has not been applied extensively. It seems quite crucial to reach their figure of 50 million metric tons—in light of present and projected deficiencies in world protein production.

Aquacultured Species

There are some 20,000 known species of fish. This does not include various aquatic plants, shellfish, and other marine life. Most of these organisms can be manipulated during at least one stage of their life before harvest, that is, manipulated for the purpose of increasing production and yield. Given current technology, most of these organisms cannot be completely controlled from spawning to market. But many commercially valuable aquaculture species likewise cannot be completely controlled by man. For example, milkfish, shrimp, and a number of other valuable aquaculture species cannot be spawned in capitivity.

Various sources list over 200 already cultured species of finfish and shellfish, and it is an incomplete list at that. Ling (1972) lists 20 species of finfish, 25 species of crustaceans, 20 mollusks, and 10 algae already being cultivated in the coastal areas of the Indo-Pacific region alone, with roughly one-third of these being cultured extensively. To Ling's list one would have to add freshwater species, which were not covered in his study, as well as numerous species cultured exclusively outside the Indo-Pacific region and hence not covered by him. It is clear that an extremely wide variety of fishes and other marine organisms might be ennced through aquaculture if the attempt were made to do so.

At present the most extensively cultured species of fish and shellfish appear to be those shown in table 7.3 (ranked in general order of economic importance). However, this list does not fully reveal aquaculture's potential role in helping to meet world protein needs. Many aquacultured species are not even produced for food. Sport fish and bait fish are not produced primarily for direct consumption as food. Moreover, most aquacultured species are produced for the luxury trade. Species emphasized in current aquaculture are often selected because they command a favorable market price, not because their cost of production per unit of protein content is low. This will be discussed in greater detail in the latter part of this chapter.

Aquaculture in the United States

Over the last ten years, intense interest in the United States has been generated in the business of fish-farming. This interest is shared by small businessmen as well as giant corporations, which are constantly on the lookout for a promising new business or a good investment for diversification. Commercial fish-farming in the United States today is being conducted on a larger scale than most people realize. This is evident not only in the raising of the more familiar species of fish, such as trout, catfish, and bait min-

Table 7.3: Fin Fish and Shellfish Species Most Extensively Cultured

Species	Estimated Production
Common Carp	
Chinese Carp	210,000 metric tons in 1965
Indian Carp	
Tilapia	N A
Shrimp and Prawns	14,298 metric tons
Oysters	710,000 metric tons
Milkfish	167,000 metric tons
Mullet	N A
Clams and Mussels	236,260 metric tons
Clarias	N A

Source: Pillay (1972, 1973); Bardach (1972).

nows, but also in the cultivation of mullet, salmon, bass, and carp. Table 7.4 shows some very crude estimates of the magnitude of the U.S. aquaculture industries (*American Fish Farmer* 1969). The five largest industries have a retail value of over $218 million, which is indeed considerable. Other species, such as *mullet, salmon, bass, bream, sunfish, crappie, carp, buffalofish,* and *pompano,* are now under some degree of cultivation (mostly experimental). They have great potential for the future. In 1969, the ex vessel value of the U.S. fishery catch from *wild stocks* (i.e., does not include artificially cultivated fish and shellfish) was $526.2 million dollars. If we use markup of three times the ex vessel price (see chapter 2), the estimated retail value would be $1.58 billion. If we use the most conservative estimated retail figures on aquacultured species shown in table 7.4, the five leading aquaculture industries may account for as much as 14 percent of the retail value obtained from the wild stock.

As will be discussed in chapter 9, the commercial fisheries in the United States today are not meeting the demand for fish. In 1950, the United States imported 25 percent of its total supply of fishery products (industrial and edible). In 1972, imports had reached 65.8 percent of the total supply. There is no lack of potential in the United States for relatively low-cost aquacultured products. In addition, farm fish are becoming more useful for recreational use, since commercial recreational fishing lakes no longer can operate successfully without the fish farmer.

Potential Role of Aquaculture as a Source of World Protein

As discussed in chapter 2, fish are a superior source of animal protein. If the amino acid pattern is considered, human beings can utilize at least 83 percent of the raw weight of fish, as compared to 80 percent of the raw weight, for example, of beef. The differ-

Table 7.4: Principal Aquaculture Industries of the U.S., 1969

Industry	Value at Retail million $	Value to Producers million $	Lbs Produced millions	Fixed Investment millions $
1. Bait Minnows	125-130	40-30	40-50	25-35
2. Oysters	45- 51	15-17	31*	10-15
3. Catfish	29- 31	9-11	25	18-20
4. Trout	18- 20	10-12	13	10-12
5. Crayfish	1.4-1.6	1-1.2	4	.8-1.0
Total (highest)	218.4	75.	113	63.

* Less shell
Source: AFF (1969).

ential between fish and grains is much greater. And, of course, animal protein is especially valuable to humans, because it contains two essential amino acids (lysine and methionine) not found in adequate amounts in vegetable protein.

Generally, a pound of fish can also be produced more cheaply than a pound of red meat. Fish, being cold-blooded, take on the temperature of their environment instead of expending calories to maintain a constant body temperature. Moreover, fish live in an environment that supports them, but land-based animals must use a good deal of energy to develop a skeletal system and in general to support themselves against gravity. For these reasons, fish are better feed converters than land-based animals. Feed conversion rates for fed fish are about one and one-half times as great as for swine or chickens and about twice as great as for cattle or sheep (Bardach, Ryther, McLarney 1972). Fish can be crowded more closely than land-based animals because to some extent the habitat of fish is three-dimensional. That is, they can utilize the water column. Thus, in well-managed environments 2,000–3,000 kg or more of fish can be produced per hectare per year, but the maximum figure for cattle is 500 to 700 kg (Delaney and Schmittou 1973). And although some fish must be fed a ration as expensive as cattle or poultry feedgrains, many varieties of fish and other marine organisms can live very well on nutrients found naturally in their aquatic environment or fostered through fertilization of their environment, mostly by the addition of *phosphates* (i.e., fertilizer).

Aquaculture often requires more labor and sometimes more capital per unit of food output than other types of farming, but in many cases the labor required for aquaculture operations can be scheduled around the peak labor demand periods for agriculture, as for example at planting and harvesting time, so that fish rearing, particularly on a small scale, can at least partly complement general farming. Aquaculture can often make use of low-value land—ravines, swampland, saltwater marsh, or mangroves—that is not well suited to other uses. And the capital construction requirements for aquaculture, though often moderately expensive, sometimes complement water conservation, provide recreation, or even help in flood control—which helps to justify the construction expense even more.

Aquaculture seems a logical alternative source for fish protein production, one to which the world increasingly will have to turn. Present-day aquaculture often is, however, not a cheap source of protein, because *much aquaculture today is devoted to the production of luxury foods.* Fish-farming is now a business, often more profitable than agriculture (Shang 1973). It is not surprising that profit-motivated fish-farming has concentrated much effort on the production of high-value output: *shrimp, prawns, eels, channel catfish, salmon, yellowtail,* and others. These aquaculture products are not the cheapest source of animal protein: they usually require the feeding of prepared foods, which together with the typically higher labor and capital requirements of aquaculture operations, pegs the final product price at a high level. To be an economical source of protein, aquaculture must concentrate on the production of species that are on the bottom of the food chain. Feeding fish to fish or even grain to fish is apt to be a losing proposition unless the produced fish has a high market value. Only where trash fish are available in large numbers would one normally want to violate the general rule and feed fish protein to fish, that is, if cheap animal protein is the goal. Trash fish as a source of fish feed may be produced in some numbers by the richer nations and by those nations whose people do not have a particular fondness for fish. Trash fish are not produced in significant numbers by nations that are short of food and whose people eat much fish. Filipinos, for example, eat nearly every species of fish, will eat even the very smallest fish, and will eat nearly every part of every fish, including fins, heads, and most of the entrails. They produce a lot of fish food but not much fish waste.

In the main, aquaculture species that could serve as economical sources of protein would have to be those that are fairly popular as food, but also that are low on the food chain, hardy, easy to culture, and fast-growing. Aquaculture should concentrate on such species as *carp, tilapia, milkfish, mullet,* and *mussels.* If it does, it is bound to make an increasingly important contribution to world protein production. Now let us consider more specific principles governing aquaculture as an economic enterprise.

Selection of Species for Aquaculture

The selection of species for aquaculture farming is extremely complex. First, it would seem logical to establish some environmental and economic criteria for selection. Second, social acceptance of fish or particular species should also be considered. Last, but by no means least, is data availability. Economic data on aquaculture enterprises throughout the world are very difficult to obtain. Bits and pieces of information must be put together in order to form a rudimentary profile of a "typical" aquaculture enterprise in any given country.

In selecting species and plants for aquaculture, the following economic feasibility criteria should be considered:

A. Environmental Conditions
    1. Species that feed low on the food chain
    2. Species that have fairly wide temperature tolerance
    3. Few breeding problems
    4. Good growth rate
    5. Fair amount of research on biological aspects of fish or plant in question
B. Socioeconomic Factors
    1. Established and profitable enterprises operating for some years
    2. High growth rate in overall production and widespread use as protein in many countries
    3. Adaptable to small-farm operations (especially in developing countries)
    4. Potentially high consumer acceptance
    5. Needs little or no feed or fertilizer
    6. Readily available cost, earnings, and factor input data
    7. High potential for foreign exchange

Generally, there is an interaction between environmental and economic factors. For example, if we hold all other factors constant, species that feed low on the food chain do not have to be fed protein, thereby reducing cost (see chapter 3 for a discussion of the food chain). Species such as oysters, shrimp, and mussels are examples of aquacultured shellfish that feed low on the food chain. Many of the social and economic criteria have to be determined by a feasibility study. Aquacultured shrimp may have a high potential for foreign exchange, but it must be shown that they can compete on the world market, which consists mostly of shrimp caught from wild stocks. These factors will be considered in a feasibility model considered below.

Bioeconomic Feasibility Model

It is fairly obvious that many aquaculture enterprises are extremely complex operations, sometimes involving the production of fry or fingerlings or both jointly. It is the purpose of this section to outline a general bioeconomic model that may be applicable to the "typical" aquaculture enterprise. The model itself is more or less a systems approach to evaluating the impact of differing factor prices and environmental variables on the economic feasibility of any aquaculture enterprise in any particular country or area. The following terms will be used:

C_n = total cost of production for a "typical" enterprise of the *n*th species
Q_n = total output (quantity) for the "typical" enterprise of the *n*th species
K_n = capital (equipment) input of the *n*th species
L_n = labor (man-hours) input of the *n*th species
D_n = land input of the *n*th species
Z_n = fertilizer input of the *n*th species (if applicable)
F_n = feed input of the *n*th species (if applicable)
S_n = stocking of (fingerlings, fry) input of the *n*th species (if applicable)
I_n = other economic inputs of the *n*th species[1]

where p_K, p_L, p_D, p_Z, p_F, p_S, and p_I are input prices (i.e., cost per unit of input) for *capital, labor, land, fertilizer, feed, stocking,* and other *inputs,* respectively, for the aquaculture farm.

$E_1, E_2, E_3, \ldots, E_d$ are necessary environmental variables for viable aquaculture operations for the *n*th species. The generalized production-input relation for the *n*th species is[2]

$$Q_n = f(K, L, D, Z, F, S, I, E_1, E_2, \ldots, E_d), \quad (7.1)$$

where the nonmarket environmental variables ($E_1, E_2, \ldots, E_d$) enter the production process and ultimately the cost of production. Examples of these variables are water temperature, rainfall, nutrients, salinity, and biological organisms. To simplify the analysis, we shall view $E_1, E_W, \ldots, E_d$ as *output shifters,* which influence the level of production. Therefore (7.1) may be respecified as

$$Q_n = f(K, L, D, Z, F, S, I)\ E_1, E_2, \ldots, E_d\ . \quad (7.2)$$

Hence, the *total cost of output unadjusted for environmental changes* for the "typical" aquaculture enterprise in any area is

$$C_n = [p_K K + p_L L + p_D D + p_Z Z + p_F F + p_S S + p_I I]\ . \quad (7.3)$$

Equation (7.3) specifies the economic dimension of the aquaculture enterprise, but it must be combined with environmental considerations.

As indicated above, there are a number of critical environmental variables that may influence the level of production (E_1, E_2, . . ., E_d). These may be combined into one net variable E. Therefore, we have the following simplified concept for species n:

$$\left(\frac{C}{EQ}\right)_n = \frac{\text{economic cost}}{\text{environmentally adjusted output}} \quad . \qquad (7.4)$$

Equation (7.4) merely specifies our estimate of the aquaculture enterprise's unit cost of production adjusted for (1) factor prices, and (2) environmental conditions. As in our discussion of opportunity cost in chapter 3, the returns to capital ($p_K K$) and labor ($p_L L$) must be adequate to hold these productive factors in the aquaculture sector or if no aquaculture sector exists, must be sufficient to attract them into it.

With respect to economic feasibility, two kinds of questions usually face the decision-maker: (1) if the industry is already established, are rates of return—to capital and labor—sufficient to remain in the aquaculture business; or (2) if the industry is not established, will the aquacultured species be competitive with fish from the wild stock (i.e., how do unit costs of production through aquaculture compare with ex vessel prices obtained from the wild stock)? A second problem often occurs with the introduction of aquacultured products in countries that do not consume such species. For example, would Americans eat carp, milkfish, or true eel? Thus, the introduction of new aquacultured products might require a marketing feasibility study to determine consumer acceptance and what people would be willing to pay for unfamiliar species.

Table 7.5 shows an economic feasibility analysis using the general model discussed above. In Hawaii, for example, the culturing of prawns or shrimp involves stocking, feed, land, maintenance, and labor costs that are approximately equal to total revenue at *current* wild stock ex vessel prices, thus showing a small profit. At current wild stock prices, there would be a slight gain to capital; therefore, this operation may be economically feasible. Remember, the lower cost of input prices (e.g., labor) in other areas of the world might make this prawn operation quite feasible, since there is a well-developed international market for shrimp. According to the feasibility analysis, oysters, clams, salmon, and prawns

Table 7.5: Economic Feasibility Analysis for Aquaculture Enterprises

Species	Stocking Number	Stocking Cost	Feed Cost	Maintenance & Land Cost	Labor Costs	Total Cost	Total Revenue	Profit or Loss
Oysters[1]	15,000	$ 5,250	$ 0	$ 42,847	$ 9,447	$ 57,544	$ 74,233	$ 16,689
Clams[2]	5,000,000	75,000	0	29,847	10.025	114,872	149,519	34,647
Salmon[3]	2,600,000	195,000	60,100	753,200	25,900	1,034,200	1,107,743	73,543
Prawns (Shrimp)[4]	924,500	5,242	17.43	404.20	11,740	17,404	19,561	2,157
Lobsters[5]	250,000	402,000	173,000	126,000	134,000	835,000	511,695	- 323,305

[1] Based upon a 24 hectare farm off the New York Coast (Long Island). Number stocked will produce 63,511 kg. of meat weight or 139,724 lbs. Present (1975) price per pound (meat weight) is $.5312. Source and Comments: John Bardach, John Ryther, and William McLarney, Aquaculture, Wiley-Interscience, New York: 1972. Maintenance and land costs are based on extensive oyster culture where seed are grown in cages at the water bottom. Feed costs are zero since the species is allowed to feed off its natural environment. Labor costs are derived by converting labor in hours per year (on a similar Japanese farm) to U.S. levels employing efficiency measures in fishing from Arrow, Chenery, Minhas, Solow, "Capital-Labor Substitution and Economic Efficiency," Review of Economics and Statistics, August, 1961.

[2] Based upon a 5 acre farm off the Virginia coast (Chesapeake Bay). Number stocked will produce 492,000 lbs. Present (1975) price per pound (meat weight) is $.3039. Source and Comments: Winston Menzel, "The Mariculture Potential of Clam Farming," American Fish Farmer, July 1971. Menzel states that the market price for clams is $.03 per clam. He says optimum planting involves 25 seed per square foot. At this rate a 5-acre farm would produce 5,000,000 clams yielding $150,000. Stocking price ranges around $.015 per clam. Labor costs involve approximately five men (three full time). Paid the minimum wage this yields labor costs of $10,025.

[3] Based upon a large farm in the Northeast (Strait of Canso). Number stocked will produce 2,590,000 lbs. Present (1975) price per pound (meat weight) is $.4277. Source and Comments: W. D. Shields and J. A. Veinot, Strait of Canso Fish Farm Feasibility Study, 1969. The above costs are based on marine cage culture which involves the use of cages and pens constructed of nylon netting and supported by styrofoam floats. Tidal flushing through the pens performs the various functions of oxygen replenishment, waste removal, and supplementary feeding in the form of planktonic organisms.

[4] Based upon a 4.05 hectare farm off the Hawaii Coast. Number stocked will produce 46,255 lbs. Present (1975) price per pound is $.4229. Source and Comments: Yung C. Shang, "Economic Feasibility of Prawn Farming in Hawaii," Economic Research Center, University of Hawaii, June, 1972. Total revenue is determined on the assumption that 20 shrimp equal one pound and that the farm has a 70% survival rate.

[5] Based upon a 3 acre farm off the Coast of Massachusetts. Number stocked will produce 500,000 lbs. Present (1975) price per pound is $1.0234. Source and Comments: Kramer, Chin, and Mayo, consulting engineers, 1969. Schematic Lobster Rearing Station Design, Seattle, Wash. The stocking number is based on the rearing of cultured lobster to the weight of two pounds and extrapolated from a total production of 500,000 lbs. Farm size is based on the rearing of lobsters in cages 36' x 36' or 1296 sq. ft. stocked with 4,000 lobsters each.

Source: Bell and Canterbery (1975)

show *positive* profits. Whether these profits are sufficient depends on the investment in equipment and capital. A very simplified approach is to look at either (1) the rate of return on capital (i.e., profits divided by capital investment) or (2) the payback period (i.e., the number of years it would take to generate enough profits to pay for the initial investment). The rate of return on capital (*ROI*), or payback period, must always be related to alternative rates of return on investment. These yields will vary from country to country. Since the five species in table 7.5 have well-developed markets, their cost of production can be compared directly to prices obtained from wild stocks.

Table 7.6 Cost of Production: Aquaculture Versus Wild Stock

	Cost of Aquaculture Production (lb)[1]	Ex Vessel Prices (lb)
1. Oysters	$ .4118	.5312
2. Clams	.2334	.3039
3. Salmon	.3993	.4277
4. Prawns (shrimp)	.3763	.4229
5. Lobsters	1.6700	1.0234

1. Excludes return to management and capital.

If the unit cost of production (excluding return to capital) through aquaculture is less than the ex vessel price, the aquacultured product will generate positive profits; therefore, an *ROI*, or payback period, must be obtained to see whether the venture is worth undertaking compared to other forms of investment, including investment in wild stock fishing. It should be remembered that aquaculture has one great advantage—*all property is privately held.* (The woes of common property resources—wild stocks—were described in some detail in chapter 3 and especially in chapter 4.) Thus, no externalities exist among aquaculture enterprises. This was shown to improve productivity.

Pillay has made a survey of various aquaculture enterprises throughout the world for various species (see table 7.7). These data on cost and earnings provide valuable information; however, they are fragmentary and do not tell us whether the aquaculture venture is meeting the opportunity cost of both capital and labor within the countries studied. The *ROIs* obtained by Pillay are fairly high; however, these profits include depreciation, which is

Table 7.7 Returns on Aquaculture Enterprises for Selected Countries

Species	Country	Year	Profit[1]	Capital Inv.	ROI (%)
Rainbow trout	Ireland	1970	12,460f	35,000f	36
Rainbow trout	Norway	1969	16,100Kr	16,700Kr	96
Common carp	Poland	1969	5,665Zy	25,326Zy	22
Indian carp	India	1971	13,078Rp	90,022Rp	15
Chinese carp	Hong Kong	1972	$8,900	$23,011	39
Milkfish	Philippines	1972	15,081Ps	177,720Ps	8
Yellowtail	Japan	1971	$1,863	$536	347
Catfish	U.S.A.	1967	$42,605	$135,390	31

1. Includes depreciation; some figures expressed in local currency.

ordinarily deducted from profits as a business expense. In addition, the data are for one farm in each country and for only one year. Thus, Pillay's earnings information may be unrepresentative of the profitability of aquaculture internationally.

In summary, the economic feasibility of an aquaculture operation should be approached in the following manner:

1. The relation among economic inputs, environmental variables, and harvest must be obtained. This is specified in equation (7.1). The fish farmer should ask the question about the optimum-size farm. That is, if inputs are doubled from some level, is harvest correspondingly doubled?
2. Current input prices, including an adequate return to capital and land (if applicable), should be obtained.
3. Combining steps 1 and 2, the fish farmer can see the relation among total cost, inputs, and their corresponding prices. In the short run, the latter are assumed to be constant for any particular area. See equation (7.3).
4. The unit cost of production for various farm sizes should be computed to obtain the least-cost farm.
5. Where the aquacultured product is in direct competition with the wild stock species, ex vessel prices should be compared with unit cost incurred through aquaculture to determine the economic feasibility of the enterprise. Any adjustments for transportation to markets should be used in making final comparisons.

Potential for Aquaculture

Bell and Canterbery (1976) have recently assessed the economic feasibility of aquaculture technology transfer, primarily from the

Indo-Pacific area, where aquaculture is flourishing, to other developing countries. Ninety developing countries were considered. Selection of species was based upon at least two of the biological characteristics that have direct economic implications: (1) the species's position on the food chain, and (2) dependency upon natural breeding. As indicated above, if a species is high on the food chain, feeding is expensive. Most aquaculture operations depend upon natural breeding (i.e., *stocking* is the opposite of natural breeding and is used more in developed countries). The following criteria were used for selection of species:

Biological	Economic
1. Feeds low on the food chain	1. Established and profitable somewhere
2. Wide temperature tolerance	2. Wide use as protein
3. Limited breeding problems	3. Potential consumer acceptance
4. High growth rate	4. Adaptable to small-farm operations
5. Well researched	5. Labor-intensive
	6. Potential for foreign exchange

Thirteen species plus a species of seaweed were selected. These are shown in table 7.8 along with necessary inputs (except capital). The price of factor inputs and environmental conditions varies from country to country. A *unit cost of production* was calculated for all ninety countries in the case of each species after adjusting for differences in input cost and differing environmental effects upon production. Of course, a *newly* introduced source of protein must be (1) able to compete on a cost basis with existing protein-producing sectors and (2) have a reasonably high potential for consumer acceptance. It was found that channel catfish, rainbow trout, mullet, milkfish, oysters, and tilapia have the highest potential for technology transfer. Table 7.10 shows the countries where these species would have (or now have) great potential for augmenting food production. According to Bell and Canterbery, suitable land availability is not a significant constraint on the expansion of aquaculture operations. The low cost of *quality-adjusted* (see chapter 3) protein from aquaculture is shown in table 7.9—the simulated and actual leading cost per kilogram (i.e., least cost) for selected countries by protein-producing sector.

FAO's indicative world plan for agricultural development (1969) considers as feasible an expansion factor of five by the year 1985. According to this, the world annual aquaculture production

Table 7.8 Factor Input Quantities for Aquaculture Products in Various Countries of the World

Species	Country of Origin	Source	Stocking (Number)	Feed (kg.)	Size of Farm (ha.)	Fertilizer (kgs.)	Labor (Man-years)
1. Indian Carp	India	Pillay	2,500,000	360	3.5	10,000	10.4
2. Channel Catfish	Georgia (U.S.A.)	Georgia	10,000	8,164.67	2.025	217.7	.075
3. Walking Catfish	Thailand	IPFC	375	1,779.94	.00152*	000.00	.06
4. Tilapia	Thailand	IPFC	60	67	.1075	4.0	.02
5. Mullet	Hong Kong	IPFC	25,000	80,906	5.0*	000.00	5.57
6. Milkfish	Philippines	Carand Darrah	1,200,000	68,038	100.00	907,185[1] 181,437[2]	6.58
7. True Eel	Taiwan	Shang Chen	120,000	472,500	1.0	000.00	2.44
8. Yellowtail	Japan	Pillay	3,110	112,379	.0126	000.00	1.057
9. Penaeus Shrimp	Thailand	Pillay	000.00	000.00	8.0	000.00	5.23
10. Oysters	Japan	Bardach & Others	000.00	000.00	1.0	000.00	1.0
11. Mussels	Thailand	Pillay	000.00	000.00	.16	000.00	3.64
12. Seaweed (Blue-Green Algae)	Africa (Chad)	Sorensen	000.00	000.00	4.0	453.59	6.00
13. Macrobrachium	Hawaii	Shang	924,500	249.12[3]	4.05	000.00	2.0
14. Rainbow Trout	Japan	IPFC	857,142	196,226	.36*	000.00	10.6

*Sea Water
1. 16-20-0 fertilizer
2. Chicken manure
3. Chicken feed

Source: Bell and Canterbery (1976)

Table 7.9 Cost Per Kilogram of Protein for Aquacultured Fish and Agricultural Products

Cameroon (Africa)	Colombia (South America)	Portugal
\$0.06 Mullet[1]	\$0.08 Mullet[1]	\$0.07 Mullet[1]
1.31 Rainbow trout[1]	0.85 Rainbow trout[1]	1.06 Shellfish
1.97 Coarse Grains	1.29 Finfish	1.42 Finfish
1.98 Wheat	2.09 Coarse Grains	1.55 Rainbow

Source: Bell and Canterbery (1976)

1. Simulated aquaculture.

by that year may be around 20-25 million tons. Bardach and Ryther (1968) estimate a tenfold increase by the year 2000, which would mean a world production of 40-50 million tons annually. If Bell and Canterbery are correct, much of the increased production may come through technology transfer as well as through expansion in areas already under cultivation and improved techniques.

Only a limited part of the sea, the shallow region near shore, is suitable for aquaculture. The total shelf area constitutes only about 3 percent of the sea's surface, and in most cases, it will be possible to farm only the shallower areas of these, probably areas

Table 7.10 Countries with Highest Potential for Aquaculture Technology Transfer

Channel Catfish	*Rainbow Trout*	*Mullet*		*Milkfish*	*Mussels*	*Oysters*	*Tilapia*
Cameroon*	Rumania	Algeria	Vietnam DR	Guatemala	Khmer*	Cuba	Cameroon
Burma	Spain	Cameroon	Vietnam R	Mexico*		Dom. Repub.	Ivory Coast
P R China*	Iceland	Egypt	Argentina	Burma		Guatemala	Madagascar
India*	PR China	Morocco	Brazil	PR China*		Mexico	Malawi
Indonesia*	Taiwan*	Nigeria*	Chile	Taiwan*		Burma	Mali
Khmer*	Yugoslavia	Tunisia*	Columbia	India*		China PR	Algeria
Malaysia*	Korea DR	Cuba	Ecuador	Indonesia*		Taiwan	Sudan
Thailand*	Korea R	Dom. Repub.	Paraguay	Khmer*		India	Tanzania
Vietnam DR*	Nepal	Guatemala	Venezuela	Malaysia		Indonesia	Tunisia
Vietnam R*	Bolivia	Mexico	Greece*	Philippines*		Khmer	Uganda
	Columbia	Burma*	Poland	Thailand*		Korea DR	Zaire
	Nicaragua	PR China*	Romania	Vietnam DR		Korea R	Zambia
	Ecuador	Taiwan*	Spain	Vietnam R*		Malaysia	Dom. Rep.
	Honduras	Hong Kong*	Yugoslavia	Brazil*		Philippines	Burma
	Peru	India*	Lib.Arab. Rep	Ecuador		Thailand	China PR
	Panama	Indonesia*	Costa Rica	Venezuela		V. Nam DR	Taiwan
	Greece	Iran	El Salvador	Nicaragua		V. Nam R	India
	Albania	Iraq	Honduras	Uruguay		Brazil	Indonesia
	Hungary	Khmer*	Nicaragua			Chile	Iran
	Poland*	Korea DR	Panama			Columbia	Iraq
		Korea R	Cyprus*			Ecuador	Khmer
		Malaysia	Israel*			Peru	Malaysia
		Pakistan*	Jordan*			Venezuela	Thailand
		Philippines*	Lebanon			Portugal	V. Nam DR
		Saudi Arab.	Albania			Costa Rica	V. Nam R
		Syr. Arab Re.	Uruguay			El Salvador	Peru
		Thailand*	Portugal			Honduras	Sierra Leone
		Turkey				Nicaragua	Cyprus
						Panama	Jordan
						Cyprus	Laos
						Albania	
						Uruguay	

Note: The following species do not have the highest or average technology transfer using the criteria discussed in the text: Indian carp, true eel, yellowtail, walking catfish, macrobrachium, blue-green algae.

*Presently cultured to varying extent.

Source: Bell and Canterbery (1976)

less than 30 meters deep. Nonetheless, enormous areas are still available for aquaculture, especially in the developing areas of the world. Many of these areas are swampy, edge-of-the-sea areas that have heretofore been regarded as wastelands. Only in a few areas have surveys been made of the potential area available for aquaculture; the total over the entire world can only be guessed. In some Asian countries about 400,000 hectares of coastal waters are under cultivation, and a total of 1,500,000 hectares are believed suitable for cultivation. This calculation does not include Japan, China, or India, three of the largest countries and those most active in fish culture. In Africa a single country, Nigeria, has at least a million hectares of mangrove area that could be made into sea farms, and it is not inconceivable that half of this could be used more profitably for aquaculture than for other uses. There are also millions of hectares along the edges of other countries in Africa. South America and many other regions also have enormous areas of coastline suitable for aquatic farms.

Limitations on Aquaculture

In some areas of the world, social problems pose greater difficulties than any others to the development of aquaculture. Shoreline property has had a higher value than other land in most parts of the world, especially in developed countries. This is partly so because human populations tend to crowd along shorelines: seven of the world's largest cities are on estuaries, and over half the world's population lives within 100 miles of the sea. In the United States one-third of the population lives in coastal counties, and one-third of the industry is located there. Aquaculture is in many regions a relative latecomer to the activities that need shoreline space and thus is often at a strong competitive disadvantage vis-à-vis commercial and sport fishing, swimming, boating, oil, other mineral exploring, and waste disposal. For example, plans to develop an ambitious operation to raise peneid shrimp in Florida was severely handicapped because of opposition from several groups of residents in the area. Commercial shrimp and mullet fishermen protested the closing of part of the coastal region to their operations. Boaters and sport fishermen claimed rights of free passage through the coastal region. These attitudes are so prevalent in the United States that the future of many kinds of coastal aquaculture is very uncertain. To a greater or lesser degree the same problems occur in other parts of the world.

A number of failures in aquaculture can be traced to the lack of trained personnel. Aquatic pollution (see chapter 5) has become a very significant limiting factor in aquaculture development, especially in industrially advanced countries. In Japan eutrophication has made 20-30 percent of the Inland Sea unfit for fish or shellfish culture. A major risk in aquaculture is loss through diseases and parasites. Thus, we must be cautious about fish farming as even a partial solution for the world food/population imbalance.

Summary

The fish farmer, or aquaculturist, differs greatly from his counterpart, the fisherman, who must *hunt,* rather than farm, fish. Iversen (1968) points out the several differences between the fisherman and the fish farmer. In this chapter we have restricted our discussion to aquaculture under controlled conditions. For example, we have not dealt with the rearing and release of young fish into open bodies of water to supplement catches of commercial fisheries—this is called *sea farming* and was discussed in chapter 4 under enhancement of the wild fishery stocks.

Table 7.11: Comparison of Activities of Commercial Sea Fishermen and Sea Farmers

Activity	Fisherman	Farmer
Obtaining stock	Relies almost completely on natural propagation, in public areas.	May obtain adults and spawn them artificially, or collects young to rear on private areas (leased or owned).
Care of stock	Provides no care	Generally provides restricted body of water, dikes, and sluice gates. Controls quality and quantity of water. Provides food and controls competition, disease, and predation.
Production increase	Relies on governmental regulations. More efficient fishing gear. Possible contribution from hatchery-reared stocks.	Increases growth and survival through feeding, artificial selection, and protection from disease, predators, and competitors.
Harvesting	Relies on searching or some means of accumulating in public areas. Carried out by any licensed or otherwise qualified fishermen, during open seasons, using legal gear, when fish schools are available.	As needed, draining, nets, or otherwise, by owner, lessee, or his employees, on private areas (leased or owned). Can be done at the most suitable economic time.

Source: Iversen (1968)

Aquaculture is now flourishing in Asia and the Far East, particularly in China. Aquaculture currently provides an estimated 10 percent of the world's water-derived protein with a harvest of between 5 and 6 million metric tons valued in excess of $2.5 billion. The growing demand for protein is likely to spur the growth of aquacultured fish species, especially in light of the widespread depletion of the ocean's wild stocks (see chapter 3).

Some countries now get more of their finfish from aquaculture than from wild stocks. For example, Hungary and Czechoslovakia—both landlocked countries—derive over 75 percent of their fishery protein (catch) from aquaculture. Indonesia and Israel derive 10 and 36 percent, respectively, of their catch from aquaculture. Although the United States derives only 1.5 percent of its catch from finfish aquaculture, indications are that the potential is great, especially for high-value species such as shrimp and lobsters.

Throughout the world, the most commonly cultured species are carp, tilapia, shrimp, oysters, milkfish, mullet, clams, and mussels. So far, the U.S. major commercial aquaculture enterprises are bait minnows, oysters, catfish, trout, and crayfish, which at the retail level constituted a $219 million industry in 1969. These five major aquaculture sectors may account for approximately 14 percent of the retail value of U.S. commercial landings from wild stock fisheries.

The land requirements for aquaculture often permit the use of low-value land—ravines, swampland, saltwater marsh, or mangrove areas—that is not well suited to other uses. Generally, a pound of fish can be produced more cheaply than a pound of red meat. That is, fish are better feed converters than land-based animals, a fact that makes aquaculture particularly attractive.

In general, species that feed low on the food chain (i.e., do not require protein as food), have a fairly wide temperature tolerance, are free of complex breeding problems, and grow rapidly have a fairly high potential for the production of low-cost protein. Of course, these species must be accepted by the consumer; therefore, marketing is important and may well present a formidable problem.

In attempting to establish an aquaculture enterprise, an economic feasibility study should be made to determine the least-cost-size farm. This is generally dependent upon the relation between harvest and economic inputs plus environmental factors. Pillay has found fairly high rates of return to capital on aquaculture enterprises throughout the world. Ultimately, aquaculture must be competitive with wild stock species as well as with other protein-producing sectors.

The world potential for aquaculture development lies in a combination of (1) expansion of areas already under cultivation, (2) increases in productivity through research and development, and (3) transfer of existing technologies to countries where aquaculture is not now extensive. The last possibility would significantly aid in increasing protein production through aquaculture. A study by Bell and Canterbery (1976) showed great potential for technology transfer, especially to developing countries, where the food/population imbalances are most pronounced. Given the demand for protein (see chapter 2), it would seem that the FAO projection of 20-25 million metric tons per year from aquaculture is certainly conservative. The Bell-Canterbery study indicates that the Bardach and Ryther projection of 40-50 million metric tons annually by the year 2000 is not unrealistic. Aquacultured species are usually for direct food consumption. If the food fish catch from the wild stocks does not increase, it is quite possible that more fish will be produced through aquaculture than through commercial fishing in all the oceans of the world. The major limitations to aquaculture development are both technical and social. Conflict over land use, especially in developed countries, and pollution may severely hamper the expansion of aquaculture.

Notes

1. Economic inputs are defined as those factors of production that have positive external price. The firm either pays for their use or some opportunity cost.

2. To simplify the analysis, we shall assume that inputs such as capital, labor, and fertilizer *cannot* be substituted for each other.

References

American Fish Farmer. 1969. Fish farming today—a rapidly expanding multimillion dollar business.

Bardach, John E., and Ryther, John H. 1968. The status and potential of aquaculture. Clearinghouse for Federal Scientific and Technical Information, P.B. 177,768. Springfield, Virginia.

Bardach, John E.; Ryther, John H.; and McLarney, William O. 1972. *Aquaculture: the farming and husbandry of freshwater and marine organisms.* New York: Wiley-Interscience.

Bell, Frederick W., and Canterbery, E. Ray. 1976. *Aquaculture for the developing countries: a feasibility study.* Cambridge, Mass.: Ballinger Publishing Co.

Caton, Douglas D.; Moss, Donovan D.; and Urano, James A. 1974. Improving food and nutrition through aquaculture in the developing countries. Unpublished manuscript, March 5, 1974.

Davidson, Jack R. 1971. Economics of aquaculture development. In *Proceedings: Fourth National Sea Grant Conference.* University of Wisconsin, Madison Sea Grant publication WIS-SG-72-112.

Delaney, Richard J., and Schmittou, Homer R. 1972. Aquaculture production project, Philippines. World Bank loan proposal, unpublished.

Food and Agriculture Organization. 1969. The prospects for world fishery developments in 1975 and 1985. *FAO Indicative World Plan.* Rome: Food and Agriculture Organization.

Furukawa, A. 1971. *Outline of the Japanese marine aquaculture.* Tokyo: Japan Fisheries Resource Conservation Association.

Iversen, E. S. 1968. *Farming the edge of the sea.* London: Fishing News (Books), Ltd.

Ling, S. W. 1972. A review of the status and problems of coastal aquaculture in the Indo-Pacific region. In *Coastal aquaculture in the Indo-Pacific region,* ed. T.V.R. Pillay. London: Fishing News (Books), Ltd.

Pillay, T.V.R. 1973. The role of aquaculture in the fishery development and management. *Journal of the Fisheries Research Board of Canada* 30: 2202-2217.

Shang, Y. C. 1973. Comparison of the economic potential of aquaculture, land animal husbandry and ocean fisheries: the case of Taiwan. *Aquaculture* 2:187-195.

8. *Underutilized Fishery Resources*

As indicated in chapter 3, the oceans' potential as a source of food for mankind has been the subject of considerable controversy. Unfortunately, most people still think that high seas fisheries are inexhaustible. We have shown that the opposite is true and that most commonly exploited species of fish are depleted, in imminent danger, or under intensive use. Much of the confusion over ocean food potential has stemmed from different scientific views over (1) the efficiency by which carbon (fixed by plant photosynthesis) is transferred to higher trophic levels in the food chain, and (2) factors that will limit utilization of the biological material produced. This was discussed in chapter 3 at some length. Through both food chain analysis and simple extrapolation, we can estimate the potential fishery yield from the world's oceans at approximately 120 million metric tons; the harvest in 1973 was about 55.3 million metric tons (see table 3.8). Obviously, not all species of fish are fully utilized. However, the basic problem is that the aggregate potential is subject to some variability and tells us little about which species and which locations are "underutilized." According to Alverson (1975), "most recent estimates associated with the biological basis [food chain] for developing conventional marine fisheries range from 100-250 million metric tons—2-4 times the existing world catch."

We should make the important distinction between *conventional* or *traditional* species and *unconventional* biological forms. The former refer to species that are widely used for food or for industrial or recreational purposes; but the latter are not utilized to any great extent. These are the so-called latent fishery resources. Although the conventional finfish species are fully utilized (and many are overfished), there is a worldwide potential for increases in such conventional species as shrimp, crabs, scallops, clams, and oysters from wild stocks. This was discussed in chapter 3 for the major conventional species. In this chapter, we shall place great emphasis upon the unconventional species as an economically feasible source of protein for a food-starved world. But first, what is an underutilized fishery resource?

Biologists have always had a predilection for assuming that if a fishery resource is not exploited at maximum sustainable yield, economic waste is the result—all those fish are going to waste! As shown in chapter 4, *economic efficiency* in fishery exploitation depends on the relation between the marginal cost of fishing and the price people are willing to pay—the demand curve. At the outset, we can now make it quite clear that underutilized fisheries (i.e., exploited not at all or below *MSY*) do *not* involve economic inefficiency. Either the demand for the product is nonexistent or relatively low, or the cost of harvesting—technology—limits the development of the fishery. For most unconventional species, both factors are usually at work. A general theme running through the fishery programs of many fishing countries is somehow to "develop" underutilized, and especially unconventional, species. Finally, this chapter restricts the discussion to the hunting phase of fishing, or wild stocks, since we have discussed the potential growth of aquaculture in chapter 7.

Unconventional Fishery Resources: Biological Potential

The most frequently mentioned unconventional form is the "shrimp-like" *Antarctic krill.* Soviet and Japanese biologists have investigated the krill, which reaches a maximum size of about 70mm (2.75 inches). Although krill are small, large concentrations are generally found in upper ocean layers to depths of 250 meters. Moiseev (1969) reports that swarming krill may attain a density of 10-16 kg per cubic meter of water.

Recent Soviet estimates of standing stock of Antarctic krill are 800-5,000 million tons (Lyubimova, Naumov, and Lagunov 1973) with an annual sustainable yield forecast of 100 million tons (Lyubimova, Naumov, and Lagunov 1973). Gulland (1971) and Omura (1973) have placed the *MSY* between 100 and 200 million tons per year. If these estimates are in any way accurate, this single species is capable of supplying as much as or more than the potential for all conventional fisheries—estimated by FAO at 120 million tons. The Soviets feel that as a matter of state policy the urgent task is to develop commercial fisheries among animals at the second trophic level, the zooplankton or, more specifically, the Antarctic krill. If feasible, this is an excellent way for centrally planned economies such as the USSR to maximize protein production. Remember our discussion in chapter 3 of demand-supply determinations in centrally planned countries. That is, sources of protein production are determined by the state; therefore, there is reasonable expectation that the Soviets will make sizable efforts to exploit the krill. So far, only the Japanese are sufficiently equipped to compete with the Soviets for this resource. We shall discuss below the krill and other unconventional species with respect to the economics of their exploitation.

Although dwarfed by the krill in terms of potential, another crustacean that reaches a moderate size is the *red crab.* Presently, the red crab is a principal food source for tuna in the eastern tropical Pacific. Longhurst (1968) has suggested that the red crab could be harvested for food at an annual rate from 30,000 to 300,000 tons off lower California and Mexico. Of course, going to lower levels on the food chain may increase potential harvest; however, it may seriously endanger the valuable tuna—trophic level 5 (see figure 3.16)!

Suda (1973) has surveyed the potential for unconventional fishes and cephalopods (squid, octopus, and others) excluding crustaceans and mollusks. He concludes that the major part of the potential increase is of small-sized species at low trophic levels. The *Falkland herring* is a latent coastal pelagic fish that appears to be abundant from southern Chile to the southernmost point of South America. The annual catch is estimated at 1 million tons. The Pacific anchovy, although under increased exploitation, has a potential *MSY* of 2 million tons. The *flying fishes* comprise about sixty species in tropical and subtropical regions including Polynesia, the Philippines, India, and Vietnam. The Japanese have begun to develop this fishery, which may yield as much as 100,000 tons from coastal areas. Marine biologists have long speculated on the potential for developing *lanternfish.* Lanternfish are covered with numerous "eyespots," or luminescent organs. They are deep-sea dwellers in offshore areas, which makes it difficult to study them and expensive to harvest them. Lanternfish are small fish from 2.5 to 25 cm (1-9.8 inches). During the day, they stay as deep as 500 meters, and at night they come up to shallower water, sometimes to the surface. Gullard (1971) estimates that these rather strange-looking fish could provide a harvest up to *100 mil-*

lion metric tons. A close cousin of the lanternfish is the *lightfish,* which feeds on plankton and is eaten by cod and rockfish. These also represent a large underutilized fishery stock.

Suda's (1973) survey of unconventional species is restricted to finfishes and cephalopods (e.g., squid). Table 8.1 shows the estimated additional catches from this source. Thus Suda estimates that 43.1–55.3 million tons of additional catch are available from *unconventional species.* As indicated above, Gulland (1971) foresees a much larger potential catch of lanternfish than Suda does. The literature in this area is understandably confusing, since not much research has been done on underutilized species; therefore, estimates vary considerably. Table 8.2 summarizes the results of many studies dealing with the potential catch from unconventional species and from some species that, although greatly underexploited, have been commercially harvested. Figure 8.1 compares selected conventional and nonconventional forms of protein from the ocean.

Except for tuna, squid, octopus, and seaweed, there is no commercial market for the species shown in table 8.2. However, the estimated potential for lanternfish, Antarctic krill, and the cephalopod group seems enormous. If we take the upper estimate of these *three* fisheries, a potential yield of 400 million metric tons is obtained. However, before sounding the trumpet of optimism regarding food from the sea, one must ask why, with a partially starving world, these resources have not been exploited.

Table 8.1: Distribution of Additional Catch of Unconventional Finfish and Cephalopods (millions of tons)

Ocean	Pelagic Fish[1]	Demersal Fish[2]	Cephalopods[3]	Total
Northern	4.2-5.7	1.8	1.7	7.7-9.2
Southern	6.8	1.8-2.3	1.8	10.4-10.9
Tropical	10.9-17.2	11.9-14.5	2.2-3.5	25.0-35.2
Total	21.9-29.7	15.5-18.6	5.7-7.0	43.1-55.3

[1]Falkland Herring; anchovy; thread herring; horsemackerel; mackerels; flyingfishes; sauries; rainbow runners; dolphenfishes; small tunas; laternfishes; lightfishes.

[2]Capelin, Argentines, sandeels, hakes; poutasson (blue whiting); nototh-enids.

[3]squid; cuttlefish and octopus

Source: Suda (1973)

Exploiting Unexploitable Species: Technological Capability

Since World War II, there has been a rapid advance in fishing technology—such as kinds of vessels, electronics, and the materials actually used in fishing. Improvements in radar, sonar, radio, navigational systems, and spotter aircraft, have greatly enhanced the development of offshore fishing. As indicated in earlier chapters, by the 1960s fishing had become a global activity, with the Japanese and Soviets leading the way. Although fishing extraction systems are in many cases very sophisticated and efficient in exploiting the fish presently being harvested, the capacity of existing technology must be altered or in some cases changed drastically in order to harvest unconventional species. Before one can design new technologies to cope with the unique features of underutilized species, studies must be made of these unique features.

Saetersdal (1973) has surveyed the various techniques for assessing unexploited ocean fishery resources. The possible methods and sources of information are of two kinds: indirect and direct. Indirect methods make use of information on the oceanography and basic productivity of sea areas and of ecological

Table 8.2: A Survey of Potential Annual Sustainable Yield from Unexploited Species (millions of metric tons)

Category	MSY	Area	Source
A. Finfish			
Demersal[1]	15.5-18.6	Mostly tropical	Suda (1973)
Pelagic (including lanternfish)[1]	21.9-29.7	Mostly tropical	Suda (1973)
Lanternfish	100.0	Ubiquitous	Gulland (1971)
Skipjack tuna	.5	Central Pacific	Matsumoto (1974)
B. Shellfish			
Antartic krill	200.0	Antarctica	Gulland (1971)
Red Crab	.3	Mostly tropical	Longhurst (1968)
C. Cephalopods (Squid, cuttlefish & octopus)	5.7-7.0	Ubiquitous	Suda (1973)
	10-100	Ubiquitous	Gulland (1971)
	8	Ubiquitous	Voss (1973)
D. Seaweed (all kinds of species)	Increase of several fold from 1 million metric tons harvested	Ubiquitous	Alverson (1975)

[1]For a listing of species in this category, see Table 8.1

Figure 8.1: Shows Conventional Fish: (1) tuna; (2) cod; (3) flounder; (4) sardine; (5) crab. Nonconventional Fish (6) euphausiid; (7) red crab; (8) squid; (9) lantern fish; and (10) seaweed.

Source: Alverson (1975)

relations. Examples are the mapping of upwelling areas (i.e., intermixing of waters, which contributes to phytoplankton growth), the charting of observations of basic (i.e., plant life) productivity, and stomach content studies. Direct methods are based on observations of any stage in the life of the resource itself, including egg, larval, and postlarval surveys, exploratory fishing, acoustic surveys, and aerial scouting. Many governments are now doing this in order to pave the way for new technology. Whether government should be involved in such activities depends upon the political organization of the country (i.e., planned vs. decentralized decision making) and internal perceived "needs."

Alverson states about what we now know, "The changes to resolve technological problems inhibiting use of some of the unconventional species appear better [than conventional]. Most of the unconventional species under consideration aggregate in a way that permits their detection and harvest with some minor modifications using traditional fishing techniques" (1975, p. 37). For example, Antarctic krill can be detected by sight at sea and can be fished by surface trawls and purse seines. However, they are not always easy to detect, since they migrate to deeper waters. Special acoustical techniques must be designed. In 1973, only 5,000 metric tons of krill were harvested, and that entirely by the Soviet Union. The next section will deal with the possible uses of krill (food vs. industrial) as well as the obstacles to consumer acceptance of unconventional species.

Numerous techniques can be used to harvest lanternfish and squid (the cephalopod group). Squid are attracted to light, and an experimental light pump system has been successfully used in Monterey Bay, California. As is well known, squid are likely to be near the bottom of the ocean during the day and catchable with bottom trawls at that time. Lux, Handwork, and Rathjen (1974) report that in experimental fishing for squid in the northwest Atlantic using trawls, the best catches were about 750 lbs per hour.[1] Commercial exploitation of lanternfish using trawls has already started off South America. In contrast to Alverson's optimism, Suda believes that the deep swimming habits and the small size of lanternfish may greatly discourage commercial utilization. In 1973, 1.029 million metric tons of squid were harvested, with Japan being the major producer. Red crab, an unconventional species, can be caught with the use of midwater trawl gear.

Finally, table 8.2 lists skipjack tuna as an underexploited species. In fact, skipjack tuna is the only species of tuna now exploited in large quantities that can withstand large increases in fishing pressure. We have included the central Pacific skipjack tuna along with more unconventional forms because of the estimated enormous potential for a species that is eaten throughout the world. We have already discussed the potential for traditional species in chapter 3. At present, tuna purse seine operations have *not* been successful in the central Pacific. Traditional purse seine fisheries for skipjack have been successful because the fishing area was characterized either by murky waters, shallow thermoclines (layers of water at different temperatures), or the presence of a marked front between warm and cold currents. In equatorial waters such as the central Pacific, waters are exceptionally clear, and the thermocline is deep enough to permit fish schools to escape by diving under the net or seine. Modifications of the purse seine are necessary for midoceanic use.

Consumer Acceptance of Unconventional Species

Krill is processed in different ways for *direct* food consumption. The Japanese have generally frozen the product after boiling it; however, they also have experimented with drying it. At this stage, krill has not won any degree of acceptance in Japanese markets. Lyubimova, Naumov, and Lagunov (1973) report more varied Soviet experiments with krill. The protein of the Antarctic krill contains amino acids indispensable to man. The Soviets have produced a shrimp paste. The method of preparing the protein paste consists of pressing the krill, heat treatment of the extracted juice, and separation of the thick coagulated paste, which is red in color. The paste is then packed in trays and frozen at -18°C. Biological studies of animals receiving krill paste as a protein component of their diet showed a higher rate of increase in weight than those receiving beef. As for krill paste as a direct food, the Soviets found that it goes well with butter and cheese. That is, recipes were formulated for "shrimp butter" and melted cheese, which was labeled "coral." Krill is also sold directly as "ocean paste." Many ready-to-eat products using krill paste have gained wide recognition in the Soviet Union (e.g., salads, stuffed eggs, and stuffed fish).

Red crabs are caught in great numbers in trawling for Pacific hake off Baja California. There are many problems associated with commercial production. The yield using current techniques (i.e., crab picking) is small—from 8 to 10 percent of body weight. The Soviets regard the red crab as a high-protein source of fish meal. Thus, this resource now has no food fish market.

Compared to other marine animals eaten by man, the squid has a larger proportion of edible parts to the whole body. With vertebrate fish, the edible portion ranges from 20 to 50 percent; shellfish range from 20 to 40 percent; and squid will yield from

60-80 percent of the weight of the animal. Squid meat is equal to fish meat in protein content and amino acid composition (see chapter 2). Squid is eaten in fresh and frozen form in Japan, the Indo-Pacific countries, and Southern European countries. At present, the only commercially prepared squid products produced in the United States are squid canned (with or without the ink) in brine, in oil, and in tomato sauce. The production of canned squid in the United States has increased from approximately 1 to 11 million pounds over the 1970-1974 period; most was probably destined for domestic ethnic markets or for export. Ampola (1974) reports that frozen breaded squid and marinated squid show definite promise as consumer items in the United States.

Finally, lanternfish represent a vast resource that at this time can only be used for the production of fish meal—not food fish. This is one of the problem areas for ocean fisheries, that is, the catch from the lower trophic levels will be utilized in various ways, sometimes as fresh fish, but mostly for fish meal.[2] If eaten directly by mankind, the marine resources can provide more protein, so every effort should be made to process fish and other aquatic species for man's direct consumption. The Soviet use of "ocean paste" is a good example.

A limiting factor to marketing unconventional fish is often their unusual appearance, such as awkward shape or body structure. The usual solution to overcome this resistance is by heading, skinning, or filleting fish and by inventing inexpensive, but appealing, packaging. Often, it is not the consumer who needs persuading to try something new; the fishing industry simply needs to be more enterprising. The development of the American hotdog from beef leftovers is a classic example of such ingenuity. Anderson (1973) has outlined some factors that bear upon developing markets for unconventional species. Many developing countries rely exclusively on fresh fish. The local population could be supplied with trawled fish from medium and distant fishing grounds. Unfortunately, this fish may initially be unacceptable, because it is completely unfamiliar and because it has to be delivered iced (i.e., not fresh). Markets have been created for unfamiliar iced and frozen fish in the Caribbean islands and West Africa. In addition, many developing countries have a poorly developed market economy; therefore, governments may have some justification for employing marketing staffs to aid industry in developing new markets. This condition is not so prevalent in developed countries. Developed countries are prone to consume frozen, canned, and convenience-prepared fish. This form of processing appears to offer considerable scope and promise for increased markets for unconventional species. The general reputation of fish must also be improved; fish is still a second choice or a once-a-week item. The public must be convinced that "fish is good for you." This is expensive, but it is a cost the fishing industry must pay for previous neglect.

Summary

There is much evidence that the ocean produces an enormous biological surplus capable of feeding a protein-starved world. For traditional or conventional species, however, many problems prevent their expansion—such as overfishing, the international wave of extended jurisdiction, and environmental deterioration. Gulland (1971) has estimated that no more than *100 million metric tons* of conventional species can be taken from the ocean on a sustainable basis. As chapter 3 indicated (table 3.8), a doubling of the present catch is possible; however, if we fail to establish effective national and international management institutions, we may never approach this figure.

For the most part, this chapter has reviewed the biological and technological potential for the exploitation of unconventional species. The projection of the world's marine resources indicates that there is an unquantified and virtually untapped reserve of lesser-known fishery stocks, which, pending the development of efficient and economical harvesting methods, may prove to be a significant addition to the world's sources of fish and protein. To a large extent such stocks are located off the shores of the developing countries in the Southern Hemisphere. Excluding crustaceans, Suda has estimated a total potential increase of unconventional pelagic, demersal, and cephalopod species of 43.1–55.3 million metric tons. Gulland is much more optimistic when he indicates that lanternfish and cephalopods may provide up to 200 million metric tons. Of course, the Antarctic krill has received much publicity, and it is estimated that this vast resource may have a potential of around 200 million metric tons. Although Alverson is rather optimistic about our ability to make slight alterations in existing technology to catch unconventional species, Suda feels that such apparently vast resources as the lanternfish may require new technology, especially in mid-ocean areas. Consumer acceptance of unknown species depends to a considerable degree upon industry's ingenuity in packaging its product. The krill has had some success in the Soviet Union as a *food* fish, and there is great potential for squid and other cephalopods as direct food fish. However, lanternfish and other pelagic fish, such as the Falkland herring and thread herring, are most likely to be used for industrial purposes.

What countries are likely to gain from these unconventional and untapped fishery resources? The demersal and pelagic resources are mostly tropical and are found at some time in their life cycle primarily off developing countries. As chapter 5 pointed out, the developing countries may gain greatly from extended fishery jurisdiction. The Antarctic krill may become the exclusive property of the Soviets and Japanese, since they are the countries best equipped to exploit such species. Lanternfish, squid, and seaweed are ubiquitous throughout the world. For example, a rapidly growing squid fishery is developing off the U.S. East Coast. Within 200 miles of the United States (including Alaska and Hawaii), there are many underutilized fishery resources, such as alewives, mackerels, flying fishes, squid, sharks, grenadiers, ocean quahog, skipjack tuna, mullet, snappers, bonito, and tanner crabs. The low level of utilization can easily be explained by our supply and demand analysis in chapter 4. In summary, considerable food resources from the ocean remain to be exploited. These resources may eventually be used, but if past trends continue, the distribution among the countries of the world will probably be uneven, and the "have-nots" may not share to any great extent in this abundance unless extended fishery jurisdiction becomes a major factor (as discussed in chapter 4). The spoils will go to enterprising nations such as the USSR and Japan. The United States is likely to play a minor role in exploiting unconventional species such as krill, lanternfish, and cephalopods, species that require deep-sea fleets. The U.S. tuna fleet is now the only U.S. fleet with such long-range capabilities.

Notes

1. This trawl productivity is not significant in itself, since it must be related to cost of trawling and prices received in order to determine economic feasibility.

2. We have not discussed fish protein concentrate (FPC) as a food fish source. Experiments in the United States and other places in the world to provide a low-cost protein supplement have failed because of the rising cost of raw materials and lack of consumer acceptance. The FPC dream is essentially dead.

References

Alverson, Dayton L. 1975. Opportunities to increase food production from the world's oceans. *Marine Technology Society* 9:33–40.

Ampola, Vincent G. 1974. Squid—its potential and status as a U.S. food resource. *Marine Fisheries Review,* paper 1110.

Anderson, A. M. 1973. Developing markets for unfamiliar species. *Journal of the Fisheries Research Board of Canada* 30:2166–2169.

Gulland, John A. 1971. *The fish resources of the ocean.* London: Fishing News (Books), Ltd.

Longhurst, A. R. 1968. The biology and mass occurrence of Galatheid crustaceans and their utilization as a fishery resource. *FAO Fish* 2: 95–110. Rome: Food and Agriculture Organization.

Lux, F. E.; Handwork, W. D.; and Rathjen, W. F. 1974. The potential for an offshore squid fishery in New England. *Marine Fisheries Review,* paper 1109.

Lyubimova, T. G.; Naumov, A. G.; and Lagunov, L. L. 1973. Prospects of the utilization of krill and other nonconventional resources of the world ocean. *Journal of the Fisheries Research Board of Canada* 30:2196–2201.

Moiseev, P. A. 1969. *The living resources of the ocean.* Translated for the National Marine Fisheries Service.

Omura, H. 1973. *Whale and Euphausia superba.* Marine Resources Research and Development Center. Tokyo.

Saetersdal, G. 1973. Assessment of unexploited resources. *Journal of the Fisheries Research Board of Canada* 30:2010–2016.

Suda, A. 1973. Development of fisheries for nonconventional species. *Journal of the Fisheries Research Board of Canada* 30:2121–2158.

Voss, G. L. 1973. *Cephalopod resources of the world.* FAO fisheries circular no. 149. Rome.

9. *The Economic Plight of the Fishing Firm and Industry*

The Market for Fishery Products and the Individual Firm

In chapter 4, we discussed how demand and supply determine prices in commercial fishing. But we did not discuss the characteristics of the market for fishery products. In market-oriented economies (as opposed to planned economies), the fishing industry generally operates under a market form that approaches what is called *pure competition.* This market form is characterized by the following: (1) homogeneous products, (2) large numbers of buyers and sellers, and (3) easy entry and exit from the industry. These characteristics have profound effects on the individual fishing firm. *First,* a fishing vessel can rarely differentiate its product or harvest from others. (Many producers attempt to differentiate their products—from toothpaste to automobiles—usually through advertising.) *Second,* it must compete with hundreds or sometimes thousands of other firms. The action of one fishing firm (i.e., how much it harvests) has no appreciable or even noticeable impact upon total supply. This would *not* be true of US Steel or General Motors, which operate as oligopolies, where there are few competitors and where each competitor has an appreciable impact on market supply. *Third,* fisheries range from inshore lobster, clam, and oyster industries, in which the capital investment to start a new fishing firm is rather low, to deep-sea fleets such as the tuna or groundfish, which require an investment of several hundred thousand dollars. Compared to the investment in a steel mill, automobile plant, or petroleum refinery, the investment in the individual fishing enterprise is small. Thus, there are no real constraints on entry into most fisheries. This was indicated quite pointedly in chapter 3, which discussed the rapid increase in the number of vessels in most fisheries of the world as the demand for fishery products increased.

We must consider one more element: the behavior of individuals or fishing enterprises operating in the market. We make the fundamental assumption that every fisherman in the market behaves rationally. Rational economic behavior implies that an individual's economic actions seek to "maximize gains and minimize losses." Critics often say that this assumption is unrealistic and point out that individuals often behave irrationally in the economic sense; they are often careless, charitable, or listless rather than vigilantly acquisitive. Even though some of these criticisms are true, an appeal to personal experience still reveals the general validity of this assumption: as producers and consumers, we seek, in general, to maximize our gains and minimize our losses.

Since the fish-harvesting market is characterized by pure competition, each fishing firm is regarded as a "price-taker" as opposed to a "price-maker." As shown in chapter 4, aggregate supply and demand determine price, and individual fishermen *have no control over this price*—they are price-takers! Figure 9.1 shows for the individual fisherman the typical relation between price and quantity *under pure competition.* In panel *a,* the forces of supply and demand determine a stable equilibrium (*AC=P* is a stable equilibrium for the fishing industry because of the yield curve and the common property nature of the resource) at a price of 50 cents per pound of fresh cod and a total market quantity sold of 200 million pounds. For any one fishing trip, the hold capacity of the

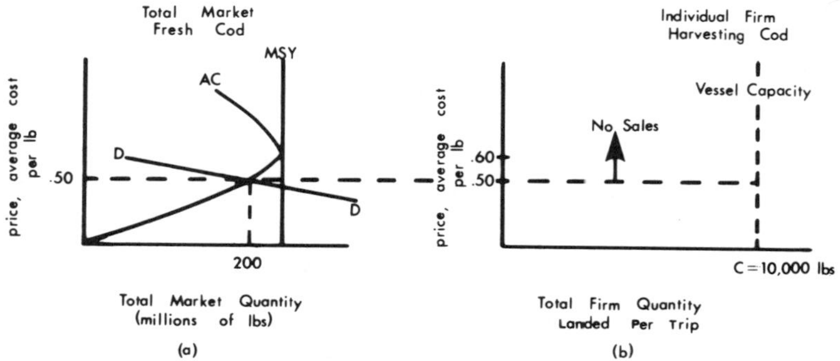

Figure 9.1: Relation Between Price and Quantity That Can Be Sold for the Individual Fisherman Under Pure Competition

Source: Hypothetical

individual cod vessel in question is 10,000 lbs, as shown in panel *b*. Why is the individual cod fishing firm a price-taker? Let us look at his options. Suppose the cod-fishing firm is not willing to accept 50 cents per pound for cod and raises its price to 60 cents. What buyer will pay 60 cents for cod if he can buy it at the market price of 50 cents? Cod is cod, or—simply—the product is homogeneous from firm to firm. If the cod-fishing firm could differentiate its product by claiming that it is superior—more fresh, more meat per pound, more nutritious, more everything—it might persuade buyers to pay a little more. However, this is unlikely, since the entire cod fleet catches its fish from the same fishing grounds, spends about the same number of days at sea, and preserves the fish with about the same amount of ice per fishing firm. Thus the individual firm can make no sales above 50 cents per pound. At 50 cents per pound or lower, the firm can sell its entire capacity of 10,000 lbs. In addition, withholding the 10,000 lbs from the market will have but a small impact on total market supplies (0.005 percent). This illustrates the second characteristic of perfect competition: large number of sellers and buyers, each having no appreciable impact on total market supply. Buyers such as processors, wholesalers, or other middlemen, must also be numerous. According to Smith, "There are millions of sea-food consumers. There are over 100,000 commercial fishermen, but there are only several hundred sea-food dealers. Some ports have as few as one dealer. In situations such as this, there is little incentive for the dealer to increase the price to fishermen when he can meet an increased demand with his existing supply and a greatly inflated price to the consumer" (1975, p. 28). This condition, when it exists, makes the market imperfect and may violate several of the federal antitrust laws. Fishermen in the United States have filed many suits alleging uncompetitive practices. Finally, ease of entry (and exit) into fishing is necessary in response to increased (decreased) demand. If not, present producers would reap abnormally high profits, or profits in excess of their opportunity cost (see chapters 3 and 4 for a discussion of this point). In sum, the individual fisherman must, in most cases, take price as a given. He can do little to influence it; however, he will indeed have to make some intelligent guesses about future prices in planning his vessel's operation.

As discussed in chapter 2, many countries are not organized as market economies in which decentralized decision making is employed. In 1970, for example, of the *total gross tonnage* of fishing vessels, factory ships, and carriers in the world, centrally planned economies such as the USSR, the Eastern European countries, and North Korea had over 57 percent (Lloyd's 1970).[1] Since the state owns the means of fishery production, managers or captains are given targets or harvest objectives. In addition, the state will determine the equipment and labor needed to accomplish these objectives. This is not to imply that there are no incentives present in a centrally planned economy; bonuses and additional rewards are often given to the fishing captain who accomplishes his objectives in the most efficient manner (e.g., obtains the target in less time than allowed; reduces the need for allocated labor and still meets his target; or overfulfills the planned catch). In the USSR, Sysoev (1970) indicates that the principle of distribution provides that everyone gives according to his abilities and receives according to his labor. Wages are dependent on the quantity and quality of labor. This is the basic principle of socialism and is called "socialist emulation." That is, personal interest depends on the welfare of the state. Sysoev states, "In the past three years (1968-70) socialist emulation throughout the country, thus including the fishing industry, was launched under the slogan of celebrating Lenin's 100th birthday in a fitting way" (p. 296). In lieu of great *monetary* rewards, the USSR gives the Order of Lenin and the Red Banner of Labor. In 1968 as many as 87 fishing industry workers were awarded the title Hero of Socialist Labor; 4000 workers were awarded orders and medals of the Soviet Union. Of course, Sysoev says nothing about those who consistently underfulfill. When looking at the world fishing industry, one must constantly keep in mind the political and economic organization of the countries exploiting the resource.

Producing Fish: The Production Function

For the individual fishing firm, we must first explore the relation between output, or pounds of fish harvested, and the determinants of this catch. The production function is a technical or engineering relation between inputs and outputs and is the basis upon which the economic theory of *supply* for the individual fishing vessel is built. The annual catch for the individual vessels is determined by the following variables:

1. Vessel size
2. Characteristics of the vessel
   a. horsepower

   b. age of vessel
   c. type of construction
   d. design (technology)
3. Crew size
4. Fishing time during any specified time period (i.e., season)
5. Skill of the captain (knowledge of fishing grounds, etc.)
6. Fishing effort by the total fleet
7. Other inputs (ice, fuel, food, insurance, etc.)

The *size* and *characteristics* of the vessel are highly correlated with the catch per vessel during any period of time (i.e., annual catch). Sometimes these two determinants are specified as capital investment and expressed in dollar terms. Crew size, or labor, is a cooperating factor of production; however, the size of the crew is subject to diminishing returns—beyond a certain point, the incremental catch gets smaller and smaller as the crew size is increased with any given vessel. Of course, a critical variable is the actual time spent fishing. For some fleets, the fishing grounds are rather close; therefore, most of their time is not spent en route to the grounds, but actually fishing. For example, the United States and Canada are relatively close to the groundfish resource of the northwest Atlantic and should have a comparative advantage over European fleets. The skill and experience of the captain are very critical to the success of the individual vessel. This, of course, is true of any firm, no matter what industry it is in. In many fleets, the captain and vessel owner are the same person. Finally, as chapter 3 pointed out, the catch of the individual vessel is subject to a technological externality. That is, the individual fishing firm has no control over the size of the fleet, which determines the average catch per unit of fishing effort—the catch per unit of fishing effort falls with increases in *total* fishing effort. In addition, other inputs such as fuel, ice, and food are necessary to the harvesting process.

Many empirical studies have been made of these catch determinants discussed above. Carlson (1973) has studied catch determinants for individual vessels for the New England trawl and the tropical tuna seine fleets. For the New England fleet, he found the most powerful explanatory catch determinant was *fishing time.* A 10 percent increase in fishing time increased the catch per vessel by 6.5 percent. The size of the vessel and horsepower were also significant determinants of annual catch. A 10 percent increase in gross registered tons (vessel size) increased catch by 4.1 percent, but a 10 percent increase in horsepower increased annual catch by less than 1 percent. As might be expected, older vessels were less productive than newer vessels, presumably because older vessels tend to have more breakdowns and equipment that is not in the best working order. For the tuna purse seine fleet, Carlson found that an increase of 10 percent by each of the following production determinants—capacity, horsepower, and days fished—increased catch by 3.7, 3.1, and 3.7 percent, respectively. Crew size was not an important factor, presumably because there is such small variation in crew per vessel in this fleet.

Studies of different technologies in catching fish are very important, since they help appraise the economic feasibility of new techniques. Bell (1966) analyzed side vs. stern ramp trawling for the New England groundfish fleet. Historically, New England fishermen used beam trawling. That is, the side of the trawler was more convenient for shooting (casting) and hauling (retrieving) the net with the large beam. A vessel must be designed initially for either side or stern trawling. With a stern trawler, a ramp must be built from the deck to the water level in the stern of the vessel so that the net can be dragged behind the vessel and hauled up the stern ramp. According to Bell's analysis, the change from side to stern ramp trawling increased landings 29 percent for a seventy-foot vessel. Dow, Bell, and Harriman (1975) found that boat size, age, number of trraps, and number of trips (i.e., fishing time) were the principal determinants of landings in the Maine lobster fishery. Comitini and Huang (1967) used a somewhat different approach in their study of the Pacific halibut fleet. First, they adjusted the market value of each vessel by days fished for halibut per year. Presumably, the market value of the vessel reflects such vessel characteristics as horsepower, age, and type of construction as well as size, or gross tonnage. For labor, they used the number of fishermen days.[2] Besides capital and labor inputs, they recognized one other factor that significantly affects the level of firm output: the "density of the fish population." The density of the fish population (i.e., size of the population for a given area) will be determined by the level of fishing effort; therefore, Comitini and Huang were testing for technological externalities. The following equation was estimated:

$$q = AK^{\alpha}L^{\beta}C^{\gamma} , \tag{9.1}$$

where

q = catch of halibut per vessel

K = market value of the vessel (adjusted for utilization)
L = fishermen days
C = catch per unit of fishing effort (proxy for population density)
A = constant .

They found that $\alpha = +0.25$, $\beta = +0.75$, and $\gamma = +0.576$. Holding C constant, increases of 10 percent in capital and labor will increase output by 10 percent. Thus, they concluded that the Pacific halibut fleet is operating under the condition of constant returns to scale; that is, proportional changes in inputs will give an identical proportional change in harvest. Finally, a test was made to determine the validity of the "good captain" hypothesis by asking an expert on the halibut fishery to give his subjective evaluation of the managerial skills of the boat captains and rank them as excellent, good, and average. Statistical tests confirmed the hypothesis that managerial ability is of great importance in fishing operation.

How does the investor select the "optimum vessel"? The "optimum vessel" is determined by finding the point where the ratio of inputs to output is minimized. For example, the Comitini and Huang study indicated that within the range of halibut vessels studied, the ratio of inputs to output did not vary with vessel size. This was referred to as constant returns to scale. Bell (1966) concluded that larger New England groundfish vessels earn more on investment than smaller vessels. That is, larger vessels generate more income (i.e., value of landings) per unit of capital (expressed in dollars) during the year. Bell's conclusion was that larger vessels generate more income per unit of capital than smaller vessels because of a greater degree of utilization during the year. Smaller vessels may be idle a large part of the year, for example, owing to adverse weather conditions, while larger vessels are free to operate. This would be an economy of scale resulting from seasonal factors. Vessel size in itself will not always yield better performance. Such factors as the skill of the captain may offset any scale effects. Furthermore, superior captains usually choose larger vessels, where financial returns are usually higher, and this may be the real reason that larger vessels exhibit better performance (i.e., better captains, not the size of the vessel in itself). Thus, the selection of the optimum vessel is a complex matter. Production function studies, financial analysis of the existing fleet, and economic feasibility studies of new technologies are the basic ingredients. For purposes of our analysis here, we shall assume the optimum vessel has been selected. Now let us look at the relation between revenue and costs.

Decision Making by the Fishing Firm

Decisions in the fishing industry are made in the short run (should the firm fish this month or year?) and in the long run (is fishing profitable enough to remain in the industry or even increase the firm's investment in the industry?). In the short run, what options does the fishing firm have? Vessel size, horsepower, age, construction, and design are fixed.[3] And, as indicated above, the fishing firm is a price-taker; therefore, it must sell at the prevailing market price. However, the vessel owner may, within limits, vary the size of his crew and spend more time fishing or decide not to fish at all! Costs of operations may be explicit (what is actually paid out by the firm) or implicit (the opportunity cost of capital and labor that are not charged as explicit costs). This topic was discussed in chapter 3. In the short run, certain explicit costs are fixed by contractual obligation, such as hull fire insurance, interest on the vessel's mortgage, and the fact that the vessel depreciates in value whether it is used for fishing or is tied up. Thus, fixed cost must be paid regardless of whether the vessel lands any fish. Variable cost consists of those expenditures that increase with catch, such as ice, fuel, nets, supplies, packing, and a "share" to the crew. The fishing industry has a long tradition of the "lay system," in which the risk of fishing success is borne by both capital and labor. This is accomplished by "profit sharing," or labor receiving a fixed percentage of the gross revenue or net revenue (after trip expenses are deducted from gross revenue). Thus, labor may be viewed as a component of variable cost. That is, the percent of gross revenue going to labor—the lay system— is a part of variable cost. In addition, we shall regard as a component of cost the opportunity cost (implicit cost) on investment that must be achieved during any time interval, say a 7.0 percent return on investment after taxes (federal and state). The opportunity cost may be considered an addition to fixed cost—a fixed profit must be obtained during any one fishing season to yield 7.0 percent on investment. Figure 9.2 illustrates these relations.

In reality, the main variable over which the vessel owner has control in the short run is vessel utilization. If the owner desires greater production or harvest during any fishing season, he can increase the number of days fished or crew size in some cases or

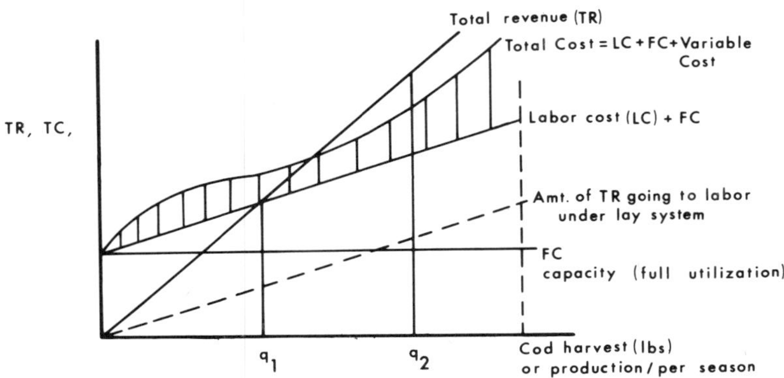

Figure 9.2: Relation of Total Revenue and Cost to Cod Harvest

both. An increase in fishing effort for the individual firm will increase revenue, as shown by the *TR* line in figure 9.2. *TR* will increase in a linear manner since price is constant (i.e. $\overline{p}q$). An increase in fishing effort will increase cost. For any one season, therefore, the higher the production, the higher the cost. In figure 9.2, we see that because of the lay system, labor costs increase in proportion to total revenue and must be added to fixed cost (*FC*). Note that at a production level of q_1, total revenue equals labor cost plus fixed cost. However, there are additional variable costs, such as vessel maintenance, nets, ice, and fuel. As a general principle, as a vessel is used more and more during the season, the total variable cost will rise slowly and then increase rapidly as 100 percent capacity utilization is achieved (i.e., the maximum number of days at sea spent fishing). With greater utilization comes more maintenance, replacement of nets, and the like. This additional variable cost is shown in figure 9.2 by the lined area. How much will be harvested? The vessel owner, in his desire to maximize profits, will harvest up to the point of maximum difference between total revenue (*TR*) and total cost (*TC*), or q_2.[4] The vessel owner is making an economic profit, because $TR > TC$ since we have already included an adequate return to the vessel (i.e., its opportunity cost—see chapters 3 and 4). As discussed in earlier chapters, this is a signal for more fishing firms to enter the industry, thereby increasing production and lowering price. Figures 9.3, 9.4, and 9.5 show positive, zero, and negative economic profits, respectively. In the long run, if negative economic profits persist, the fishing firm may decide to leave the industry. A negative economic profit reduces the normal rate of return (opportunity cost). In addition, a decrease in price lowers labor's *absolute* share (not percentage) of total revenue, and less income will be spread over a crew of the same size. Thus, annual wages per deckhand will decline and provide a motive for leaving the industry should better economic opportunities be available elsewhere.

Thus, one can clearly see the plight of the individual fishing firm. In the short run, the firm is at the mercy of so many factors beyond its control: prices, cost of variable inputs, fixed costs, level of total fleet fishing effort, fixed vessel size and characteristics, and others. This is in addition to the wide fluctuations in fishery abundance demonstrated in chapter 3. The *CPUE* cannot be precisely predicted from year to year. For example, a drop in *CPUE* due to oceanographic conditions will lower the total revenue curve, possibly causing negative economic profits or even zero returns to the vessel. Fishermen may be unwilling to work for such low wages or may be laid off. They may ask for welfare or unemployment compensation. If adverse returns persist, labor and capital may not leave the industry immediately. The reasons why will be discussed using the example of Atlantic groundfish.

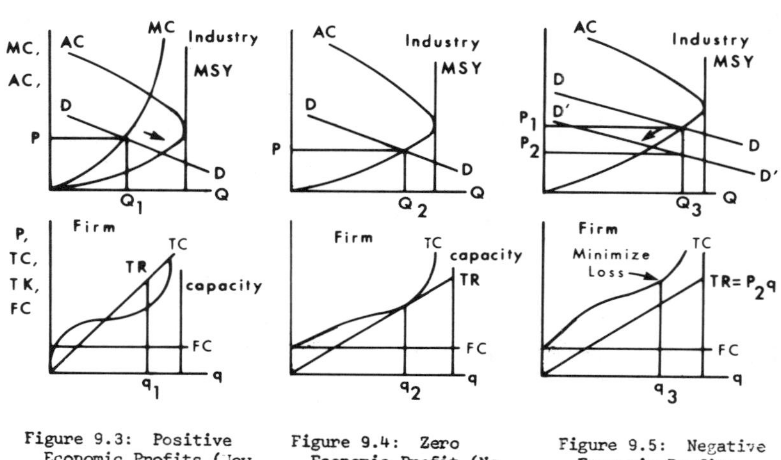

Figure 9.3: Positive Economic Profits (New firms will enter)

Figure 9.4: Zero Economic Profit (No entry or exit)

Figure 9.5: Negative Economic Profits (Pressure to exit)

Financial Profitability of the U.S. Fishing Fleet

The U.S. fishing industry is really many industries, each with different economic circumstances. In this section, we shall analyze some selected U.S. fisheries with respect to returns to capital and labor. This will be in the context of the theoretical discussion of individual fishing firm's behavior.

Atlantic Groundfish

For 1974, Bell (1976) estimated the average annual share per crewman at $12,958 compared to $9,958 in 1965, a 30 percent increase. During this same period, the consumer price index increased 55 percent, and real income therefore declined 25 percent. In addition, we must adjust wages in fishing because of either a longer year spent fishing or a shorter year in seasonal fisheries compared to shore-based employment. Workers in manufacturing, for example, work approximately 230 days at eight hours per day, or 1840 hours per year. A "full-time" Boston fleet fisherman spends 248 days at sea at approximately ten hours per day, or 2480 man-hours per year, to generate his annual wage. If the Boston commercial fishermen spent 1840 hours, or 74 percent of their time, at sea, they would earn an estimated $9,278 (adjusted annual wage) in 1974 compared to the average wage in manufacturing in Massachusetts of $8,794. Compared to manufacturing, fishing is (1) more hazardous, (2) more risky because of the "lay system," (3) provides fewer fringe benefits, and is (4) probably more arduous. Therefore, the differential in favor of fishing may hardly be sufficient to cover these differences.

An alternative way of judging whether a firm (capital) is earning economic profits is to compare its accounting profits (total revenue less explicit cost) after taxes as a percent of capital invested to that *ROI*, or rate of return, it could yield in other sectors. Bell (1976) reports a *ROI* for the average Boston and New Bedford groundfish vessel of 3.1 and –1.8 percent in 1974. Presumably, the same capital invested in U.S. manufacturing could earn an average of 7.0 percent. Table 9.1 illustrates these calculations and is consistent with figure 9.5. Noetzel (1977) uses a different sample but also reports relatively low or negative returns on capital for Atlantic groundfish in the same year. Although returns can vary from year to year, this fishery shows a persistent tendency toward declining real and relative returns to fishermen and vessels. Thus, negative economic profits have acted to force labor and capital from the industry. Noetzel (1972) indicated a

Table 9.1: Cost and Earnings--Average Vessel-Atlantic Groundfish (Boston)* 1974

Gross Receipts		$441,360
Explicit Costs		
Variable		
Food	$ 25,320	
Fuel and Lube	60,779	
Ice and Icing	9,265	
Maintenance	46,747	
Crew Share	169,049	
Captain's Commission	17,654	
Miscellaneous	13,644	
	$ 342,458	
Fixed		
Depreciation	$ 15,000	
Insurance	37,152	
Miscellaneous	24,155	
	$ 66,307	
Total Explicit Cost		408,765
Accounting Profit (before tax)		$ 32,595
Less Federal and State Corporation Income Tax		10,595
Accounting Profits (after tax)		$ 22,000
Implicit Cost		
Opportunity Cost of $720,602 investment @ 7%		50,442
Economic Profit		(28,442)

ROI (Accounting Profits after Tax) = 3.1%
ROI (Economic Profits) =-3.9%

*Sample of 19 vessels averaging 112 ft. in length making 28 trips with 16 crewmen.

Source: Bell (1976)

17 percent decline in the number of fishermen over the last decade.

Why has this occurred? First, Bell (1966) and Van Meir (1969) showed that in the late 1950s and early 1960s, the New England groundfish industry had to compete with frozen groundfish imports from Canada and Iceland, which were 33 percent cheaper than locally produced groundfish. The foreign advantage is based upon (1) higher vessel productivity, (2) lower labor cost, and (3) an array of subsidy programs in exporting countries. By the late 1960s the New England industry "survived" by selling fresh fish as opposed to frozen. By 1974, nearly 80 percent of the U.S. supply of groundfish came from foreign imports. Reduced to the fresh fish market, the U.S. industry has encountered a price-cost squeeze in recent years. Insurance rates and other costs have increased by 79 percent over the last decade, but revenue has increased by only 61 percent. As discussed in chapter 3, dwindling productivity of the resource base increases the unit cost of production. The New England fishing industry was the prime mover toward extended jurisdiction. Extended jurisdiction (see chapter 4) can increase vessel productivity if stocks are rebuilt; however, the problems facing the industry are not confined to the state of the resource alone. Tables 9.2 and 9.3 show the relative returns to labor and capital for selected U.S. fisheries.

Gulf Shrimp

As indicated in chapter 3, the U.S. Gulf shrimp fleet over the 1962–1973 period increased its fishing effort by 89 percent. The number of shrimp vessels increased from 2,193 in 1950 to 3,569 in 1969, or 63 percent.[5] The number of fishermen on these vessels increased by 64 percent. During the 1950s and 1960s, shrimp vessels earned positive economic profits. Average share per crewman increased from $4,659 in 1964 to $7,952 in 1974, or 71 percent, compared to an increase in the CPI of 61 percent for an estimated increase of 10 percent in real income. In 1974 the "adjusted" annual wage in the Gulf shrimp fishery was $7,952; wages in manufacturing ranged from $6,737 (Mississippi) to $8,846 (Texas), or an average of $7,792 (see table 9.1). Hence, relative wages in the Gulf shrimp fishery are very competitive with manufacturing. However, according to Griffin and Nichols (1976), "Lower shrimp prices coupled with rapidly escalating prices for fuel and other input items have brought about a cost-price squeeze that has severely affected vessel owners." Although the Gulf shrimp fishery shows an *ROI* of –15.9 percent for 1974,

Table 9.2: A Comparison of Estimated Wages Per Crewman for U.S. Fisheries with Wages in U.S. Manufacturing Industries (1974)

	Commercial Fishery	Average Wage Per Crewman (from fishing only)	Average Wage Per Crewman At Annual Rate	Average Wage in Manufacturing (State)[1]	Average Wage in Agriculture (State)[2]	Average Wage for for all Industries (State)	
1.	Atlantic Squid	$25,718*	$26,436	$ 7,544	$5,692	$ 7,827	(RI)
2.	Pacific Yellowfin-Skipjack-Bluefin Tuna	27,635	19,261	10,054	6,317	9,444	(CAL)
3.	Atlantic Surf Clams	21,207	17,602	10,023	6,403	9,687	(NJ)
4.	Pacific Groundfish	16,964	15,212	10,943	5,250	9,322	(WA)
5.	Wash - Oreg Salmon Seiner	4,601	13,932	10,943	5,250	9,322	(WA)
6.	Gulf Industrial Croaker	27,040*	12,438	8,110	4,580	8,079	(LA)
7.	Pacific Halibut	12,375	12,375	13,458	5,250	9,322	(WA)
8.	Alaska Salmon Seiner	2,690 (4,410)	10,760	13,458	9,067	11,949	(AK)
9.	Rhode Island Groundfish	11,724	9,965	7,544	5,692	7,827	(RI)
10.	Alaska King Crab	6,113 (9,860)	9,860	13,458	9,067	11,949	(AK)
11.	Boston Groundfish	12,958	9,589	8,794	6,860	8,786	(MASS)
12.	New Bedford Groundfish	11,238	9,573	8,794	6,860	8,786	(MASS)
13.	Gulf Food Croaker	16,218*	7,974	8,110	4,319	7,871	(AL)
14.	Gulf Shrimp	7,952	7,952	8,029	6,410	8,201	(FL)
15.	Gulf Spiny Lobster Vessels	6,159	6,159	8,029	6,410	8,201	(FL)
16.	Atlantic Sea Scallops	7,463	6,045	8,794	6,860	8,786	(MASS)
17.	Gulf Snapper-Grouper (Large Vessels)	5,548	5,289	8,029	6,410	8,201	(FL)
18.	Alaska Salmon Gillnetter	1,294 (1,505)	5,176	13,458	9,067	11,949	(AK)
19.	Alaska Salmon Troll	1,257 (1,632)	5,028	13,458	9,067	11,949	(AK)
20.	Gulf Snapper-Grouper (Small Vessels)	5,289	4,961	8,029	6,410	8,201	(FL)
21.	Gulf Spiny Lobster Boats	4,535	4,535	8,029	9,067	11,949	(AK)
22.	Wash - Oreg Salmon Troller	730	3,540	10,943	5,250	9,322	(WA)

*Highliner vessels (i.e., vessels in top 5 - 10 percent of fleet in earning capacity)

[1]State was selected on the basis of highest relative volume of landing for the particular fleet

[2]Data unadjusted for seasonal factor present in agriculture.

numbers in parentheses indicate wages from all fishing during year

Source: Bell (1976)

Table 9.3: A Comparison of Estimated Rate of Return (after tax) on Investment for U.S. Commercial Fisheries with the Rate of Return in U.S. Manufacturing 1974

	Commercial Fleet	ROI[1]
A.	Positive Return	
	1. Gulf Snapper - Grouper (Large Vessels)	24.8
	2. Gulf Snapper - Grouper (Small Vessels)	16.2
	3. Alaska Salmon Seiners	9.4 (15.4)
	4. Atlantic Squid	8.6
	5. Pacific Halibut	8.3
	6. Gulf Food Croaker	7.1
	7. All U. S. Manufacturing	7.0 (1.73 - 12.73)[2]
	8. Alaska Salmon Gillnetters	6.2 (7.2)
	9. Gulf Industrial Croaker	5.7
	10. Gulf Spiny Lobster Vessels	5.2
	11. Atlantic (Rhode Island) Groundfish	5.0
	12. Alaska King Crab	4.3 (6.9)
	13. Atlantic Surf Clam	3.5
	14. Pacific Groundfish	3.2
	15. Atlantic (Boston) Groundfish	3.1
	16. Gulf Spiny Lobster Boats	2.8
	17. Alaska Salmon Trollers	1.0 (1.3)
B.	Negative Return	
	18. Wash - Oreg Slamon Seiners	- .02 (?)
	19. Pacific Yellowfin-Skipjack-Bluefin Tuna	- .9
	20. Atlantic (New Bedford) Groundfish	- 1.8
	21. Wash - Oreg Salmon Trollers	- 3.7 (?)
	22. Atlantic Sea Scallops	- 6.5
	23. Gulf Shrimp	-15.9

[1]profits (after taxes) ÷ total investment (total assets)

[2]range of 24 manufacturing industries.

numbers in parentheses indicate rate of return from a combination of fisheries. Question mark indicates that data are not available on other fishing to obtain annual rate of return from all fishing.

[3]Federal Trade Commission.

Source: Bell (1976)

we do not believe this figure is a long-run true reflection of returns in this fishery. Noetzel (1977) also shows a negative *ROI* for 1974. Positive accounting profits were earned in 1971 and 1973. In 1974, however, prices fell over 15 percent, and operating cost increased by 14 percent. Although costs (especially fuel) will probably continue to increase, prices will probably recover if the overall world economy expands toward full employment. We believe that the major factors operating in this fishery are vessel productivity, which will continue to decline unless catch limitations are imposed and unless prices revert to prior rates of change.

However, landings of shrimp from the Gulf of Mexico by the U.S. fleet have averaged around 200 million pounds over the last two decades. Therefore, imports have increased steadily (they constituted 54 percent of U.S. consumption in 1974), despite increases in domestic production from the Pacific Ocean and New

England. (Remember our discussion in chapter 3 as to why the U.S. demand for shrimp has grown so rapidly.)

Finally, the Gulf shrimp industry will lose some of its catch through Mexican extended jurisdiction, and it must pay for the right to fish off Brazil. Recent events make it unlikely that even with substantial price recovery the great positive economic profits of the 1950s and 1960s will reappear and produce further overcapitalization.

Pacific Tropical Tuna

The analysis here is based on tuna purse seine vessels operating principally out of Southern California ports. Over the 1960–1974 period, the tropical tuna fishery increased rapidly in both aggregate tonnage and average size. The increasing size and productivity of tuna purse seine vessels increased annual wages per crewman from $11,842 in 1962 to $34,178 in 1971, an increase of 189 percent compared to a 34 percent increase in the CPI. The increase in return to both labor and capital coupled with extensive construction subsidies from the federal government (see chapter 10 for an extended discussion of the government role) finally led to an overcapitalized fleet relative to a resource that is presently exploited at about its maximum sustainable yield. Since 1971, the decline in vessel productivity has more than offset a rising ex vessel price, resulting in a decline in vessel revenue (about 2.2 percent over the 1971–1974 period). Meanwhile, trip, repair, fuel, maintenance, and fixed costs increased by 50 percent. Average crew wages plummeted to $27,635 in 1974. Adjusted wages for 1974 were $19,261, which still compares very favorably with the $10,054 earned in California manufacturing.

For 1974 Bell (1976) estimated the *ROI* for this fishery at -0.9 percent. Projections for 1975 indicate further losses. Studies by Noetzel (1977) and Altrogge (1976) show positive, but relatively low, rates of return on investment. Economic profits are surely negative. Altrogge states that "the results of this analysis argue more convincingly that U.S. purse seining for tuna will represent a serious financial risk in 1976" (p. 6). Labor income is still competitive, but it is not adjusted for the hardships of being at sea for forty-five days at a time, for example. From 1971, the decline in vessel productivity was offset by rising tuna prices until 1974. Revenue per vessel remained flat over this 1974–1975 period, but operating costs increased by over 64 percent. It would appear that the tuna fishery will (and should be) be very unattractive for capital investment. In chapter 10, we shall discuss the economic impact of many government programs, including the Marine Mammal Act, on the tuna fishery.

The Recent Financial Condition of U.S. Fisheries

According to tables 9.2 and 9.3, most fisheries are yielding returns to capital and labor that are competitive with U.S. manufacturing.

With respect to wages, we cannot overemphasize the intangible differences between fishing and other lines of work. Fishing is generally considered more hazardous, risky, and arduous than other jobs. On the other hand, the independence of small fishermen and the "love for the sea" are equally difficult to evaluate and quantify. Furthermore, returns to capital are blurred by the lack of a clear-cut distinction between capital and labor income, especially for small fishing operations.

Table 9.4 shows Bell's (1976) general ratings for the fisheries in his study. Atlantic groundfish, Atlantic sea scallops, and Alaska salmon gillnetters and trollers were not considered competitive with alternative employment of capital and labor if the economy were at full employment (i.e., tighter capital and labor markets). The Alaska fishermen derive much of their income from nonfishery sources; therefore, we are less confident about our conclusion regarding salmon fisheries in general. On the capital (vessels) side, two of our largest fisheries are experiencing difficulty: Gulf shrimp and Pacific tuna. A vigorous recovery of shrimp and tuna prices with some reduction in inflation of cost may help these fisheries return to positive returns on investment. However, the overcapitalization problem must be solved by the Fishery Conservation and Management Act. Again, the reader should be aware of the volatility of profits from year to year.

Labor Productivity Trends in U.S. Fisheries

The most comprehensive study of labor productivity in the fisheries has been made by Bell and Kinoshita (1973). The growth in productivity, or annual landings per fisherman, is an important determinant of the economic welfare of the U.S. fishing industry. Small or negative productivity gains in a fishery are often associated with lagging profits, wages, and employment, because U.S. fishermen must compete with foreign fishery imports and other protein substitutes—where productivity is a main ingredient of competitive advantage. Moreover, rising productivity in the fishery sector has

Table 9.4: An Estimate of the Relative Competitiveness of U.S. Commercial Fisheries for Labor and Capital in a Full Employment Economy

Commercial Fishery	Labor		Capital	
	Competitive	Not Competitive	Competitive	Not Competitive
1. Atlantic Groundfish				
(a) Boston		X		X
(b) New Bedford		X		X
(c) Rhode Island		X		X
2. Atlantic Squid	X		X	
3. Atlantic Sea Scallops		X		X
4. Atlantic Surf Clams	X		X	
5. Gulf Croaker				
(a) Food	X		X	
(b) Industrial	X		X	
6. Gulf Snapper - Grouper				
(a) Small Vessels	X[1]		X	
(b) Large Vessels	X[1]		X	
7. Gulf Shrimp	X			X[2]
8. Gulf Spiny Lobster				
(a) Vessels	X[1]		X[1]	
(b) Boats	X[1]		X[1]	
9. Pacific Yellowfin, Skipjack and Bluefin Tuna	X			X[3]
10. Pacific Groundfish	X		X	
11. Pacific Halibut	X		X	
12. Alaska King Crab	X		X	
13. Pacific Salmon				
(a) Wash-Oreg Seiners[4]	X			X[5]
(b) Wash-Oreg Trollers[4]	X			X[5]
(c) Alaska Seiners	X		X	
(d) Alaska Gillnetters		X[6]		X
(e) Alaska Trollers		X[6]		X

*These ratings assume (1) a fairly tight labor market; (2) intersector mobility of labor within a labor market area. Presently, unemployment conditions throughout the country help maintain labor in fishing. In addition, the average age of fishermen in many fisheries is over 45 years; therefore, their mobility intersectorally is very limited (i.e., manufacturing industries prefer younger men).

[1]We feel that wages may be competitive with rural or farm employment so we have placed this fishery in the competitive category.

[2]This may be the result of the current recession; since, the unprofitable operations are largely due to a decline in price. Hence, this may be only a temporary classification or possibly long term in nature. See 3.

[3]Both tuna and shrimp fisheries are greatly overcapitalized. They will not attract additional capital unless ex vessel prices rise extremely rapidly.

[4]Data on other fishery incomes are not available to form a complete judgment; however, these fisheries were given a provisional competitive rating for returns to labor.

[5]Although losses (profits) were shown for 1974, the reader should keep in mind the lack of distinction between capital and labor income due to the size of the fishery operation.

[6]Rated not competitive if they base entire yearly income on salmon.

Source: Bell (1976)

helped reduce inflationary tendencies, which have been most prevalent in meat and fish products. Productivity gains, in the long run, raise standards of living or reduce the amount of time we must work to produce a pound of fish, or a television set, or an automobile.

In general, gains in productivity are determined by the increasing efficiency of our vessels and gear; the education, training, experience, and morale of our fishermen; and, of course, the condition of the fishery stock and other environmental factors.

Table 9.5 shows the compound annual growth rate of labor productivity for seventeen of the nation's major fisheries over the 1950–1969 period. The Gulf of Mexico blue crab, Atlantic clam, and Gulf of Mexico menhaden fisheries all had rates of productivity over 5 percent. Unfortunately, some of our largest fisheries, such as the Gulf of Mexico shrimp, Atlantic sea scallop, Atlantic and Gulf of Mexico oysters, and Alaskan salmon, exhibited negative trends in productivity.

Although the performance of individual fisheries is important, we do want some summary measure to tell us how the entire fishery sector is doing with respect to the rate of growth in labor productivity. We can then compare this summary measure with other important sectors in the U.S. economy. Fortunately, we can construct an aggregate index of labor productivity. The construction of this index is rather technical in nature and will not be discussed in detail here. Suffice it to say that when constructing an aggregate index over a period of time, we cannot add the total pounds of fish landed in the United States and divide by the number of fishermen employed. This is because there may be appreciable shifts in the production of various species with differing absolute productivity, thereby biasing the index. That is, the constructed index controls product mix.

If we construct an index based on the seventeen fisheries shown in table 9.5, we find that aggregate productivity grew at an annual rate of 0.7 percent. In order to obtain a more representative figure for all fisheries, we added an eighteenth fishery, which represents all the U.S. fisheries not included in the original seventeen. The aggregate index showed productivity growth at an annual rate of 2.5 percent over the 1950–1969 period. However, the growth rate of fishermen's productivity noticeably tended to decline over this period; i.e., the annual growth rate over 1950–1959 was 4.7 percent, but it slackened to 0.5 percent in the last ten years. This was probably the result of increasing fishing pressure in established fisheries (see below on factors behind productivity advances). This index is plotted in figure 9.6. On the average, the American fisherman has been able to raise his productivity over the last nineteen years. This is especially encouraging when we realize that the fishermen, unlike their counterparts in manufacturing and service industries, must exploit a resource that has a fixed biological maximum whose tendency is to depress labor productivity (see below).

Figure 9.7 compares the growth of labor productivity over

Table 9.5: Ranking of Fisheries by the Rate of Growth in Output per Fisherman, 1950-69

	Fishery	Rate of Growth[1]
1.	Gulf of Mexico blue crab pot fishery	+7.8[2]
2.	North-Middle Atlantic and Chesapeake Bay dredge clam fishery	+7.0[2]
3.	Gulf of Mexico menhaden	+6.8[2]
4.	Pacific yellowfin-skipjack tuna	+4.5[2]
5.	Pacific halibut	+3.8[2]
6.	North Pacific groundfish	+3.1[2]
7.	Atlantic menahden	+2.4[2]
8.	Atlantic blue crab pot fishery	+1.3[2]
9.	Pacific albacore	+0.8
10.	Atlantic shrimp	+0.7
11.	North Atlantic groundfish	+0.5
12.	Pacific (excluding Alaska) Dungeness crab	-0.4
13.	Inshore American lobster	-0.5
14.	Gulf of Mexico shrimp	-1.3[2]
15.	Atlantic sea scallop (subarea 5Z)	-1.5
16.	Atlantic and Gulf of Mexico oyster	-2.0[2]
17.	Alaska salmon	-3.1[2]

[1]Linear least squares trends of the logarithms of output per fisherman.

[2]Trend was statistically significant at the 5% level.

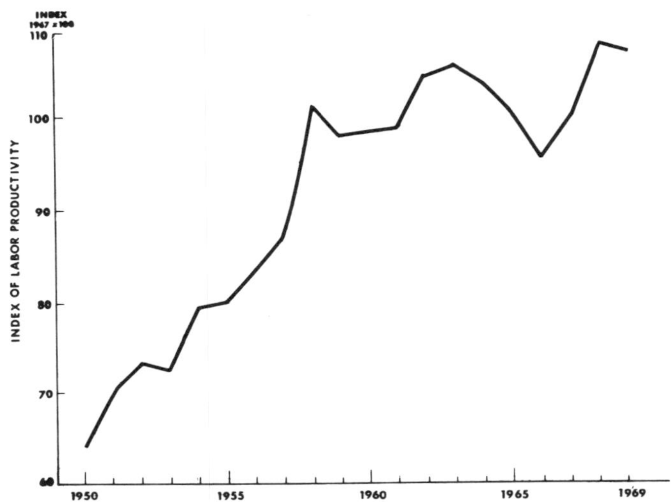

Figure 9.6: Index of Labor Productivity for the Fishing Sector, 1950-69 (Productivity index is based on 17 individual fisheries and 18th residual category.)

Source: Bell and Kinoshita (1973)

the 1950–1969 period in the total economy and in specific categories encompassing all agriculture, meat, poultry, nonagriculture, and fishing industries. The rate of growth in fishing was less than that for the U.S. economy as a whole. In contrast, the rate of growth of labor productivity in agriculture was nearly twice that of the entire economy. Of special significance, the growth of productivity in fishing lagged considerably behind that of the poultry industry and over one percentage point (per annum) behind that of the meat industry. Since labor productivity is a prime ingredient in relative price changes, it may be concluded that these trends were generally adverse to the fishing industry. That is, the more rapid advance in agriculture (including meat and poultry) lowered the price ratio of agricultural to fishery products. Remember our discussion of cross elasticities in chapter 3. For example, the annual rate of growth (1950–1969) in the wholesale price index of processed finfish was 3.9 percent, but the wholesale price index for processed foods and feeds was 0.9 percent, which partially reflects the differential gains in productivity. The consumer may then substitute the less expensive agricultural products for fishery products, and the share of

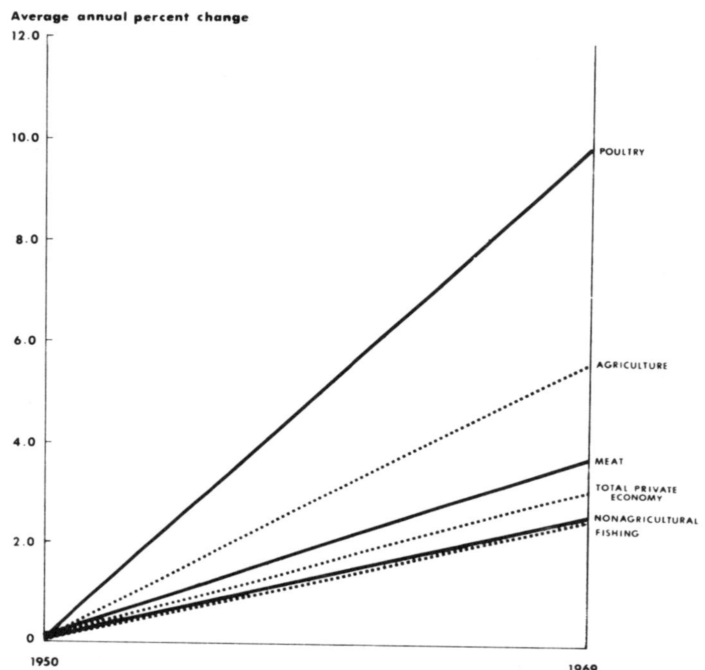

Figure 9.7: A Comparison of the Rates of Growth in Labor Productivity for the Total Private Economy, Agriculture, Meat, Poultry, Nonagricultural, and Fishing Industries, 1950-69. (Source: U.S. Department of Labor, Bureau of Labor Statistics; U.S. Department of Agriculture, "Agricultural Statistics"; and Economic Research Division.)

Source: Taken from Bell and Kinoshita (1973).

the total food markets will decline for fish. This is reflected in the data that show a 0.8 and 3.6 percent increase in per capita consumption of meat and poultry, respectively, while the per capita consumption of fish remained constant over the 1960–1969 period.

Why has labor productivity increased at a lower rate in fisheries than in competing sectors such as meat and poultry? Have fishermen been technologically backward, or are they not working harder? As indicated earlier, many factors influence trends in the productivity of fishermen. Probably, there are two important opposing forces. First, fishermen attempt to improve their technology, training, and experience so that their capability of catching fish will be enhanced. This tends to raise productivity. Second, fishermen, unlike their counterparts in agriculture, are characterized by finite limitations to production. The buildup of aggregate fishing effort (i.e., vessels, gear, and fishermen) tends to lower the productivity (catch per unit of effort) of those fishing the resource—more people are then sharing a fixed pie. This is a paradoxical result: improvements in technology increase gear efficiency but also increase effective fishing effort, which in turn depresses the catch per unit of effort. Unless the level of effective fishing effort is controlled (e.g., through limited entry and not merely through making gear less efficient or through maintaining constant gear efficiency [see chapter 4]), the fishermen will always remain on a treadmill—attempting to balance changes in technology against the finite productivity of the resource. This is why fishery management may be one of the more important solutions to the problem. In addition, other factors influence labor productivity in the fisheries, such as changes in the environment and institutional changes (see chapter 4 on regulated inefficiency).

The following factors are important determinants of labor productivity for each fishery:

1. Aggregate fishing effort
2. Fishing effort per fisherman
3. Secular time trend
4. Environmental factors
5. Institutional or regulatory changes

It is hypothesized that increases in aggregate fishing will depress productivity; that increases in fishing effort per worker (e.g., traps fished per fisherman, standard days fished per fisherman, or other gear used per fisherman) will increase productivity; that a secular time trend represents all other factors, such as changes in technology that may raise productivity; that environmental change may either raise or lower productivity depending on individual factors; and that regulatory changes may raise productivity. Let us look at one example: the eastern tropical Pacific yellowfin and skipjack tuna. The fishery for tropical tuna in the eastern Pacific Ocean developed shortly after the turn of the century. The exploitation increased steadily as the U.S. fleet, which lands the major portion of the catch, grew and as the fleets of Latin America and Japan developed. Before 1959, yellowfin and skipjack tuna from the eastern tropical Pacific Ocean were taken by bait fishing vessels that use live bait and pole. After 1959, many fishermen converted their bait vessels to purse seiners, which have subsequently proved to be more efficient fishing vessels. Over the 1935–1969 period, annual landings per tuna fisherman showed an upward time trend, growing at a rate of 2.1 percent per year.

To analyze the growth in labor productivity in the eastern tropical Pacific tuna fishery, we specified the following explanatory variables: fishing effort, or the aggregate number of standard fishing days; fishing effort per worker (i.e., standard units of fishing effort expended per worker); secular trend variable; crew size; and a variable to reflect any residual increase in labor productivity because of the switch from bait fishing to purse seining. As expected, the statistical analysis revealed that the buildup in fishing effort had a negative impact on labor productivity; that fishing effort per worker had a positive influence on labor productivity; and that the other factors were not statistically important. The Inter-American Tropical Tuna Commission apparently did a good job in adjusting its effort series for the switch in technology over the 1960–1967 period. Therefore, it must be concluded that the switch in technology is primarily reflected in the effort-per-worker variable. A look at the effort-per-worker series reveals that it increased from approximately thirteen to twenty standard units of effort per worker from 1959 to 1960. Before 1959, the standard unit of fishing effort per worker increased gradually, presumably because of more efficient use of labor in searching for and catching tuna. Although fishing effort increased appreciably over the period, its negative effect was greatly offset by increases in effort per fisherman, resulting in an annual growth rate of 2.1 percent over 1935–1969. The actual and computed (using a statistical equation) yellowfin landings per fisherman are shown in figure 9.8.

Summary

In market-oriented economies, most fishing sectors are characterized by pure competition. Each firm must accept the price determined by supply and demand. Fishing firms tend to be relatively small and in the short run have little flexibility in decision making except to vary fishing time or crew size. Nevertheless, fishing firms attempt to maximize profits. Positive economic profits have acted to entice more vessels into fishing. As discussed in chapter 4, these profits (i.e., economic rents) are the catalyst that, without proper management controls, produces

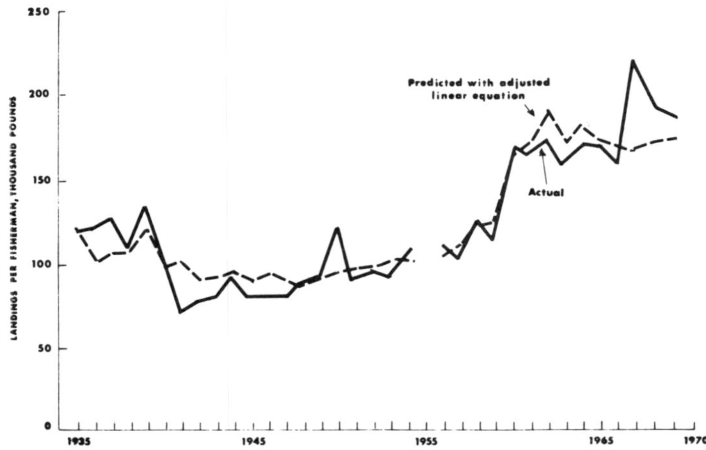

Figure 9.8: Observed and Predicted Labor Productivity (Annual Landings per Fisherman) for the Eastern Tropical Pacific Tuna Fishery, 1935-54 and 1956-69. Estimating equation: Q/L = -50615 - 2.524E + 7406E/L + 13501L/K. Variables: E = effort in fishing days; E/L = effort per fisherman in days; L/K = crew size. R^2 = 0.82; D-W = 1.22; t values-E = 2.81; E/L = 8.04; L/K = 1.76. Annual compound rate of growth = 2.1%. Data source: Inter-American Tropical Tuna Commission.

Source: Taken from Bell and Kinoshita (1973).

overcapitalization (e.g., Gulf shrimp) and overfishing (e.g., Atlantic menhaden).

In physical terms, a vessel's catch is determined during any time period principally by the amount and kind of capital used (e.g., vessel size, age, horsepower, technology), crew size, degree of fishing time (i.e., vessel utilization), technological externalities, and the managerial skills of the captain. Once the owner decides on the least-cost or optimal vessel, he must relate total revenue to total cost (which includes the opportunity cost of capital); that is, he varies the level of catch through deciding on the number of days at sea during any season.[6] As a vessel approaches full utilization, diminishing returns, or lower productivity, will occur, and total cost will rise rapidly. Zero economic profits (yet an adequate return to capital) may result at less than full utilization of the vessel during any given fishing season (see figure 9.4). If the total costs of fishing rise in a linear fashion (and if the slope is less than the total revenue curve), full vessel utilization will be the owner's objective. This will maximize profits.

Despite a sizable increase in imported fish, our survey indicates that many sectors of the U.S. fishing industry are competitive for capital and labor, except for Atlantic groundfish and scallops. The Alaskan salmon fishery is also a marginal operation. These three fisheries have had persistent secular difficulties. In general, most of the major U.S. fisheries are overcapitalized—they have too many vessels and fishermen. This is especially true of the Gulf shrimp and eastern tropical tuna fisheries, which have shown adverse trends in recent years. Returns to capital in these fisheries have dropped to uncompetitive levels, which on the one hand deters further overcapitalization, but creates hardship for those presently in the fishery.

Finally, labor productivity gains in the U.S. fisheries have been remarkable given the depressing effect of the resource limitation (i.e., technological externalities). In 1950–1959, labor productivity grew by 4.7 percent but slackened to 0.5 percent over the 1959–1969 period. The evidence would indicate that many sectors of the industry are competitive and are able to increase productivity. However, the common property nature of the fishery resource will ultimately bring about overcapitalization and eventually overfishing. Thus, as indicated in chapter 4, proper management of the fisheries is critical.

Notes

1. Vessel counted if 100 gross tons or more.

2. In other studies, the number of days fished was a separate catch determinant; however, Comitini and Huang chose to adjust capital and labor directly for degree of utilization in comparing vessels.

3. The captain's skill is fixed unless the owner would like to change captains.

4. Economists sometimes discuss marginal or incremental cost and revenue in arriving at profit maximization. At q_2, the price will just be equal to the marginal cost of the last pound of fish harvested. Production beyond this point would be unwise, since the marginal cost of harvesting a pound of fish would be greater than the given price.

5. A "vessel" is defined by the NMFS as having a capacity of five net tons or over. Thus, these figures exclude boats under five net tons.

6. Deciding on the least-cost vessel requires an economic feasibility study. See Bell (1966) for techniques used.

References

Altrogge, Phyllis D. 1976. *Further analysis of the estimated 1976 financial condition of the American tuna purse seine fleet.* National Marine Fisheries Service.

Bell, Frederick W. 1966. *The economics of the New England fishing industry: the role of technological change and government aid.* Boston: Federal Reserve Bank research report no. 31.

——. 1967. The relation of the production function to the yield on capital for the fishing industry. In *Recent developments and research in fisheries economics,* ed. Frederick W. Bell and J. E. Hazleton. Dobbs Ferry, N.Y.: Oceana Publications.

——. 1976. A survey of earnings of fishermen and vessels for selected U.S. commercial fisheries. Manuscript submitted to the U.S., Congress, Office of Technology Assessment.

Bell, Frederick W., and Kinoshita, Richard K. 1973. Productivity gains in U.S. fisheries. *Fishery Bulletin* 71.

Carlson, Ernest W. 1973. Cross section production functions for North Atlantic groundfish and tropical tuna seine fisheries. In *Ocean fishery management: discussions and research,* ed. A. A. Sokoloski. National Oceanic and Atmospheric Administration technical report, NMFS CIRC-371.

Comitini, Salvadore, and Huang, David S. 1967. A study of production and factor shares in the halibut fishing industry. *Journal of Political Economy* 75:366–372.

Dow, Robert L.; Bell, Frederick W.; and Harriman, Donald H. 1975. *Bioeconomic relationships for the Maine lobster fishery with consideration of alternative management schemes.* National Oceanic and Atmospheric Administration technical report, NMFS SSRF-683.

Griffin, Wade L., and Nichols, John P. 1976. An analysis of increasing costs to Gulf of Mexico shrimp vessel owners: 1971–75. *Marine Fisheries Review,* no. 1178.

Lloyd's Register of Shipping Statistical Tables. 1970.

Noetzel, Bruno G. 1972. New England Trawlermen's struggle for survival. *Marine Fisheries Review,* no. 943.

——. 1977. *Revenues, costs, and returns for vessel operation in major U.S. fisheries.* National Marine Fisheries Service.

Smith, Frederick J. 1975. *The fisherman's business guide.* Camden, Maine: International Marine Publishing Co.

Sysoev, N. P. 1970. *Economics of the Soviet fishing industries.* Translated for the National Marine Fisheries Service.

U.S. Department of Interior. 1969. *The effect of imports on the United States groundfish industry.*

Van Meir, Lawrence. 1969. *An economic analysis of alternatives for managing the Georges Bank haddock fishery.* National Marine Fisheries Service working paper 21.

10. The Role of Government in the Fisheries

The Pros and Cons of Government Intervention

Throughout this book, it has been quite obvious that the government is a principal participant in the economic affairs of both commercial and recreational fisheries. William M. Terry (1972), a U.S. fisheries administrator, once asked, "Why should government treat the fishing industry any differently than it treats lots of other industries, the hula hoop industry for example? That is, why should the general revenue be spent on the fishing industry?" Fishery agencies have argued that the fishing industry produces necessary food, offsets imports (the balance-of-payments problem), offers opportunities for employment in rural areas, and even that a decline in the fisheries will damage U.S. prestige among the major fishing powers (see chapter 1 for an analysis of this assertion). At best, U.S. per capita utilization (not direct consumption) of fish was 53.7 pounds in 1976 (only 12.9 pounds of direct consumption) compared to overall direct food consumption of 1500–2000 pounds. The flow of fishery imports to the U.S. market represents a bargain for consumers and foreign exchange to countries such as Japan and Canada, countries that purchase machinery and other products from the United States. There is no question that unemployment and lack of alternative job opportunities are characteristic of rural areas. It is now true of many U.S. central cities. Should the government aid the steel, construction, and automobile industries?

Even the most conservative economist, one who believes in minimum government participation in the free marketplace, will admit that Adam Smith's "invisible hand" (see chapter 4) will result in market failure in the form of overcapitalization and eventually the demise of the fishery resource. Free fish (the common property problem) means ultimately no fish if the demand for protein is increasing as it is today. Thus, government intervention is necessary to repair the breakdown in fishery markets, which has manifested itself in the resource crisis discussed in chapter 3. As we look around the world fisheries, we find that there are two traditions. One is the tradition of the common property resource, and with this tradition we often find two sets of circumstances, either a more less permanently subsidized fishing industry, or a sick or unprofitable fishing industry. The other tradition is that of creating property rights in effect, a tradition practiced to some extent in Japan and in the USSR, and other countries more centrally controlled than the United States and Western Europe. Extended fishery jurisdiction is an attempt to reduce the freedom of the seas—which means too many fishing vessels from too many countries working on the resource at the same time, with the threat of depletion and the corresponding economic disadvantages.

A second market failure associated with the fishing industry is environmental deterioration. Changes in water quality are a direct result, as discussed in chapter 5, of the free use of this common property resource. The external diseconomies of adverse spillover effects destroy fishery habitats or result in direct fish kills. The failure of polluters to internalize the full social cost of production results in the closing of areas to fishing because of high bacterial counts, contamination of fish with anything from kepone to DDT, or an immediate and massive fish kill due to phosphate slime. There is little debate within the economics profession: government must intervene to correct these two failures of the private market. Thus, expenditures from the general revenue fund are warranted. However, the methods of correcting these market failures are hotly debated. Some believe that protection of the resource through an overall quota will solve the market failure resulting from "free" or common property fish. In chapter 4, we argued against this remedy: it does not really deal with the property rights problem.

A third, and sometimes overlooked, market failure is the allocation of the fishery resource between recreation and commercial use. Since there are as a rule no user charges for recreational fishing, the proper balance between the recreational and commercial use of a fishery resource cannot be attained by the private market. This was discussed in chapter 6.

Many of the arguments for government intervention in the fisheries are based upon those made for the agricultural sector. The two sectors have many common characteristics:

1. the price elasticities for both sectors as a whole are inelastic;
2. resources (labor and capital) are relatively immobile;
3. great price fluctuations are produced by natural phenomena such as oceanographic changes.

However, agriculture is based upon a private property resource (i.e., land) and has had a rapid increase in technological change (output per man-hour); fishing is based on common property and has had slow technological change.[1] Furthermore, recreational and commercial fishing deliver quite different "commodities" to the consumer. In comparing agriculture to fishing, we are primarily talking about the sea as a source of food. Given the inelastic demand for farm products, the increase in supply—due, for example, to technological change—has created persistent tendencies for farm income to fall. Of course, farm incomes over time have not fallen absolutely, but statistics indicate that farm incomes have clearly lagged behind the nonfarm sector of the economy. The immobility of agricultural labor and capital has produced chronic overcapitalization in farming, although there has been a long-run tendency of farmers to move out of the agricultural sector to the nonfarm sector. Since technological change has not been as rapid in fishing and since overfishing has been prevalent, supplies of fishery products are not currently keeping pace with agricultural production in the United States. Thus, the price inelasticity tends to keep fishing income relatively high. Evidence for this fact was presented in chapter 9. Because of supply fluctuations from year to year, however, both the agricultural and fishing sectors are subject to violent fluctuations in income. Although for different reasons, both agriculture and fishing are overcapitalized.

From statistics available, it is not a clear-cut conclusion that both farm and fishing income per worker have lagged behind their nonfarm or nonfishing counterparts. For example, both the Gulf of Mexico shrimp and eastern Pacific tropical tuna fisheries became increasingly overcapitalized during the 1960s and early 1970s; however, returns to capital and labor compared favorably to other sectors. This is illustrated in table 10.1. As indicated in chapter 9, serious economic problems have had long-term depressing influences on earnings to both capital and labor in the Atlantic groundfish, Atlantic sea scallops, and northwest and Alaskan salmon fisheries. In the United States, the failure of the NMFS to develop detailed earnings data on a consistent basis makes it difficult to argue that many fishermen are below the poverty line. Poverty in the United States is to a large extent a problem related to rural communities and farms. Table 10.2 indicates that at the county level, where fisheries-related activities are significant, per capita income tends to be lower than in the rest of the state. For example, twenty-eight of the thirty-four counties fall into this category. In eighteen of the counties with fishery-related activities, the per capita income was less than 90 percent of the state's per capita income. Fishery-related activities include the wholesaling and processing of fishery products. The national average annual pay for wholesalers of nondurable goods was 30 percent higher than for workers in seafood wholesaling establishments in 1972. In processing, seafood workers were paid about 71 percent of that paid in all processing establishments. The figures in table 10.2, therefore, may reflect the wholesaling and processing of seafood rather than the earnings of producers or fishermen. Although the agricultural sector has many similarities to fishing, many programs and policies employed in agriculture would clearly be inappropriate in fishing. However, if one sector such as farming is given special privileges (subsidies, advisory services, and the like) to the disadvantage of fishing, this is clearly anticompetitive (see the cross elasticities of demand between fish and farm products in chap-

Table 10.1: Comparison of Real Wages in Various Fishing Sectors to Those in Manufacturing* (dollars)

Fishery	1962	1963	1964	1965	1966	1967	1968	1969	1970
1. Gulf Shrimp	-	-	5,016	6,053	7,150	8,013	7,937	6,040	7,200
Texas Manufacturing	-	-	6,349	6,427	6,457	6,601	6,786	6,797	6,810
2. Tropical Tuna	14,418	11,672	14,183	14,361	12,825	-	-	24,978	-
California Manufacturing	7,412	7,568	7,844	7,942	7,988	-	-	8,120	-

*Fishing wages include food; all wages deflated by CPI where 1967 = 100

Source: Basic Economic Indicators (1973)

Table 10.2: 1973 Economic Condition of Counties with Significant Fisheries-Related Employment (note a)

County	Total personal income	Per capita personal income	Fisheries-related employment as a percent of total county employment
	(millions)		
Alaska	$ 1,958	$5,926	
Cordova-McCarthy	14	7,135	14.9
Kenai-Cook Inlet	73	5,109	3.4
Ketchikan	69	6,592	7.1
Kodiak	53	6,399	30.7
Outer Ketchikan	9	5,482	15.7
Wrangel-Petersburg	33	6,171	12.9
California	113,746	5,508	
Del Norte	73	4,881	1.6
Florida	37,799	4,880	
Monroe	240	4,729	1.5
Nassau	89	3,643	1.2
Georgia	20,928	4,343	
Glynn	229	4,386	8.5
Mcintcsh	20	2,385	5.5
Maine	4,196	4,040	
Hancock	149	3,973	5.6
Knox	120	3,877	7.7
Washington	97	3,046	9.0
Maryland	22,185	5,446	
Queen Anne's	93	4,791	3.2
Somerset	77	4,095	10.5
Dorchester	137	4,648	3.4
Talbot	141	5,645	4.8
Massachusetts	30,551	5,268	
Bristol	2,045	4,451	1.3
Essex	3,380	5,228	1.0
New Jersey	43,026	5,874	
Cape May	330	4,944	2.3
Cumberland	606	4,673	1.5
North Carolina	22,577	4,258	
Dare	32	3,981	2.9
Oregon	10,753	4,845	
Clatsop	137	4,734	9.4
Lincoln	113	4,159	1.2
Texas	53,912	4,558	
Aransas	40	3,920	9.6
Cameron	472	2,970	6.9
Matagorda	114	4,147	1.4
San Patricia	177	3,515	4.4
Virginia	23,579	4,868	
Accomock	113	3,793	3.6
Lancaster	40	4,265	13.9
Mathews	29	3,791	11.9
Hampton City	570	4,449	1.3
Washington	17,674	5,151	
Pacific	75	4,687	10.9

a/Fisheries-related employment is considered significant when it is greater than or equal to 1 percent of total employment.

Source: USGAO (1976).

ter 3). This general thesis about government intervention should also apply to international trade—where other countries violate U.S. trade laws that are designed to keep the international marketplace as competitive as possible. Thus, under free markets, government has a distinct role in the fisheries in the following areas:

1. *Market Failure 1:* Overfishing and overcapitalization resulting from common property fishery resources;

2. *Market Failure 2:* Environmental deterioration of fishery resources because polluters do not internalize the total social cost of production;
3. *Market Failure 3:* Misallocation of fishery resources between recreational and commercial use because of no user charges for sport fishing;
4. *Market Failure 4:* Anticompetitive practices either between sectors within a country or between countries.

If satisfactorily handled, the correction of these market failures should raise earnings (if depressed) in the commercial and recreational fishing industries to competitive levels, but more importantly it should promote a more efficient use of capital, labor, and the fishery resource. Government should not under ordinary circumstances prop up an inefficient fishing firm or industry that is about to go bankrupt. If the firm is a victim of a market failure, the government is not functioning adequately. In this case, industry and government may come into direct conflict. Finally, we do not mean to confine ourselves in this chapter to government intervention where only market failures are present; we also wish to evaluate other issues which may be expressed as "societal preferences" (e.g., the Marine Mammals Act). We shall first consider the role of government in the international fishery commission discussed in chapter 4. How effective have fishery negotiators and administrators been? Second, we shall consider for selected countries the public choice about the extent of government participation in the fisheries. Emphasis will be placed upon the experience of the United States. Finally, we shall consider the role of politics in determining the government role in both recreational and commercial fishing.

International Fishery Commissions

In chapter 4, we discussed international fishery commissions very briefly. Our survey showed that both the perceived and actual function of these commissions is to study the fishery stocks in question and to adopt conservation measures to protect the fishery resource. The commissioners by and large have been attempting to resolve the market failure resulting from the common property nature of the resource—overfishing. However, the historical experience has been one economic disaster after another as international commissions have attempted to "protect the fish." The failure of many commissions can be traced either to inaction or to misunderstanding of the economics of resource use. In addition, these commissions are a little like international cartels: governments make agreements, but it pays to cheat on the agreement. Let us consider a few examples

The International Whaling Commission (IWC) was established in 1937. Although there was enough biological information to recognize that many stocks of baleen whales had been decimated in various parts of the world, particularly in the Antarctic, the IWC was powerless to halt its decline: its terms required unanimity of decision and precluded assigning national quotas. It took until 1972, or thirty-five years later, to correct these weaknesses, and in the meantime the blue and humpback whale were rendered "commercially extinct" (see figure 3.2). D. G. Chapman deemed the IWC a "horrible example of the results of inaction" (1973, p. 2419).

The International Commission for the Northwest Atlantic (ICNAF) came into existence in 1950. The catch increased from 1.7 million metric tons in 1950 to 3.9 million metrics tons by 1968, after which it declined. Again, ICNAF was given the responsibility for protecting and maintaining the fishery stocks at maximum sustainable yield. As discussed in chapter 3, many fishery stocks—haddock, yellowtail flounder, and herring—became overfished while ICNAF argued about methods of management. According to Elliot,

> A learned friend in fisheries administration once propounded his own law of conservation which is roughly that no conservation measures are ever applied until the species which they are intended to protect has been exterminated. . . . Everyone is enthusiastic, there is gold-rush fever in the air, aided and abetted by government agencies whose job it is to promote new investment by lavish grants and loans. Even when the euphoria has gone and it has become manifest that there are too many boats and plants in the industry, there is a long lag before action is taken (1973, p. 2489).

The Inter-American Tropical Tuna Commission (IATTC) has a much better record than most commissions—the problem of overfishing was recognized in 1960, and overall catch quotas were implemented in 1968. The evils of overall catch quotas will eventually lead to a waste of capital and labor because overcapitalization occurs even if there is no overfishing. The record would indicate that international commissions have largely been failures.

Extended fishery jurisdiction is a testimony to this fact. The United States has already withdrawn from many international commissions and will probably withdraw from more, except a few dealing with pelagic (i.e., tunas) and anadromous fishes and marine mammals. The failure of international commissions is in part attributable to the difficult job of persuading people or countries to agree, especially where the force of law is relatively nonexistent. Countries can be "bribed" into agreement by proper economic incentives. However, commissions have never really been regarded as tools of economic policy. The United States has been especially guilty of relegating economic analysis to an insignificant role in fishery policy, especially in international negotiations. As is so often seen, the goal of the latest NMFS plan is "to restore, maintain, enhance and utilize in a rational manner fisheries resources of importance to the United States" (1975). We again see the emphasis on protecting fish. According to the USGAO, the Canadian government has shifted policy: "First, the guiding principle in fishery management no longer will be the protection of fish, but the best use of society's resources. 'Best use' is defined as the sum of net social benefits, such as personal income, occupational opportunity and consumer satisfaction" (1976, p. 293).

In sum, the governments of the world have been especially ineffective in working together to solve the market failure associated with the common property nature of the resource. A recent USGAO report (1976) indicated that many governments share this conclusion. West German officials feel that the fish stock protection measures taken by international fishery commissions have generally been too weak and too late. U.K. officials believe that international quota agreements have not worked. Canadian officials complain that ICNAF has not effectively conserved fish stocks.

The Extent and Magnitude of Government Involvement in the Fisheries

The USGAO (1976) made a recent survey of the role of government in the fisheries for seven countries in addition to the United States. This role ranged from complete government involvement in the Soviet Union, where industry and government are one and the same, to the relatively hands-off approach of the Danish government.[2] Government funds are used in essentially four areas: (1) basic biological research; (2) biological and economic research as it relates to fishery management; (3) financial assistance to the fishing industry; and (4) various forms of advisory service to industry, including marketing, gear development, and business management. Let us consider the U.S. experience first.

As of June 30, 1976, the U.S. National Marine Fisheries Service had 1,762 permanent employees, of which 958, or 54 percent, were in fishery laboratories. In 1976, it received appropriations of $69 million dollars. This agency is the major federal policy-making body for fisheries and is subject to review by the Department of Commerce (of which it is a part). However, fisheries research is also done by the Office of Sea Grant and seven other identifiable federal agencies ranging from the Smithsonian Institution to the Defense Department. The Sea Grant Program funds research projects through universities. A recent study by Hollomon et al. indicated "that Sea Grant has produced results with significant commercial potential. The bulk of these are concentrated in a few projects" (1977, p. 33). Hollomon et al. estimate that their sample of seventy-seven Sea Grant projects have sales potential estimated at $122 million annually by 1980. Because there are no figures on the costs of these projects, it is difficult to evaluate the results in a benefit-cost framework. Most of the Sea Grant projects were not oriented toward the traditional market failures. Government intervention to aid industry is characteristic of the agricultural sector as discussed above. If enough tax money is used to aid the fishing industry, some income and job creation will doubtless occur. (When we use the term *aid,* we are excluding funds to solve basic market failures.) The more fundamental question is why more money in consumers' hands (as opposed to government) does not stimulate the fishing industry (processors and producers) to innovate and thereby to capture this increased purchasing power. As with other governments (see below), this has been a subject of continuing debate. Each of the states also engages in a wide variety of fishery activities—from management to basic biological research. Most of the federal activity is not well coordinated, and much of it has been challenged repeatedly by examiners from the Office of Management and Budget. The USGAO (1976) admitted that no one knows the exact (or even approximate) number of tax dollars (federal, state, and local) allocated to the fisheries. It is beyond the scope of this book to evaluate the effectiveness of basic biological research, which is certainly necessary in formulating fishery management programs. In addition, we have discussed traditional methods of fishery management in chapter 4 and found them to be economically destructive to the fishing industry in the long run. Therefore, we shall concentrate on aid to the fishing industry as direct or indirect financial assistance. The role of government in international trade and other prominent areas will also be discussed.

In an attempt to help the harvesting sector of the U.S. fishing industry maintain its competitive position in the domestic and international marketplace, the federal government has since 1956 provided both direct and indirect monetary support through four basic programs. The fishery loan fund program is designed to upgrade and maintain commercial fishing vessels by providing reasonable financial assistance not otherwise available to commercial fishermen. Loans were granted for purchasing, constructing, equipping, maintaining, and repairing or operating new or used vessels. Table 10.3 indicates that for the major fisheries, almost $25 million was loaned, primarily to the shrimp, tuna, crab, groundfish, and salmon fisheries over the 1960-1972 period. According to the USGAO,

> Loans from the Fisheries Loan Fund were often made to refinance existing mortgages, pay existing debts and provide operating funds. Although authorized by Statute, many of these loans allowed the continued use of unefficient vessels rather than improving vessels and equipment for more efficient and profitable fishing . . . vessels have been maintained in, or added to, segments of the fishing industry which service (NMFS) officials consider to have excess but not necessarily efficient, harvesting capacity (1973, p. 1).

The USGAO further criticized the NMFS for not establishing priorities for directing program funds to the building of more efficient vessels in fisheries not characterized by overcapitalization. The NMFS no longer offers this program.

The guaranteed loan fund provides up to 100 percent federal government guarantees for the repayment to private financial institutions for credit given to individual fisherman. For the fisheries listed in table 10.3, approximately $25.6 million in loans was guaranteed over the 1960-1972 period, with heavy concentration in the shrimp, tuna, American lobster, groundfish, and crab fisheries. Although still in existence, this program was criticized by the USGAO (1973) for many of the same reasons it criticized the fisheries loan fund.

In 1960, Congress passed the Fishing Fleet Improvement Act, which was designed in its initial version to assist certain depressed

Table 10.3: Financial Assistance to Selected U.S. Commercial Fishermen (dollars)

Fishery	FFI	GL	FLF	CCF[5]
1. Atlantic Groundfish[1]	10,012,208	1,997,178	2,492,264	2,580,766
2. Pacific Groundfish[1]	5,978,474	-	816,819	9,739,150
3. American Lobster[1]	62,495	2,302,828	297,238	4,248,398
4. Atlantic Sea Scallops[2]	1,526,721	907,500	536,613	-
5. Oysters	-	-	-	-
6. Blue Crab	-	-	-	-
7. Clams[3]	-	-	307,005	3,482,500
8. Shrimp[1]	-	12,615,968	5,178,626	35,363,336
9. Tuna[1]	14,884,456	5,527,675	5,888,532	174,578,300
10. Pacific Halibut[1]	-	-	991,814	474,000
11. King and Dungeness Crab[4]	788,663	2,029,657	4,261,765	41,524,455
12. Salmon[1]	-	266,059	4,087,376	26,343,750
13. Menhaden[1]	343,911	-	-	22,215,779
	$ 33,605,928	25,646,865	24,858,052	320,550,042

[1]1960-72; [2]1960-69; [3]1968-73; [4]1960-71 [5]1970-1975

FFI = Fishing Fleet Improvement Act, 1960, 1964. (Vessel construction)

GL = Guaranteed Loan (Government insured loan)

FLF = Fisheries Loan Fund (Loans for construction, purchase and repair of fishing vessels)

CCF = Capital Construction Fund (Deferred taxes to induce the building of fishing vessels)

Source: Basic Economic Indicators (1973-75); NMFS, Financial Assistance Division

segments of the fishing industry—where it could be shown that they were adversely affected by foreign imports. The initial act did try to address itself to a market failure. A 1793 law prohibits U.S. fishermen from purchasing fishing vessels outside the United States. Thus, the relatively high-cost U.S. shipbuilding industry has a captive market: Americans who wish to build fishing vessels. This, of course, violates free market conditions and protects U.S. shipyards against foreign competition. Foreign fisherman have the advantage of less expensive vessels; therefore, these savings can be used to produce less expensive fish which were and are still making massive penetrations into many U.S. markets for fishery products, especially for groundfish and sea scallops. The 1964 version of the Fishing Fleet Improvement Act eliminated the depressed segment provision of the 1960 act; its main purpose was to correct inequities in the construction of U.S. fishing vessels. The 1964 act stipulated that the vessels must be of "advanced design," which indicated that the intent of Congress was improved efficiency.

Under the act, the government would pay up to 50 percent of the cost of construction of new fishing vessels. Over the 1960–1972 period, about $33.6 million was spent under the provisions of this act for the major fisheries shown in table 10.3. The tuna and groundfish fisheries were the overwhelming recipients of so-called construction subsidies. As with the other two programs, the USGAO (1973) criticized the administering of this program—on similar grounds. The program was terminated after the construction of the Seafreeze Atlantic and Seafreeze Pacific, two large groundfish factory ships, which turned out to be financial disasters.

Finally, as part of the Merchant Marine Act of 1970, a capital construction program was established. The program provides tax deferral incentives to encourage and assist fishermen in establishing cash reserves (from profits, depreciation, etc.) for the future purchase of fishing vessels. This, in effect, provides a government interest-free loan equal to the federal taxes that would otherwise have to be paid on vessel income. Deferred taxes are eventually recaptured through a reduction in the basis for depreciation of vessels constructed with tax-deferred funds. The "interest-free loan" continues until the vessel has used its total depreciation allowance. This program is very popular with profitable segments of the fishing industry, as indicated in table 10.3. Profits and depreciation deposited into the capital construction fund over 1970–1975 have been enormous. Almost $174 million dollars came from the tuna industry alone, and the crab, shrimp, and salmon industries have each deposited at least $25 million. The ability of these overcapitalized sectors to generate profits for the fund over the 1970–1975 period indicates that our conclusions about the fairly high profitability in many fishing sectors are warranted. As table 10.3 indicates, over $320 million has been deposited to the fund for the major sectors of the U.S. fishing industry.

How important have these U.S. financial assistance programs been? In one evaluation, Living Marine Resources (1973) indicated that only about 7 percent of the U.S. vessels added to the commercial fishing fleet from 1957 to 1971 received some type of financial assistance under NMFS loan, subsidy, or guaranteed insurance programs. This is shown in table 10.4. This does not include the capital construction fund, which is enjoying increasing use.

Historically, these financial assistance programs have been counterproductive from the point of view of both the general

Table 10.4: Number of Vessels Added to the United States Fishing Fleet, and Vessels Receiving Financial Assistance 1957 - 1971

Item			Number
Vessels added to fleet[1]			10,222
Vessels receiving NMFS financial assistance[2]			
Loans "for new and replacement vessels and conversions"[3]		415	
Subsidies		45	
Mortgage Insurance			
All	247		
Less subsidy vessels	19		
		228	
Total receiving assistance		688	688
Proportion of vessels added receiving financial assistance			0.0678 (6.8%)

[1]Calendar year data from NMFS (formerly BCF), Fisheries of the United States, annual editions for 1957-71 (Washington, D.C.: NMFS)

[2]Fiscal year data, NMFS Financial Assistance Division.

[3]Total, all categories, 1,380 "vessels" (counting loans instead of vessels, 1956-1959).

Source: Financial Assistance for the Fisheries - An Issue Paper, National Marine Fisheries Service.

welfare and the fishing industry. They have accelerated overcapitalization in the tuna, shrimp, salmon, and groundfish industries. The temptation of a "free lunch" from Uncle Sam coupled with congressional pressure to "help the fishing industry" has made the National Marine Fisheries Service a partner in the overcapitalization process, despite repeated urgings from NMFS economists to establish overall economic criteria for administering these programs. During a period of enlightenment, the American Tunaboat Association requested that entry into the Capital Construction Fund be temporarily suspended in order to prevent the present overbuilding trend in the industry from being accentuated. The NMFS now establishes "conditional fisheries," which are characterized by excess capacity relative to potential harvest, so that financial assistance may be denied. Given the massive capital construction fund buildups in fishing sectors that are already overcapitalized, one might become rather skeptical of the operational effectiveness of the conditional fishery concept. The hard lesson we have been preaching in this book is that building more and even better fishing vessels will fail as long as the institutional problem of property rights—which is so often ignored for political expediency—remains unresolved. All of the major fisheries were either fully capitalized or overcapitalized before the advent of most of these financial assistance programs. Where property rights are private, as in agriculture, Griliches (1964) has shown that for every public dollar spent on research and education, 13 dollars of output per year are obtained. However, the Department of Agriculture relies greatly on its Economic Research Service for guidance. This is in sharp contrast to the NMFS, which recently abolished its Economic Research Division. As we have already noted, very little economic analysis went into the National Fishery Plan, which is allegedly an integral part of making extended fishery jurisdiction work (see chapter 4). Finally, economic efficiency sometimes gives way to emotions, especially in the political arena. The Marine Mammal Act of 1972 prohibits the catching of porpoise. Porpoise are taken along with the tuna catch, and many are unintentionally killed in the process. Except for research and commercial display, porpoises have no economic value. The population dynamics of the porpoise have not been well studied; however, since they are not pursued for economic gain, the porpoise are probably not overfished. Several environmental groups have, without considering the economic impact, brought U.S. tuna fishing to a halt as of 1977; of course, other countries that use a less efficient technology (i.e., long-lining) will step into the gap. We have a classic situation: the cost of "saving the porpoise" will be lost jobs in catching and processing and higher prices for tuna. Ironically, mammals such as the porpoise and fish such as the shark feed high on the food chain, and their elimination might increase the available supply of edible fish. It is not the environmentalists, but the American people who should decide whether they are willing to pay the price. Since tuna is a relatively insignificant part of the average American's diet, the environmentalist will undoubtedly have his way. If beef were involved, the environmentalist would be overwhelmed by the public outcry. On the other side, the fishing industry has used the "principle of being unimportant" to get legislation that favors their objectives.

What is the role of government in fisheries in other countries? In Canada, the fishing industry provides job opportunities in regions offering few alternatives, supplies protein to the consumer, and is an important source of foreign exchange. The Canadian government has provided subsidies for the construction of fishing vessels since 1944. Loan insurance is also provided. In contrast to the United States, Canada provides price supports to protect fishermen against sharp declines in prices. As in the United States, subsidies have contributed to an overinvestment in fishing and processing capabilities (USGAO 1976).

In Denmark, the fishing industry supplies enough fish to make the country self-sufficient (i.e., exports equal imports) at a per capita edible consumption of 70.2 pounds a year, about 50 percent higher than the average for Western Europe. Unlike the U.S. and Canadian governments, the Danish government does not normally interfere directly in the fishing industry. However, indirect Danish subsidies do exist in the form of government-supported bank loans. As of late, however, fishing industry influence has resulted in fuel subsidies and subsidies for the exportation of cod and saithe.

In Japan, we find widespread subsidization of the fishing industry. This takes in port facilities, aquaculture ventures, vessel insurance, price stabilization, fuel costs, and promotion. These subsidization programs amounted to almost half a billion dollars in 1976. Despite a rigorous limited entry program, Japan's fleet has overcapacity problems, and the situation may get worse as the wave of extended jurisdiction increases (see chapter 4).

Within the framework of extended fishery jurisdiction, Mexico is engaged in a massive effort to expand its fisheries through government investment in modernizing the fleet and ports and through the establishment of fishery technical schools and joint government-private ventures. Moreover, price supports are offered to the shrimp industry, and subsidies are offered to fishermen's cooperatives for the purchase of fuel and oil.

As a centrally planned economy, the USSR invests $1 billion annually in the fishing industry, which supplied over 8 percent of its food supply in 1975. The United Kingdom has encountered increased overfishing and increased quota restrictions. In 1975, large subsidies were given to the U.K. fishing fleet to avoid radical reductions in income and employment. Similarly, in West Germany, loans, interest reductions, and other benefits are available to the fishing industry.

One general theme is prominent in our analysis of the role of government in fisheries throughout the world: there is an inordinate amount of financial assistance, often no more than a direct subsidy. Although the fishing industry is usually less than 1 percent of gross national product, most countries regard their fishing fleets as important in supplying food. Despite increasing overcapitalization, governments repeatedly engage in shipbuilding programs. Fishery programs are often designed to keep people employed, even though they are not really needed. Thus, there is a strong social objective to maintain employment. Interestingly enough, wages received by fishermen in Canada, Japan, Mexico, the USSR, and the United Kingdom were well above the average wage paid in their respective economies. It should also be pointed out that by and large the U.S. fisherman relies less on government than fishermen in the overwhelming majority of fishing powers do. This was documented in a recent study of fishing industry support for the year 1968-1969. Table 10.5 shows the results of an Organization for Economic Cooperation and Development (OECD 1971) study for eighteen countries that have varying degrees of free market economies. Table 10.6 shows the ratio of financial assistance to the value of landings for all eighteen countries. For this one year, which may be biased, the "subsidy ratio" of the United States ranked third from last. It would seem that the political economy of food from the sea is based upon a vicious cycle of government assistance: it induces overcapitalization, which in turn requires more financial assistance to support fleets to provide both employment and food. This vicious cycle is hardly conducive to effective utilization of the sea as a source of food and recreation.

Summary

Although there is considerable debate over the role and extent of government intervention in the private sector among market-oriented, as opposed to centrally planned, economies, most would agree that the recreational and commercial fishing industries are subject to at least four market failures. That is, the working of the private and decentralized marketplace in the fisheries does not deliver products to the consumer at the lowest possible price—in other words, it does not make the most efficient use of capital, labor, and the fishery resource. The four distinct market failures are (1) overcapitalization and overfishing attributable to the common property nature of the resource; (2) environmental deterioration of fishery habitats resulting from the indiscriminant use of water—a common property resource—by polluters; (3) misallocation of fishery resources among commercial and recreational users, since the latter have no market price or user charge imposed;

Table 10.5: Financial Support to Fishermen in OECD Countries, 1968-69

Country	Price supports	Vessel equip-ment subsidies	Loans	Loan guarantees	Interest rebates	Other operating cost subsidies	Estimated total subsidization 1/
	$1,000						
Belgium	112	75	1,595	-	114	48	509
Canada	2/	6,372	2/	1,022	-	997	7,369
Denmark	-	-	6,066	-	-	-	607
Faroe Islands	1,427	2/	628	2/	-	2/	1,490
Greenland	-	35	1,147	-	-	-	150
Finland	48	-	-	-	48	104	200
France	-	2,072	1,441	-	238	-	2,454
Germany	-	3,743	1,740	-	119	320	4,356
Iceland	-	-	2,465	466	-	-	246
Ireland	-	41	83	-	-	-	124
Italy	-	64	622	-	118	9,973	10,777
Japan	-	- 2/	36,472	103	-	5,825	9,472
Netherlands	335	328	-	-	-	524	1,187
Norway 3/	23,191	2,360	13,891	-	-	1,190	28,130
Portugal	-	-	4,953	-	-	-	495
Spain	-	1,829	1,829	-	-	2/	2,839
Sweden 4/	-	13	2,515	-	-	-	265
Turkey	-	-	352	-	-	2/	35
United Kingdom	132 2/	1,538	1,068	-	-	1,982	3,759
United States	-	3,058	2,274	3,893	-	-	3,285

1/ Totals include items as shown, except loans (0.10 of amount shown) and loan guarantees (not counted).
2/ Assistance is given, but no amount or only a portion could be determined from source.
3/ Government purchases and sales guarantees for stockfish in 1968, due to sales contract failures related to the Nigerian civil war, are excluded.
4/ To support men who intended to leave the fisheries, $580,000 was allocated in the 1969-70 budget.

Source: OECD, Financial Support to the Fishing Industry, (Paris: 1971). U.S. data from NMFS.

Table 10.6: Fishery Catch and Assistance Data, 1969[1]

Countries	Subsidy ratio 1/	Estimated total subsidization	Value of catch	Quantity of catch (round)
		- - - - - - $1,000-	- - - - -	1,000 metric tons
Norway	.190	28,130	148,300	2,206
Italy	.056	10,777	191,200	250
Germany	.047	4,356	91,900	633
Canada	.044	7,369	168,800	1,239
Belgium	.030	509	16,766	51
United Kingdom	.024	3,759	158,300	1,062
Denmark	.023	2,247	99,558	1,275
Netherlands	.018	1,187	66,800	269
Ireland	.017	124	7,100	62
France	.013	2,454	192,400	690
Finland	.011	200	18,300	87
Spain	.009	2,839	319,000	1,482
Portugal	.007	495	73,800	334
Sweden	.007	265	39,600	268
United States	.006	3,285	518,500	1,947
Japan	.004	9,472	2,419,000	8,623
Iceland	.002	246	104,000	689

1/ Countries are ranked by the ratio of estimated total subsidization to the value of catch.

Source: FAO, Yearbook of Fishery Statistics, 1969, Vol. 28, (Rome: 1970). OECD, "Draft Review of Fisheries, 1970," Part I (Paris: 1971).

and (4) anticompetitive practices such as massive subsidies granted to one country's industry, but not to others—where each have to compete in a world market. Of course, government intervention is sometimes predicated on slogans such as "save our fishing industry," which has an emotional appeal—the public might perceive that "our" resources are being lost to foreigners.

Because of the principle of freedom of the seas, fishery resources have been common property to all countries. Countries have attempted to deal with the problem of "conserving fish" through international fishery commissions. Generally, little if any, economic analysis has been applied to the problem. Unfortunately, the problem has been viewed almost exclusively within a biological context. The historical experience of international fishery commissions is one economic disaster after another. Some of these overfishing disasters stem from acting too late; others are rooted in a failure to limit entry by economic means (see chapter 4), that is, by using overall quota systems or regulated inefficiency as "conservation measures." These conclusions are shared by many governments involved in fishing and certainly have had much to do with the current wave of extended fishery jurisdiction.

Government has greatly extended its involvement in fisheries well beyond these four market failures. For example, the United States has provided financial assistance in the form of construction subsidies, loans, loan guarantees, and deferrals of federal income tax to those wishing to construct, acquire, or reconstruct fishing vessels. In the main, this assistance has gone to the groundfish, shrimp, tuna, crab, and salmon fisheries, all of which are greatly overcapitalized. Thus, this assistance has contributed to overcapitalization. However, only 7 percent of the U.S. vessels added to the commercial fishing fleet from 1957 to 1971 received some kind of financial assistance. It has been primarily the uneven distribution of this assistance and the size of the vessels added (e.g., Seafreezes Atlantic and Pacific) that have created overcapitalization problems. The NMFS's lack of an economic policy on U.S. fisheries prompted the U.S. Office of Management and Budget to be highly critical of all phases of NMFS programs. The NMFS would rather sidestep this issue. In sharp contrast to the Department of Agriculture, which relies heavily upon its Economic Research Service, the NMFS abolished its Economic Research Division in 1976.

Our survey of other countries has indicated that such market-oriented economies as Canada, Japan, the United Kingdom, and West Germany actually offer more financial support to their respective fishing industries than the United States does. For example, the recent formation of OPEC and the accompanying higher prices for fuel has prompted many governments to offer fuel subsidies to their fishing industries. Price supports and construction loans are substantial for foreign fishing fleets. The motives for this financial assistance are diverse. Some countries desire to maintain a fishing industry for food, foreign exchange, and employment. However, these relatively large subsidies are in violation of U.S. trade laws (i.e., a market failure due to reduction in competition), and the U.S. government has rarely imposed sanctions against offending governments. Thus, various fishing sectors in the United States have suffered from government inaction.

In chapter 3, we pointed out the extent of overfishing and the depletion of fishery resources throughout the world. Governments have not only been ineffective in dealing with this problem, but have also gone to great lengths to "out-subsidize" competing nations, thereby contributing even more to the problem. Extended fishery jurisdiction is not a total solution, since heavily subsidized imports will still flow into the United States.

In conclusion, we are very pessimistic about the future of food from the sea. Governments are openly pursuing policies to diminish the resource base through overcapitalization. International commissions are ineffective. Of great importance, the fishery resources of the world are becoming more and more valuable (see chapter 1) and will be enjoyed, as in the past, by the more affluent societies. In addition, many of the underexploited species (krill, lanternfish) may be used more often than not for reduction purposes, to serve as feed for higher-value protein. As for the less-developed countries, extended fishery jurisdiction does offer them a chance to collect rents for the use of their fishery resources and thus to earn foreign exchange. It offers less chance to improve their direct food consumption.

Notes

1. See the discussion of technological change in fishing and agriculture in chapter 9.

2. These countries were selected by GAO since they either fished off the U.S. coastline, have economic and political systems similar to the United States, have large, efficient, and profitable fishing industries, or are neighboring countries that will affect the U.S. extended fishery jurisdiction zone.

References

Chapman, D. G. 1973. Management of international whaling and North Pacific fur seals: implications for fisheries management. *Journal of the Fisheries Research Board of Canada* 30:2419–2426.

Elliot, G. H. 1973. Problems confronting fishing industries relative to management policies adopted by governments. *Journal of the Fisheries Research Board of Canada* 30:2486–2489.

Griliches, Zvi. 1964. Agricultural production function. *American Economic Review* 54:961–974.

Hollomon, J. Herbert; Utterback, James M.; McGugan, Blair M.; and Kim, Linsu. 1977. An analysis of the potential commercial and foreign trade impact of the sea grant program. National Oceanic and Atmospheric Administration, Office of Sea Grant.

Living Marine Resources, Inc. 1973. *An evaluation of financial assistance programs of the National Marine Fisheries Service.* Contract 3-35127.

National Marine Fisheries Service. 1973. *Basic Economic Indicators, Tuna, 1947–72,* Current Fisheries Statistics 6130; and *Shrimp,* Current Fisheries Statistics 6131.

——. 1975. National plan for marine fisheries. Final draft, October 1975.

——. 1976. *A marine fisheries program for the nation.*

Organization for Economic Co-operation and Development. 1971. *Financial support to the fishing industry.* Paris.

Terry, William M. 1972. Fisheries and the national interest. In *World fisheries policy: multidisciplinary views,* ed. Brian J. Rothschild. Seattle and London: University of Washington Press.

U.S. General Accounting Office. 1973. *Need to establish priorities and criteria for managing assistance programs for U.S. fishing-vessel operators.* Report to the Congress by the comptroller general of the U.S., B-17704.

——. 1976. *The U.S. fishing industry—present condition and future of marine fisheries.* Report to the Congress by the comptroller general of the U.S., vols. 1-2, CED-76-130.

Appendix

Table A.1 Current Fishery Statistics

World[1]	1974	1975	1976
Catch (Mil. Metric tons)	70.49	69.73	73.49
United States: Commercial Fisheries			
Catch (Mil. Metric tons)[1]	2.77	2.80	3.00
Value (millions)	$ 898	$ 971	$ 1,353
Value Per Metric Ton	$ 324	$ 347	$ 451
Supply[2]			
Domestic (mil)	$ 898 (37%)	$ 971 (38%)	$ 1,353 (37%)
Imports (mil)	$ 1,571 (63%)	$ 1,566 (62%)	$ 2,277 (63%)
Total	$ 2,469 (100%)	$ 2,537 (100%)	$ 3,630 (100%)

Consumption Per Capita (1976)[3]: Total: 12.9 lbs; Tuna: 2.8 lbs; Groundfish and other species fillets and steaks: 2.6 lbs; Shrimp: 1.5 lbs; Salmon: .4 lbs; Sardines: .3 lbs.

Per Capita Utilization (1976)[4]: Total: 53.7 lbs; Edible: 34.3 lbs; Industrial: 19.4 lbs.

Inputs (1973): Vessels[5]: 15,396; Boats: 72,362; Fishermen: 148,884.

Foreign Trade (edible): See Figure A.1

United States: Recreational Fisheries (1975)[6]

Measure	Salt[7]	Fresh[8]	Total
Participants[9]	16,307	37,622	53,929
Expenditures[9]	$ 3,450,358	$ 11,756,211	$ 15,206,569
Recreational Days[9]	207,212	1,127,685	1,334,897
Expenditures Per Fisherman	$ 213	$ 312	$ 282
Expenditures Per Recreational Day ($)	16.65	10.43	11.40
Recreational Days Per Fisherman	12.71	30.00	24.75

1. Round or life weight; 2. Includes food and industrial fish; 3. Edible weight; 4. Round weight; 5. Craft over 5 net tons; 6. Data cannot be directly compared with data in chapter 6 due to changes in definition; 7. Oceanic waters only; 8. Includes cold water, sea-run and warm-water fishing; 9. Expressed in thousands.

Sources: FAO Year Book of Fishery Statistics; Fisheries of the United States (U.S.D.C.) and 1975 National Survey of Hunting, Fishing and Wildlife-Associated Recreation (U.S.D.I.).

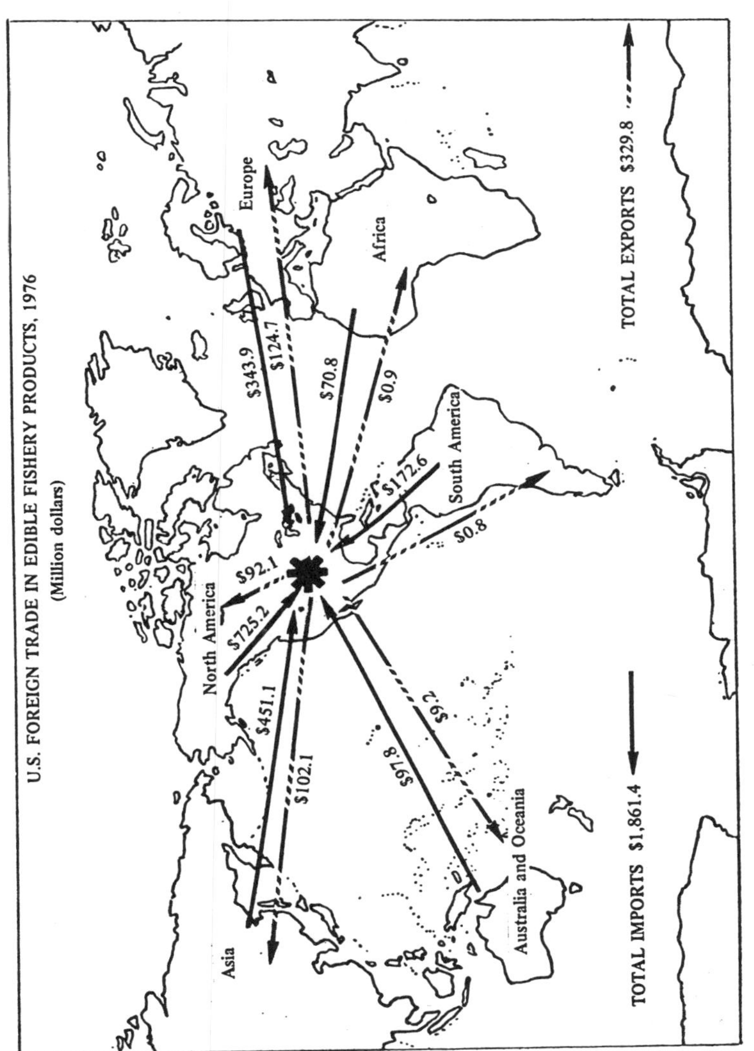
U.S. FOREIGN TRADE IN EDIBLE FISHERY PRODUCTS, 1976
(Million dollars)
Europe
Africa
North America
South America
Asia
Australia and Oceania
$343.9
$124.7
$70.8
$0.9
$172.6
$0.8
$92.1
$725.2
$451.1
$102.1
$97.8
$9.2
TOTAL IMPORTS $1,861.4
TOTAL EXPORTS $329.8

Figure A. 1

Study Questions

Study Questions

Chapter 1

1. In 1971, Peru caught 10.6 million metric tons of fish and was considered the leading fish-catching nation in the world. The U.S. catch was only 2.8 million metric tons. From an economic standpoint, evaluate this kind of comparison (i.e., Peru versus U.S.).

2. Despite a large increase in world population and a need for protein, the fishery catch from the world's oceans has shown no upward trend over the 1969–1975 period. Why do you think this has occurred? What, in general, will be the economic implications (i.e., for developed versus developing countries, etc.)?

3. The *real* value per metric ton of fish has increased 3.9 percent annually on a world basis over the 1963–1973 period. Discuss the economic implications of this trend from the viewpoint of producers (i.e., fishermen) and consumers. Is this trend related to question 2?

Chapter 2

1. Discuss the importance of fishery production (i.e., calories, protein) within the context of the total world food supply. Be sure to discuss the role of industrial as opposed to food fish.

2. The halibut catch is 20 million lbs and returns an ex vessel (dockside) price of $1.00 per pound. The next catch is 25 million lbs due to very favorable environmental conditions. The price elasticity for halibut is -5.0. What would you predict the price to be with the increased catch? What will happen to halibut sales (revenue) due to the price change?

3. List and briefly explain some *unique* factors influencing fish consumption in the U.S. and throughout the world. What are the standard demand determinants?

4. Contrast demand determination for fishery products in the U.S. and the Soviet Union.

5. Since 1953, have fishery products (production) been keeping pace with world population, a major demand determinant? Are there exceptions? In light of question 2, chapter 1, do you expect these trends to continue?

Chapter 3

1. If the *MSY* for haddock is 50,000 metric tons in an area of the Northwest Atlantic while ICNAF has placed a quota of only 6,000 metric tons, what does this imply about (a) the history of the fishery; (b) status of the stock; and (c) economic implications?

2. Define *depleted, imminent danger,* and *intensive use* of a fishery stock using the Schaefer yield function. Give an example fishery for each case.

3. Discuss the derivation and interpretation of the Schaefer yield function for a fishery resource. Define *MSY* as it relates to renewable resources.

4. Using a Schaefer yield function and the fact that total industry cost is a linear function of fishing effort, show how bioeconomic supply curves, both marginal (MC) and average (AC), can be derived. How do you interpret each, and why does the AC bend backward while the MC does not?

5. Discuss one case history of stock assessment and regulatory action if applicable (e.g., Peruvian anchoveta).

6. Discuss the use of "food chain analysis" in determining the biological potential and catch from the sea. What is the currently accepted *MSY* from the world's oceans (i.e., all fish)? Critique this figure (see chapter 8 to help with critique).

Chapter 4

1. Using graphs, show the integration of cost curves (MC, AC) with a demand curve for a fishery and show why MC=P is *not* a stable market equilibrium if the fishery is common property.

2. Why is MC=P socially optimal? Define a stable market equilibrium for a fishery (i.e., either regional or interventional) if common property.

3. Explain by the use of a graph the fundamental reason why most fisheries are eventually overfished or depleted. Also explain verbally.

4. Critique *regulated inefficiency* and the *overall quota* systems as methods to prevent overfishing or restore a depleted stock.

5. Critique the following three limited entry schemes: (a) licensing; (b) boat quotas; and (c) taxes or license fees. Select a fishery with which you are familiar and give your recommendation as to which management scheme would accomplish the objectives postulated by Crutchfield and Zellner (i.e., Limited Entry 3 in text).

6. "The U.S. 200-mile fishing zone is the solution to all our fishing

problems." Critique this statement. Why is the old law of the sea breaking down in favor of unilateral extended fishery jurisdiction? Will this new wave create a more efficient use of the world's fishery resources from the sea (e.g., increase potential catch, etc.)?

7. Discuss the future of world fisheries in the context of the FAO and Bell et al. projections of world demand and supply of fishery resources.

Chapter 5

1. How is the term *external diseconomy* related to environmental deterioration and fishery resources?

2. Distinguish between degradable versus nondegradable waterborne pollutants.

3. Outline and discuss the sources of economic losses to the fisheries through water pollution. What empirical data (e.g., fish kills, shellfish areas closed, etc.) support these theoretical sources of losses? Explain.

4. Discuss the various solutions to the pollution problem including the substance of the Federal Water Pollution Control Act.

Chapter 6

1. Compare the value of the U.S. *commercial* fishery catch (see chapter 1) ex vessel and retail (see general mark-up, chapter 2) with gross expenditures on salt and fresh water recreational fisheries. Given the great emphasis on commercial fisheries, why do you think recreation fisheries are a "sleeping giant"?

2. "As a recreation fisherman [i.e., one who does not sell fish for profit], my main motive is to obtain good protein for my dinner table." Critique this statement in light of material presented in this chapter.

3. The *value* of the recreational fishery resource (e.g., salmon) can be measured either by expenditures on equipment plus travel cost to the site or merely by pounds caught multiplied by the retail value of the fish if sold in stores. Explain why neither of these valuation procedures is correct from an economic point of view.

4. Explain what data you would need to collect and how you would use it if the Clawson technique of recreational fishery valuation were used.

5. Assume you have derived a demand curve for a fishery resource, what is the resource's "value" if (a) the state charges no fees and (b) the state acts as a monopolist?

6. "I do not care what the fishery resource is worth to me as a recreationalist! After all, it's free, isn't it?" Explain the economic implications of this statement with respect to resource allocation. *Hint:* Someone wants to

dredge and fill the recreational area where the fishery resource exists to build several seashore condominiums.

7. Cicchetti et al. expect recreational fishing days to increase by 37 percent over the 1965–2000 period (1 percent a year); growth of recreational fishing days increased 94 percent over the 1955–1970 period (over 6 percent a year). Why do you think the authors are so pessimistic about future growth? *Hint:* See sport-commercial conflicts, declining catch per angler, etc.

Chapter 7

1. What are the major aquacultured species, and where in the world is aquaculture concentrated? Is the U.S. a major producer of aquacultured products? If not, why?

2. Compare the costs (or inputs) that are necessary to aquaculture fish versus catching the same fish at sea. Does the property rights system make aquaculture different from hunting fish on the high seas? What are the economic implications of this difference?

3. According to Bell and Canterbery, what critical biological and economic criteria must be considered in technological transfer of aquaculture within the developing world? Explain.

4. Some believe that by the year 2000, aquaculture will be as important in producing fish protein as the catch from wild stocks. What factors might limit these optimistic projections?

Chapter 8

1. Give at least *two* good reasons why the Antarctic krill is not exploited by the U.S.

2. In many areas of the world, people are starving while such protein as the lanternfish and squid is in great supply (i.e., large resource). Explain this apparent inconsistency in economic terms.

3. Explain the positions of Alverson and Suda with respect to the technological capabilities of harvesting underutilized species.

4. What are some of the main reasons consumers tend to reject underutilized species? Is industry (producers, processors, etc.) at fault?

5. Contrast the individual *MSY* estimates for underutilized species with the "food chain analysis" of the total biological potential (i.e., catch) from the sea (see chapter 3).

Chapter 9

1. Economists indicate that fishing firms operate in a market characterized by pure competition. Describe how pure competition influences

fishing firm decision making.

2. Describe the main vessel characteristics and other factors that determine the size of a vessel's fishery harvest. Define *economies of sale* (see Pacific halibut in this chapter).

3. In the short run, for example, about one year, what factor(s) determine vessel utilization? Or, how does the firm try to maximize profits in the short run in the fishing industry?

4. Like any other industry, the fishing industry must compete for capital and labor. Explain, using the Gulf shrimp fishery, whether it is currently competitive with other sectors of the fishing industry. What have been the main forces at work in determining this fishing fleet's competitiveness?

5. Overall, is the U.S. fishing industry competitive for capital and labor? Explain the main reasons that the U.S. fishing industry is uncompetitive in some sectors.

6. "Labor productivity or technological change is a two-edged sword in the fishing industry based upon open access (i.e., common property resources)." Explain. Why is productivity in fishing like "running on a treadmill" just to keep a given level of labor productivity? Looking at productivity trends in the fishing industry, does the industry really deserve to be termed "technologically backward"?

Chapter 10

1. Why should government get involved in the fishing industry? Should the U.S. attempt "fishery independence" as has been promised through extended jurisdiction? Or are there fundamental reasons why government intervention is warranted in a free enterprise, capitalistic system? Explain.

2. The U.S. and other governments have worked through international fishery commissions. Using the material in chapter 4 and this chapter, evaluate the performance of these organizations as government tools for management. Be specific on performance criteria.

3. Summarize and evaluate U.S. financial assistance to domestic fishermen. Have these programs reduced or created economic problems?

4. Given the present trend of international subsidization of domestic fishing fleets, do you believe that this accomplishes nationalistic objectives such as creation of employment or reduction in balance of payments problems? In other words, are other nations dealing with "market failure" or offsetting each other's activities in the fisheries? Discuss the economic results of this government competition. Should the U.S. government join the competition or abstain? If the U.S. abstains, could there be advantages? To whom?

Index

Index